The
School of
Hard
Knocks

NUMBER TWELVE:
*C. A. Brannen
Series*

The School of Hard Knocks

Combat Leadership in the American Expeditionary Forces

Richard S. Faulkner

Texas A&M University Press

COLLEGE STATION

Library of Congress Cataloging-in-Publication Data

Faulkner, Richard Shawn.
The school of hard knocks : combat leadership in the
American Expeditionary Forces / Richard S. Faulkner. — 1st ed.
p. cm. — (C. A. Brannen series ; no. 12)
Originally presented as: Thesis (Ph.D.)—Kansas State University, 2008.
Includes bibliographical references and index.
ISBN-13: 978-1-60344-297-8 (cloth : alk. paper)
ISBN-10: 1-60344-297-9 (cloth)
ISBN-13: 978-1-60344-698-3 (e-book)
ISBN-10: 1-60344-698-2 (e-book)
1. Command of troops—History—20th century. 2. United States. Army—Officers—
Training of. 3. United States. Army—Non-commissioned officers—Training of.
4. United States. Army. American Expeditionary Forces—Unit cohesion. 5. United States.
Army. American Expeditionary Forces—Mobilization. 6. Military morale—United States.
7. United States. Army. American Expeditionary Forces—Operational readiness.
8. United States. Army. American Expeditionary Forces—History. 9. World War,
1914–1918—United States. I. Title. II. Title: Combat leadership in the
American Expeditionary Forces. III. Series: C. A. Brannen series ; no. 12.
D570.2.F38 2012
940.4′1273—dc23
2011025679

On the cover: Machine gun platoon from the 80th Division attacking
in the Meuse-Argonne between La Chalade and Le Claon, 29 October 1918
(US Army Signal Corps photo 31991)

For
Laura,
Connor,
Brenna,
and
Shelby

Contents

Illustrations, Maps, Tables

Acknowledgments

G ROWING UP IN NORTHERN G EORGIA on land crisscrossed   **xi**
with the trenches of the Atlanta Campaign of 1864, I, perhaps under-
standably, developed an interest in the Great War. Although I cannot
remember when I first became fascinated with the First World War, I
know that my love of history came from my parents, Larry Faulkner
and Gloria Nan Faulkner. To them I owe the deepest thanks for my
abiding passion for the military and the study of the past.

I, and this work, have also been shaped by twenty-three years of
service in the United States Army. I enlisted in the Army Reserve as a
sophomore in college and thus was able to gain a personal familiarity
with the challenges and rewards of training and leading citizen soldiers. This time also gave me the rich first-hand experience of being
on the receiving end of leadership as a back ranker. The Regular Army
that I joined as an armor officer in the mid-1980s was a force with a
solid tactical doctrine, organization, training focus, and weapons base.
In short, "my" army generally had all the benefits that the American
Expeditionary Forces (AEF) lacked, including the most important
advantage: the time to correct its shortcomings. I have tried to keep
the differences between my army and the AEF from influencing my
view of the latter, but my service as an officer, to include commanding a tank company in combat, unquestionably informed my understanding of the realms of the possible and impossible when it came to
developing officers and noncommissioned officers (NCOs), training
units for war, and the challenges of leading soldiers in battle.

Over the course of my military career I had the privilege of learning the art and craft of combat leadership from some of the republic's finest commanders and NCOs. I am in the debt of Cols. Richard
Geier, Edward Kane, Thomas Wallace, and Michael Alexander for
mentoring and teaching me how to be an officer. I am also humbled
and indebted to the soldiers and NCOs with whom I have served
for schooling me in the skills of leading the American soldier. I must
especially single out M.Sgt. Thomas Thacker and Sgts. Maj. Robert
Hayden and Samuel Colter for my thanks for exemplifying the highest standards of noncommissioned officer professionalism.

Acknowledgments

      This work would not be possible if not for the assistance of a number of friends, colleagues, and talented scholars. I am grateful to Emory Thomas, John Morrow, Michael Ramsay, Mark Parillo, and Charles Sanders and my other professors at the University of Georgia and Kansas State University for their efforts to shape me into a scholar and, ultimately, my thoughts into this book. While serving as a history instructor at the United States Military Academy, I had the privilege of serving with Robert Doughty, Lee Wyatt, Gian Gentile, Evan Hoffer, Ty Seidule, and the other talented members of the history department. These gifted soldier-scholars offered me sage advice, access to sources, and friendship that materially aided and fueled my quest to be a historian. I owe my comrades at the Department of Military History, United States Army Command and General Staff College at Fort Leavenworth, most notably Scott Stephenson, Marlyn Pierce, John Suprin, Chris Gabel, and Sean Kalic, my deepest thanks for their support, wise counsel, and friendship. Due to their efforts and dedication, DMH is truly a fun and rewarding place in which to work. I give special gratitude to Jim Willbanks for encouraging me in my efforts to obtain a PhD and for granting me the sabbatical needed to complete the work that led to this book.

      This book also rests upon the hard work and assistance of numerous dedicated librarians, archivists, and researchers. The efforts, understanding, and kindness shown by Timothy Nenninger and Mitch Yockelson of the National Archives; Mike Browne, Kathy Buker, and Elizabeth Merrifield of the Fort Leavenworth Combined Arm Research Library; Alan Aimone of the West Point Library Archives; Sandra Reddish and Robert Smith of the Fort Riley Cavalry Museum; Jonathan Casey of the Liberty Memorial and National World War I Museum; and David Keough of the Military History Institute at Carlisle Barracks were essential to the crafting of this work. Their commitments to helping this novice researcher navigate the records of the Great War have won my undying gratitude and admiration. I must also thank Mary Lenn Dixon and the staff of the Texas A&M University Press for their patience, understanding, and hard work in bringing this work to fruition. Given the talents of all those listed above, any errors in this book are purely my own.

      Most of all I must thank my wife, Laura, and my children, Connor, Brenna, and Shelby, for their love and support during my education, writing, and military service. Laura has untiringly worked and sacrificed to be my closest companion, best editor, and most able sounding board. For the promise of adventure and good French meals she has been my loving camp follower and fellow traveler on numerous military moves, deployments, and treks across the battlefields of the Great War. Laura has also graciously acquiesced to my eBay-fueled addiction to purchasing World War I letters, photos, and other military ephemera. In the words of the great American philosopher Ralph Kramden, "Baby, you're the greatest." Although Laura somewhat knew what she was getting herself into when she left hearth and home to follow the drum, my kids had no say in being born into a military family in which one parent also suffered from the psychosis of studying history. My oldest two children had the pleasure of living in seven different houses and learning in school in four different districts before they were twelve years old. Despite these sacrifices, they have remained steadfast in their backing of their father, and I may even have convinced "The Boy" that "they don't call it the Great War for nothing."

The
School of
Hard
Knocks

I   Combat Leadership in the AEF
A Tale of Alvin and Charles

THE DAY WAS NOT GOING WELL for the 82nd Division's 328th Infantry. As the regiment attempted to seize the Decauville railroad in the early morning of 8 October 1918, German riflemen, snipers, and machine gunners on Hill 233 and Champrocher Ridge caught the Americans in a vicious crossfire that quickly halted the momentum of the assault as the doughboys sought shelter from the defenders' remorseless firepower. The assault itself had been rather clumsy and ill-coordinated. It was a frontal attack into an open valley, and the attack had been preceded by little effort to suppress the German defenders with artillery or machine gun fire.

The American attack degenerated into a confused effort by individuals and small clumps of doughboys to move forward by running from shell hole to shell hole. Part of the confusion was caused by a general shortage of leaders within the regiment. The two previous days of fighting had already taken a heavy toll on the regiment's commissioned and noncommissioned officers (NCOs). The loss of these key leaders meant that many of the unit's companies and platoons were being led by junior officers and NCOs with little or no experience at the levels of command into which they had been thrust.

As the 328th Infantry's attack stalled, its casualties began to mount. Along with the rest of the 328th, G Company found itself beset by enemy fire, heavy losses, and the inability of its officers to bring order to the chaos reigning on the valley floor. It was not that the company's leaders were not trying to reorganize the advance. In fact, 2nd Lt. K. P. Stewart was killed by machine gun fire while exhorting his men to move forward, but the officers and NCOs faced a situation that little in their training or previous experience had prepared them to meet. Heavy enemy fire had isolated the company commander from much of his unit, and his span of control extended merely to those soldiers in adjoining shell holes that were within the range of his voice.

The 1st Platoon leader, Sgt. Harry M. Parsons, realized that his unit's position was precarious and that something had to be done to reduce the enemy's fire. He ordered corporal turned acting sergeant Bernard Early to take three squads and attempt to flank the German position and silence their machine guns. Early's small command of three corporals and thirteen privates succeeded in surprising and capturing a number of Germans, but a more alert group of enemy machine gunners discovered the American detachment and pinned it down with accurate fires.

After the German fire killed one corporal and severely wounded Early, the command of the detachment devolved onto Cpl. Alvin York. York ordered the surviving squad members to remain under cover and guard their prisoners while he worked himself into a location where he could enfilade the German positions. York managed to kill fifteen to twenty of the German defenders and then led his detachment back to the American lines, forcing the surrender of additional German units as he went. The American detachment ultimately returned with 132 German prisoners, completing a mission that gained York the Medal of Honor and the distinction of being the most decorated soldier in the American Expeditionary Forces (AEF).[1]

While York was winning his honors on the morning of 8 October, a different drama was being played out less than two kilometers away. Pvt. Charles Clement, a scout with the intelligence section of 2nd Battalion, 328th Infantry, had been moving throughout the morning delivering messages from the battalion headquarters and reporting on the conditions of the unit's scattered companies. Clement repeatedly braved the enemy fire that had done such grave damage to G Company and the rest of the 328th Infantry. As Clement rushed forward from crater to crater with A Company, a German sniper on the ridge shot him in the forehead, killing him instantly.[2]

Although Clement was only one of scores of the regiment's privates to die that day, his story was unique. This twenty-six-year-old private had in fact previously been a captain in command of the battalion's E Company. Clement's fall from grace was a sad tale of the systemic problems associated with the nation's hasty mobilization and the pressures of command and leadership in modern war. On the surface, Clement was everything that the army wanted of its new officers. He was a 1912 graduate of Mercer University, a respected teacher at Atlanta's Boys High School, and a man whose peacetime college athletics and work with the YMCA embodied "muscular Christianity" and adherence to Theodore Roosevelt's "strenuous life" that supposedly marked the alpha male of the Progressive Era. After the army established a series of Officer Training Camps (OTCs) to provide officers for the new draftee divisions, Clement signed up for the first iteration of the training to be held at Fort MacPherson, Georgia, in April 1917. After his assignment to the 2nd Battalion,

Charles Clement shortly before he joined the army.
(Photograph courtesy of Benjamin Byrnes.)

328th Infantry, his commander, Maj. Edward Buxton, praised Clement as an "indefatigable student" of war, "one of the hardest working men in the regiment," and a person marked by "his higher ideals of helpfulness toward military service."[3] Unfortunately, Clement's hasty training and the burden of making life-and-death decisions left him ill-prepared to exercise effective combat leadership.

In late June 1918, the 82nd Division's regiments rotated into the trenches of the quiet Langney sector of the French line to receive their first combat seasoning. On the night of 1 July, shortly after his company had occupied their section of trench line, Clement ordered a small ambush patrol into no-man's-land. Shortly before they went out, Clement informed the patrol leader, Sergeant Cunningham, that he would accompany him on the mission. Cunningham noted that the captain was in "an intoxicated condition" and, with the aid of a lieutenant, tried to dissuade him from coming with the patrol. Despite their efforts, Clement insisted on going. The patrol was only in no-man's-land for a short time before Cunningham returned to the American trenches, bodily carrying the stupefied captain. Cunningham reported that "Captain Clement made so much noise that he thought it was foolish to stay out there."[4] Because of that behavior, Clement's battalion commander felt that he had no other option but to bring the young captain before a court-martial for violation of Article 85 of the Article of War, an offense dealing with an officer found drunk on duty.

During his court-martial Clement readily admitted his guilt. He noted that he normally abstained from alcohol (a fact supported by his fellow officers) and was unsure exactly why he had drunk heavily from a bottle of Scotch on the night of the patrol. In his last statement before the court, Clement declared, "I am guilty, but if the verdict of this court be the death penalty, I have nothing to say. If it be, however, dismissal from the service, I have. In "No Man's Land" I disgraced myself, my uniform, and my country, and in "No Man's Land" I would like to have the opportunity of redeeming myself, at least partially. Take my commission away from me, but allow me to go to my own company as a private and allow me to serve there shoulder to shoulder with the men that I have commanded."[5] Although the board sentenced him to be cashiered and confined at hard labor for five years, the members of the court-martial unanimously signed a plea for clemency for the disgraced captain. Ultimately, the appealing authority granted Clement's request, and he was allowed to enlist as a private in the 2nd Battalion's Headquarters Company.

One of Clement's friends, Pvt. Ernesto Bisogno, noted that the busted captain frequently maintained that the drinking incident was not due to cowardice, but he never went further in explaining the cause of the binge. Although he sometimes expressed bitterness at his fall, Private Clement remained true to his promise and repeatedly volunteered for patrols and other hazardous missions. When he was last seen alive, Clement was forward of one of the 328th Infantry's lead companies, apparently seeking a blood sacrifice to redeem his lost honor.[6]

The experiences of the officers, NCOs, and soldiers of the 328th Infantry on 8 October 1918, and the specific cases of Alvin York and Charles Clement, offer a window into the overall experiences of the AEF. In a microcosm, these events and people highlight one of the American military's greatest challenges in World War I: how to build a cadre of combat leaders at the company level and below who were capable of fighting a modern industrial war without sustaining prohibitively high casualties. The 8 October attack reveals some of the realities that made combat leadership such a challenge for the US Army in World War I. These included inadequately or inappropriately trained officers and NCOs, leaders inexperienced with the tools and techniques of modern war, organizations too ungainly to be controlled by novice leaders, the innate challenges of command and control in a pre–radio communications army, and the impact on unit effectiveness of heavy losses in leader cadres.

In the actions of Sergeants Parsons and Early we see a battlefield initiative born more of desperation and a sense of survival rather than the cool and deliberate actions of battle-wise and professional NCOs. Parsons later admitted, "It was an awful responsibility for a non-commissioned officer to order his men to go to what looked to be a certain death. But I figured it had to be done. I figured that they had a slight chance of getting the machine guns." In York's case, his actions reflected more of his individual prewar marksmanship and stalking skills than of his strength as a leader. He admitted that during the battle, "I hadn't time to give orders nohow," and that he limited his actions to targeting "them there Germans machine gunners and [giving] them the best I had."[7] In the tragic case of Charles Clement, we see the psychological toll the burden of command took on individual leaders when the coping mechanisms of training and experience had not adequately prepared and armored them for the realities of combat. Clement was everything that the army sought in selecting its new officers, yet he clearly recognized his own limitations and made a mistake that could have been costly to the men under his command.

In the end, the 82nd Division's 8 October attack achieved its goal. However, this success cost the 328th Infantry twenty-eight officers and 718 soldiers. One is left to wonder if these meager gains were worth the losses spent to achieve them. This question is tied to the larger issue of the AEF's overall combat effectiveness. For the last twenty-five years, historians studying the AEF have explored the larger issues of the army's senior leadership and their attempts to build an effective tactical doctrine. Much of that scholarship has been critical of Pershing's leadership and the AEF's operational effectiveness. For example, James Rainey's "Ambivalent Warfare: The Tactical Doctrine of the AEF in World War I" (*Parameters,* 1983) notes that the AEF's problems on the battlefield resulted from Pershing's inability to transform his nebulous concept of "open warfare" into a sound doctrine that could be used by battlefield commanders. In a similar vein, Timothy Nenninger has likewise taken a critical view of the AEF's performance. In "Tactical Dysfunction in the AEF, 1917–1918" (*Military Review,* October 1987), he argues that the Americans' disdain for "European" methods along with their own flawed training and personnel practices prevented the AEF from becoming an effective fighting force. Nenninger expands this argument in Allan Millett and Williamson Murray's *Military Effectiveness: The First World War.* In his chapter "American Military Effectiveness in the First World War," Nenninger concludes that while the United States was strong in the political-strategic arena, the nation's overall lack of readiness to fight a modern war, and its subsequent rapid mobilization, undermined the AEF's operational and tactical efficiency.

More recently, Mark Grotelueschen has argued that while Nenninger and Rainey were correct to note the bankruptcy of Pershing's open warfare doctrine, the true impact of that doctrine was much less

5

dramatic than previous historians had maintained. By studying the training and actions of the 1st, 2nd, 26th, and 77th Divisions, Grotelueschen accurately argues that leaders at the division and brigade level disregarded the directives from the AEF's General Headquarters (GHQ) and took a more pragmatic approach to operations than those prescribed by Pershing. One of the key lessons that these divisions learned was to substitute the firepower of artillery and other supporting weapons for that of the individual rifleman. This embrace of a firepower-centric doctrine ultimately allowed these divisions to achieve their missions without prohibitive casualties He also maintains that the Americans' training with the Allies was much more effective in preparing the AEF for the realities of combat than the AEF GHQ admitted.

Of all the books and articles about the American army, none directly address the topic of the competency of junior leadership in the AEF. Although Nenninger, Rainey, and Grotelueschen all examine the tactical level of war, they give little attention to small-unit leadership or its pervasive effect on the AEF's operations and overall effectiveness. For example, by focusing on the division and brigade levels, Grotelueschen seems to have overlooked the baleful influence of inadequate and poorly focused training on operations at the company level and below. The ill-coordinated and costly frontal attacks that continued to characterize American infantry operations from Soissons through much of the Meuse-Argonne seem to belie the assertions that the skills of the AEF improved over time.

This book will examine how the training, professional development, and expectations of the AEF's junior officers and NCOs influenced the American army's effectiveness in World War I. It goes further than Nenninger, Rainey, and Grotelueschen in examining the tactical world in which these leaders operated and the factors that undercut their ability to build cohesive units capable of accomplishing their missions without prohibitive casualties. This focus on the lowest tactical levels not only furthers our understanding of the American experience at "the sharp end" of the war, but also provides insights into how the Great War was a watershed moment in the development of modern warfare.

World War I was a catalyst for the emergence of radically new conceptions of combat leadership and tactics, for it fundamentally changed the role and importance of junior officers and NCOs in battle. As machine guns, rifles, and improvements in artillery forced armies to spread and thin out their tactical formations and seek protective cover in field fortifications, the traditional centralized discipline of the eighteenth and nineteenth centuries was no longer possible. Commanders at the regimental level and above could not see, much less directly command, their scattered soldiers. Combat on the expansive and deadly battlefields of the First World War placed much greater demands on the initiative, aggressiveness, and motivation of the junior leader and the individual soldier than had been required of them in earlier times. Thus, while Pershing and his corps commanders could plan operations and order their execution, in the end it was the sergeants, lieutenants, and captains, far removed from Chaumont, who determined what would be accomplished on the battlefield. When the junior leaders failed, faltered, and bungled, the AEF's battles became confused and uncoordinated slugging matches that confounded the plans and expectations of the army's senior leaders.

Although the US Army was somewhat slow in accepting this shift in the nature of junior combat leadership, the question of who would lead the legions was of paramount concern to the Regular Army's leadership from the earliest days of the American entry into World War I. When the United States declared war on Germany in April 1917, the army immediately faced the problem of how to obtain, train, and develop a corps of officers and NCOs for an expeditionary force to fight in France. By June the army was rapidly expanding from a peacetime strength of 209,000 officers and men (including the National Guard) to a force that would eventually reach four million men. Within weeks of the declaration of war, army planners estimated that the nation would need to commission an additional 200,000 officers. To understand the magnitude of the problem, one should keep in mind that on 30 June 1916 the Regular Army had only 4,843 officers on its active rolls and could call on just 3,199 additional officers from the National Guard.[8] The army could, in theory, produce adequately trained riflemen within a few months; officers and sergeants, however, usually required years to master the tactical, technical, and leadership responsibilities of their positions. To add to this problem, the Americans had to change their doctrinal outlook from that of a small professional constabulary force to that of a mass, technical army. To put it bluntly, it was a pre-1914 army thrust into the ugly realities of a 1917 industrial total war.

Under the stress of time, the army's short-term solution to the shortage of leaders was to graduate the West Point class of 1917 early, recall as many retired officers as possible, and federalize all fit and competent National Guard officers. The army's long-term solution was to establish three-month-long OTCs to commission new captains and lieutenants. The army also chose to select and promote sergeants directly from the ranks of the Regular Army and selected draftees. Although these solutions filled the required positions in the expanding National Army divisions, the abilities of company-level leaders remained questionable.

Although some commentators in the army's professional journals had long stressed the need for the service to develop junior leaders able to cope with the demands of modern war, the Regular Army's own efforts to address this issue had been haphazard at best. While the United States pulled off a minor miracle by deploying a two-million-man army to France in nineteen months, this was accomplished only through a series of shortcuts and mistakes that ultimately carried severe consequences in terms of human life.

Leadership is the bedrock of all armies. It is the leader who forms the discipline and cohesion of the unit and directs it toward a collective effort to achieve the unit's assigned mission. In many ways, combat leadership is a rather nebulous thing to classify. Perhaps the most simple and accurate definition of combat leadership is that it is the art of getting soldiers to do more or less willingly what instinct and society have programmed them not to do: to place themselves at mortal risk and to kill others while engaged in battle. War is about fear, mental and physical exhaustion, killing, and dying. Junior officers and NCOs are the key element that mitigates and channels the full range of human emotions unleashed by combat. Combat leadership is also based on a social contract between the leader and the led. In return for their subordinates' obedience, battlefield leaders must show a genuine concern for their welfare and demonstrate a level of tactical competency that assures soldiers that their lives will not be placed at unnecessary risk.

Ultimately, battlefield leadership rests on a foundation of mutual trust and confidence between the soldier and his superiors. The cornerstone of that confidence is the subordinates' faith that their leaders have mastered the technical and tactical aspects of their jobs and that the leader can provide the basic requirements of food, clothing, shelter, ammunition, and medical aid that hold together the body and soul. The last point should not be downplayed, for it is part of the social contract of leadership and a foundation of unit cohesion and effectiveness. The soldiers expected that their leaders could do these routine things routinely, and when leaders proved unable to produce those basic necessities, unit discipline and cohesion suffered.

Although this book will touch on officers from a large array of commissioning sources, it will concentrate mainly on the officers commissioned through the OTCs. The OTC graduates comprised over

74 percent of the officers commissioned during the war and more than two-thirds of the army's line officers. National Guard officers comprised only 9 percent of the commissioned ranks, while officers of the Regular Army accounted for only 5 percent of the officer corps during the war.[9] Also, while the study will cite the experiences of leaders in all the combat arms, it will focus mainly on the leadership in infantry and machine gun units. Infantry officers and soldiers constituted the bulk of the troops in the AEF, and the army leadership from Pershing downward considered the infantry to be the core element of the army. Leadership, both good and bad, is also generally more discernable in the historical records of infantry and machine gun officers and NCOs. All of the caveats applied to the officer corps will also apply to the examination of the AEF's NCOs. Throughout the work, the term "junior leadership" is meant to encompass all ranks within a company from captain to corporal squad leader.

The book's examination of the leadership of African American officers in the AEF will be brief. African American officers faced problems in leadership training and professional development that were largely identical to those of their white peers, but the endemic racism of the period added a dimension of complexity to the subject that can not be adequately addressed here. Without in any way slighting the contribution of African American officers in the war, it should be noted that they made up less than 0.07 percent of the wartime officer corps.[10]

In the final analysis, the US Army's own shortsightedness, institutional uncertainty, and administrative missteps greatly contributed to the AEF's costly and awkward tactical performance in the last six months of the war. Nowhere was this more evident than in the selection and training of its cadre of junior leaders. The systemic problems associated with mass mobilization, poor personnel policies, and incomplete or ill-focused training meant that the AEF's companies where led by officers and NCOs who did not understand how to employ the new weapons introduced in the war, lacked basic skills such as map reading, and were largely unable to employ basic tactics. This lack of leader know-how resulted in the formation of companies and battalions that often lacked a strong cohesiveness and were frequently incapable of executing offensive tactics beyond costly frontal attacks. Leadership at the most basic tactical levels was one of the most important factors in determining how Americans fought their Great War and why the AEF never truly lived up to its full operational potential.

2 "To Be Instructed in the Dark Art and Mystery of Managing Men"
Junior Officers in the Old Army

IN HIS 1888 SHORT STORY "ONLY A SUBALTERN," Rudyard Kipling wrote of his young subject: "He became an officer *and* a gentleman, which is an enviable thing." In the story Kipling also noted that the subaltern was expected to sit at the feet of his veteran captain "to be instructed in the dark art and mystery of managing men."[1] To understand the First World War US Army's conception of company-level leadership, we must first appreciate the prewar Regular Army's expectations for its junior leaders and its system and traditions for passing on to its young officers and NCOs those leadership traits and skills that would allow them to command in combat.

In the first decade of the twentieth century the US Army's conception of leadership was a complex mixture of long-held attitudes and assumptions and more recent and novel ideas about military professionalism. This was most apparent in the Regular Army's often paradoxical attitudes and assumptions toward its junior leaders. On one hand, the army continued to follow eighteenth-century ideas of leadership that focused on gentility, noblesse oblige, paternalism, deference, and an apprenticeship approach to officership. On the other hand, the turn-of-the-twentieth-century army also embraced professionalism and specialist education, meritocracy, and a drive for scientific management and efficiency. It was from this conflicted crucible that the officers and NCOs of the Regular Army developed their expectations of the attributes, proper background, and experiences needed for their wartime junior leaders and the role that the army expected them to play in combat in World War I.

The army of the World War I era had no set doctrine to define, codify, or explain the organization's views on leadership. In fact, the *Field Service Regulations of 1913 (FSR)*, the army's definitive doctrinal work at the beginning of World War I, made only vague and passing references to the command and management of soldiers in combat.[2] This oversight was not lost on certain members of the officer corps. In 1911 Robert Bullard, the future commander of the AEF's Second Army, noted, "As far as I know, hardly a suggestion is contained in the whole West Point curriculum of the need or value to a young officer of knowing or understanding either his soldiers or his fellow countrymen." Another regular officer complained: "We have lectures and manuals and treatises and textbooks on all sorts of technical subjects. On the subject of how to manage men, the most important subject of all, the young officer will find pretty nearly a barren field. A few paragraphs in Army Regulations, a few scattered magazine articles, and a general order or two compose the literature available. Neither at West Point, or our service schools, has this subject received the attention that it deserves." But while the prewar army lacked a set leadership doctrine, it still understood the centrality of leadership to combat operations and had developed its own institutional norms to define its expectations of officers and NCOs.[3]

While they lived in a world of massive technological, economic, and social change, it is interesting to note the degree to which the Regular Army's prewar junior officers and NCOs were shaped by concepts of leadership based on paternalism, noblesse oblige, and social deference that dated back to the founding of the Republic. For example, Frank McCoy, who would rise to command a brigade during World War I, reminded young lieutenants that "by law an officer is set down as an officer and a gentleman and with that high privilege there goes *noblesse oblige.*"[4] This concept of paternalism rested upon the assumption that if the leader looked after the welfare and comfort of his soldiers, they in turn would reciprocate with loyalty and obedience.

One also finds these institutional norms of paternalism and deference propagated in the semiofficial manuals for NCOs and junior officers written by veteran Regular Army officers for commercial sale. Since the army lacked its own doctrinal guides and manuals to instruct its company-level leaders in the art of leadership, these books became a key source for passing on the military's culture and institutional wisdom to its corporals, sergeants, and lieutenants. These works, such as James A. Moss' *Officer's Manual,* O. O. Ellis and E. B. Garey's *The Plattsburg Manual,* and M. B. Stewart's *Handbook for Noncommissioned Officers of Infantry,* transmitted the army's internalized, if somewhat officially unspecified, views of "correct" leadership and subordination to the generation of regulars who would raise, train, and lead the soldiers of the Great War.

Without a set doctrine or a process for young officers to learn the army's institutional norms of paternalistic leadership, the Regular Army officer corps expected lieutenants to learn how to lead soldiers during an extended apprenticeship in their first units. As Maj. Gen. David Shanks recalled in 1917, "Before our entry into the present war . . . a second lieutenant was assigned to a company, and he had

11

the benefit of learning by observation and experience. His captain was generally an officer who had received a certain amount of seasoning. The green subaltern had abundant opportunity to become acquainted with his profession gradually." An army study in 1908 estimated that an officer would spend nearly seven years as a second lieutenant and nine years as a first lieutenant before being promoted to captain. Thus, junior officers had the opportunity to learn the skills and traits of military leadership long before the novices would actually be allowed to command a company.[5]

This system of apprenticeship worked well in a small professional army in which junior officers could expect to serve for long periods of time at the same post. Even when the regulars faced a massive influx of enlisted men, civilians, and citizen soldiers into the army's commissioned ranks between 1898 and 1905, this reliance on on-the-job training managed to continue virtually unchanged into the Great War. During this period of expansion, the apprenticeship system also served as a means for inculcating the non–West Point "others" into the institutional culture of officership.

Although the Regular Army followed a traditional apprenticeship approach to training officers, it still had institutional gates in place to control who entered the profession. One of the greatest discriminators that the Regular Army used to evaluate the character and abilities of those seeking a commission was education. To the regulars, the proper education for officers generally meant attendance or graduation from a college or university. Gen. John Schofield acknowledged this predilection for college men by noting, "It is a feeling, and a very strong one, in favor of *education,* of qualification in all respects for the service which may be required."[6] The Regular Army's emphasis on college education stemmed both from the preeminence of the Military Academy in providing much of the prewar army's officer corps and from its underlying assumption that the best approach to warfare was one grounded in scientific examination and problem solving. The army's desire for college-educated officers also stemmed from social prejudices and a desire to keep "the wrong sort" from obtaining commissions and undermining the efficiency and exclusivity of the officer corps. Army leaders maintained that college graduates generally possessed the proper attributes, gentility, and character required for good combat leadership.

While seeking educated men to fill its officer ranks, the army also worked to improve and codify its professional standards. To answer congressional concerns that the army's seniority system promoted officers regardless of individual physical ability and professional competency, the army instituted a requirement in 1890 that all lieutenants and captains eligible for promotion first had to pass a rigorous series of physical and professional examinations. A year later, the secretary of war ordered that all officers below the rank of colonel be given efficiency reports by their superiors. The professional examinations tested the officers' tactical and technical competency through both a written exam and their hands-on ability to maneuver and deploy a company of soldiers. The examination boards for company-grade infantry and cavalry officers focused on the officers' knowledge of administration and *Army Regulations,* drill and *FSR,* fire discipline, basic military field engineering, military law, minor tactics, and basic topography. It was not unusual for the officers being examined to write over one hundred pages in answering the questions posed by the board.[7]

As did other professional organizations in the Progressive Era, the army tried to systematize its institutional process for ensuring competency and mastery of specialty knowledge through a system of postgraduate education. Beginning in the 1880s, the army began to improve professionalism in the service through a system of post, branch, and advanced tactical schools. These efforts to educate and professionalize the officer corps were further spurred by the army's lackluster performance in the Spanish American War.

In November 1901, Elihu Root ordered the army to establish a coherent and progressive school system designed to educate an officer from his precommissioning training to his ascension to senior field grade. Although Root's educational reforms never overcame the systemic problems associated with instruction at post schools, nor reached as many officers as he had hoped, he and military reformers such as Eben Swift, William Harding Carter, Arthur L. Wagner, and Franklin Bell were able to instill a deep and abiding professional ethos within the army's culture. It is little surprise that this "cult of professionalism" arose. To regular officers, professionalization would cement the regulars' position as the nation's preeminent source of military leadership and know-how and lay to rest decades of debate over the superiority of regulars over citizen soldiers. In the cult of professionalism, specialist knowledge, attained through education and honed during an apprenticeship overseen by expert practitioners, was the true and only path to professional enlightenment and salvation.

The drive for professionalism did not come without a cost. In the decade prior to the American entry into World War I, one-fifth of the infantry officers were absent from their commands because of details, staff assignments, and attendance at army schools. During the same

13

time period, up to one-quarter of cavalry and field artillery officers were absent from their commands for similar reasons. In all of the years between 1908 and 1916, the vast majority of those officers on detached service were company-grade officers. In 1910, for example, 85.4 percent of detached officers were captains or lieutenants. Despite complaints from field commanders over the absence of these officers, the reformers were not going to backslide on their commitments to establish coherent education and staff systems and to purge the army's ranks of those failing to adhere to the cult of professionalism. In fact, Secretary of War William Taft rebuked those in the army who grumbled over the number of officers on detailed service. In 1905 he noted, "The remark is sometimes heard in the Army that an officer's time now is completely taken up in going to school. With due allowance for exaggeration in this statement, it is well for the Army and for the country if it be true."[8]

The tapering off of criticism of the educational and staff systems between 1908 and 1916 reflected the army's general acceptance of the cult of professionalism. Senior officers accepted the premise that it was tolerable for junior leaders to be gone from their units for long periods of time to achieve their required education and that the frequent rotation of officers from positions of command was an unfortunate but necessary effect of this policy. Thus, the Regular Army entered the Great War with a reverence for professional schools, a willingness to pull officers from their units in the name of education, and an intolerance for those unable to meet the standards of know-how and efficiency demanded by the cult of professionalism. These beliefs would later have a powerful, and sometimes baleful, influence on combat leadership in the AEF.

As with the officers, to understand the AEF's approach to the leadership of its corporals and sergeants we must first appreciate the prewar Regular Army's expectations for its NCOs. Undoubtedly, the Regular Army paid much less attention to the education, selection, training, and development of its NCOs than it had lavished on its prewar officers. Thus, while James Moss would state that "the noncommissioned officers are the backbone of an army" and declare that "experience has shown that the efficiency, discipline, and reputation of a command depend to a great extent on its noncommissioned officers," those truisms were based more on tradition and serendipity than a concerted effort to professionalize the Regular Army's NCO corps from 1900 to 1917.[9]

As an institution, the Regular Army had long understood the correlation between having a strong cadre of NCOs and the smooth running of its units. It expected the NCO corps to train new recruits, administer the routine details of army life, oversee the "good order and discipline" of its soldiers, and assist its officers in the command and direction of its soldiers in combat. Company and regimental commanders generally expected their NCOs to identify and solve most of their soldier's everyday problems or lapses in discipline before their officers were even aware that such issues existed. At one lecture in 1916, an officer informed his NCOs that they were "valuable in their capacity as instructors, as disseminators of technical information, but they are doubly valuable as leaders, to whom the men look for moral, social, and intellectual inspiration."[10] However, despite the recognized importance of NCOs, from 1865 through 1916 the Regular Army had no formalized method for selecting NCOs for its infantry and cavalry units.

Historian Ernest Fisher Jr. argues that all NCOs within the company were "commander's men" who were selected by and served at the pleasure of their company commanders. This system of the company commanders' "owning" their NCOs was enshrined in Article XXX of Army Regulations, which stated that NCOs were to be appointed by the regimental commander "on the recommendation of their company commanders." Army Regulations also aided commanders in selecting NCOs by allowing them to "test the capacity of privates for the duties of noncommissioned officers" by temporarily appointing select enlisted men to the position of an NCO. James Moss admitted that this system deliberately set out to throw the newly minted NCO "into deep water" to find out if he would "sink or . . . swim," but believed that it was still the best way to determine if the soldier would "make good" as an enlisted leader. The goal of these regulations was to build harmonious units in which a company commander selected NCOs who had ability and would best complement the captain's leadership style. This also meant that while NCOs were given their warrants from the regimental commanders, their company commanders still managed to exercise a great deal of control over their NCOs by their power to demote those corporals and sergeants who failed to live up to their expectations. In other words, those who the company commander promoted, he could just as quickly demote.[11]

In the prewar regulars, this idiosyncratic method for selecting NCOs was not a great hindrance. Company commanders were usually mature veterans with much experience and command time within their units. This allowed the long-service captains to gain a deeper appreciation for the abilities of their soldiers and to develop a discerning eye for nascent leadership talent in their pool of potential NCOs. Some officers, however, recognized the inherent drawbacks to this system. In 1908 the commander of the Department of the Columbia remarked in his report to the secretary of war, "There is no carefully thought out system of promotion for noncommissioned officers. In one regiment it is done one way, in the next another, and even in the same regiment there may be a dozen ways of appointing and promoting noncommissioned officers, depending on the ideas of individual company commanders."[12] The same year, Maj. William Burnham recommended that the army solve this problem by establishing a three-year course of instruction for all NCOs and that "appointment to the grade of corporal be made after a written competitive examination by a board of officers."[13] None of these recommendations gained traction within the War Department, and the ad hoc system for the selection and promotion of NCOs remained in force when the army entered World War I. Unfortunately, this trial-by-doing approach, which worked so well in the small professional Regular Army, was not a system that was well suited to wartime conditions requiring a rapid expansion of a mass army and its NCO corps.

Although the Regular Army touted the NCO as being the backbone of the army, it was somewhat imprecise in actually defining its institutional expectations of its corporals and sergeants. For example, the *FSR,* the army's only definitive prewar doctrinal source, mentions the duty of NCOs only three times within its pages; all were for relatively unimportant tasks. As was the case with so many leadership subjects between 1890 and 1916, the authors of the era's many semi-official publications filled a void that the War Department itself had failed to address or codify. In his *Noncommissioned Officer's Manual,* Moss maintained that the role of the company NCO was to correct errors, prevent the commission of offenses, enforce order in quarters, suppress disorderly conduct, ensure the respect and obedience of his soldiers to officers and NCOs, and "assist company commanders in carrying out orders."[14] To accomplish this, Moss argued, the NCO must exhibit efficiency, initiative, promptness, obedience, and loyalty. An NCO's authority would also rest on his superior tactical and technical skills.

An examination of Moss's definitions of initiative and obedience reveals the Regular Army's somewhat ambiguous expectations of its NCOs. Moss defined initiative as doing things without being told to and obedience as obeying "promptly and fully all orders and instructions received from superiors." In these definitions Moss seems to wrestle with the acceptable limits of NCO leadership. To Moss, initiative rested upon the NCO's understanding of the wishes of his su-

17

Maj. James A. Moss was a prolific writer of semiofficial military manuals. His *Officer's Manual* was one of the most popular publications bought by novice officers during the Great War. *(Photograph from the author's collection.)*

periors and acting within those limits without constant supervision. In explaining obedience, Moss warned, "Whether or not you like the order is neither here nor there. Your business as a soldier . . . is to obey all orders, and to do so willingly, faithfully and promptly, *without excuses or explanations.*" He went on to scold, "Remember that nine times out of ten the superior giving the order knows more about the matter than you do and is probably in possession of information that you know nothing about. If a superior makes a mistake in giving an order, it is his lookout, not yours." Moss makes clear that while the NCO was a critical member of the unit, his officers expected him to be a doer rather than a thinker. Regrettably, these expectations did not mesh well with the evolving realities of the modern battlefield.[15]

The issue of how the Regular Army selected and trained its junior leaders, and its expectations of the attributes and skills these officers and NCOs needed to possess, mattered exactly because the realities of combat radically changed between 1865 and 1914. During that period the armies of the United States and Europe all struggled to understand the effects of a vast array of new weapons upon their doctrine, tactics, and training. All the major powers were consumed by the central issue of how to achieve the decisiveness of the offensive without suffering prohibitive casualties. The central dilemma of this "men against fire" debate centered on the issues of mass and command and control. The close-order regimental formations used from 1700 through the mid-1800s could be controlled with relative ease by the voice and drum commands of field-grade officers. Junior officers and NCOs merely assisted the field-grade officers in maintaining the cohesion, discipline, and momentum of the unit in battle. Company-level leaders seldom had the chance or need to exercise independent command.

As demonstrated at Cold Harbor, Gravelotte–Saint-Privat, Plevna, and other battles of the late 1800s, the withering firepower of defenders made short work of massed regimental formations. This reality prompted the major western armies to experiment with thinning and spreading out their attack formations and to use tactics that moved their attacks forward in brief bounds using the terrain as cover. The idea was to present the smallest possible target to the defender until the attacker's last rush would overwhelm the defenses. The seemingly easy solution to the attacker's quandary was unfortunately wrecked on the rock of command and control. As the attacking commander thinned and spread his formations, he quickly found it nearly impossible to direct the fire and maneuver of his subordinates or to mass either the firepower needed to suppress the defender or the numerical superiority needed to overpower the defender during the final assault.

As the expanse and deadliness of the modern battlefield grew, the new warfare began to change the demands that armies placed on their junior combat leaders. The need to disperse combat formations exacerbated existing problems with battlefield command and control. This change demanded captains, lieutenants, and NCOs able to exercise initiative and a mounting level of independent action. A number of American officers clearly understood these new realities. As early as 1882, Capt. Francis Greene acknowledged that the open-order infantry formation "throws great responsibility upon the senior captains" and that success in battle now depended "upon the skill of the commanders of small units." Similarly, in 1909 Capt. George Baltzell observed, "It is self evident, that . . . the display of judgment among the minor leaders . . . is one of the demands of modern warfare."[16]

Although American officers may have recognized the increasing importance of junior leaders in combat, one of their greatest challenges was how to reconcile the need for independence and initiative in officers and NCOs with the senior commanders' demand for unified action, control, and proper subordination. Reflecting this tension, the *Infantry Drill Regulations* (*IDR*) stated: "When circumstances render it impracticable to consult the authority issuing an

18

order, officers should not hesitate to vary from such order when it is clearly based upon an incorrect view of the situation, is impossible of execution, or has been rendered impracticable on account of changes which have occurred since its promulgation." However, the *IDR* offered senior commanders the ability to have the best of both worlds by also declaring, "In the application of this rule the responsibility for mistakes rests upon the subordinate, but unwillingness to assume responsibility on proper occasions is indicative of weakness."[17]

The inherent conflict between subordinate's initiative and superior's control meant that the army consistently sent its junior leaders mixed messages on its institutional expectations of combat leadership. On one hand, J. Franklin Bell decried officers "trained in the old tradition of prompt and unquestioning obedience, without even a desire to understand," and argued that subordinates must be given more latitude in commanding their units. On the other hand, O. O. Ellis and E. B. Garey lectured officer candidates, "As soon as you obey properly, promptly, and, at times, unconsciously, the commands of your officers, as soon as you cheerfully give up the pleasures and personal privileges that conflict with the new order of life to which you have submitted, you will then have become a disciplined man." In truth, effective combat units had to be able to meld the two together. Soldiers and leaders had to obey orders to accomplish their assigned missions, and also they had to be willing and able to make snap decisions based on changing battlefield realities.[18]

As an institution, the army did a poor job of reconciling these competing demands within both its doctrine and its leadership training. The senior officers' demand for control and predictability all too often trumped their subordinate leaders' efforts to exercise their initiative. In 1912, another infantry officer argued, "The essential point needed in our infantry today is more independence for organization commanders and a direct holding of them responsible for results attained. To do this they must be given the freedom of method and not have every moment of the day mapped out by superior authority. This only kills all initiative so essential to war." In 1916, a major with the Mexican Punitive Expedition bewailed, "Experience has led me to believe that the with-holding of initiative from subordinates is one of the very worst faults to be attributed to the field officers of our service. . . . Many an officer who has been an excellent troop commander, makes an indifferent major, because he interferes too much with the initiative of his subordinates."[19]

Although junior officers chafed under the smothering control of their superior officers, they seem to have been just as guilty of this offense when it came to their NCOs and soldiers. In 1916 the editor of the *Infantry Journal* lamented, "It cannot be denied that our noncommissioned officers are given too little scope for the development of their initiative as group leaders in the average company of the regular service." Maj. James Chester chastised company commanders who "spend too much of their time with their companies" and thus set the condition that within their units "they are everything and their noncommissioned officers are nothing."[20]

The tension between initiative and control was a central theme in the American army's tactical doctrine. As were their European counterparts, American officers were bedeviled by the need to address the challenge of the mass industrial battlefield. The US Army's solution was to place its faith in infantry firepower and forceful offensive action. This focus on the offense meant that the army stressed the need for an aggressive spirit in leaders and soldiers, the ability to use infantry firepower to set the right conditions for the assault, and the skill and willingness to force the decision by use of the bayonet. Although he viewed the bayonet as the weapon of the tactical coup de grace, Capt. Charles Crawford confidently proclaimed, "The end of all warfare is attained by breaking up and destroying the enemy's forces in battle, and the chief instrument used is the small arm rifle." In a lecture at the Army War College, Lt. Col. R. K. Evans amplified this argument by noting, "Fire action is the controlling factor in deciding battles," and he declared, "Over 80 per cent of the men that fall in battle go down under infantry fire."[21]

The army's fixation with infantry firepower superiority exacerbated the tensions between control and initiative. The weight that army doctrine placed on gaining and maintaining mass small arms fire dictated very strict direction by field officers of the targeting and orchestration of the firefight. It is only a slight exaggeration to argue that in the American infantry doctrine the chief role of maneuver was merely to move units to a place that allowed them to gain a firepower advantage and to set the conditions for the final charge. The dictates of fire superiority tended to push officers at all levels toward micromanaging their subordinates in an effort to maintain strict control of the combat action.

This obsession with controlling fires was deeply embedded in the *IDR* and forcefully argued in the professional writings of regular officers. The *IDR* delineated the roles and responsibilities for leaders at each echelon and provided detailed instructions on the conduct of the firefight. The *IDR* made clear the paramount importance of making the fire of the individual soldier an extension of the will of the

20

commander. It unequivocally stated, "The best troops are those that submit longest to fire control. Loss of control is an evil which robs success of its greatest results. To avoid or delay such loss should be the constant aim of all."[22]

Instructors at army schools and writers in the army's professional journals constantly harped on this theme. Capt. Henry Eames warned his students at Fort Leavenworth, "Fire Discipline is different from any other kind of discipline and it is vastly more important, and much more difficult to instill into the soldier." Its goal was to get the soldiers to perform "without any conscious mental activity" so the "very muscles may instinctively obey the word of command."[23] Although all military training basically seeks to instill a reflexive response that allows soldiers to overcome fear and to act rapidly and instinctively to battlefield events, the army of the turn of the twentieth century took this Pavlovian concept to an extreme.

For all the talk by some officers of the need for leaders to understand the psychology of their soldiers, to appeal to their men's reason, and to admit that the modern battlefield prevented the direct supervision of individuals, the army's compulsive obsession with control actually seemed to be a throwback to the rigid automaton discipline of the earlier era. This reality encouraged leaders to attempt to micromanage their subordinates. In 1909 Capt. George Baltzell wrote that in the firefight the main duty of lieutenants and NCOs was "not a mere perfunctory repetition of commands, but a close supervision of their execution" and that this close control should be taken "even to the point of examining the sight adjustments [of individual soldiers]."[24] Despite the dispersion of the modern battlefield and subsequent increase in the individual infantryman's freedom of action, the army still expected its junior leaders to adhere closely to the will of their superiors and, in the process, exercise a strict and close control of their subordinates.

To produce the control of action that the army expected, it placed much emphasis on close-order drill. Close-order drill has long been a tried and true method for training new recruits in automatic response to orders and to act as part of a unit collective. The turn-of-the-century army, however, continued to stress close-order drill of its units above and beyond the dictates of recruit training. In his 1914 semiofficial manual *Fundamentals of Military Service,* Lincoln Andrews stated that the purpose of continual and constant close-order drill was for "*training your minds and bodies into HABITS of precise unhesitating obedience to the will of your leader. . . .* Then when the stress of battle comes, and men's faculties are paralyzed by the un-

21

wonted roar and loss of life and straining fear, they may still be controlled because H A B I T *has made obedience automatic and the easiest line of action.*"[25]

George Baltzell also justified the focus of close-order drill on control over the individual soldier by noting, "In the heat of battle, the average man fires on in an almost cataleptic state, his mind incapable of ordinary obedience. Obedience under such conditions must, therefore, be instinctive and the result of long continued habit. This habit can only be learned by strict close order drill on the parade ground. . . . Any carelessness allowed on the parade ground will bear ugly fruit on the battlefield where we require that under whatsoever stress of circumstances, danger and death, when the soldier hears the word of command his muscles if not his mind shall instinctively obey." This belief that close control by leaders and the conditioned response of soldiers could somehow overcome, or at least mitigate, the fog and friction of combat was a touchstone of army training practices prior to the Great War. Soon after the United States entered the war, Frank McCoy admitted, "The Great War has been terribly hard on the textbooks. About the only military china that has not been smashed are the primary functions of discipline and its drill, *drill,* D R I L L ." This approach to training (especially close-order drill) guided the wartime instruction of the AEF's leaders and soldiers.[26]

The unfolding events of the Great War did little to change the American army's doctrinal assumptions or expectations of small-unit leaders. From the start of the Great War in 1914 it is clear that the US Army closely followed the military developments on the Western Front. The War Department received a constant flow of reports from its military attachés in Europe, and its professional journals published a steady stream of accounts of the conflict from American observers and European participants. Despite this flow of information about the war and its larger tactical, organizational, and technological implications, there is little indication that these reports led to any substantive changes to the US Army's doctrine or patterns of military thought before 1918. Although fiscal, operational, and political constraints prevented the army from fielding new weapons or experimenting with new organizational models during that period, there were no tangible restraints on the officer corps' ability to grapple intellectually with the doctrinal challenges being evidenced on the Western Front. The saddest indictment of the prewar officer corps was that it continued to cling to doctrinal assumptions long after the evidence from Europe had proven them false.

From the very outbreak of the war, American military attachés reported the military situation and doctrinal developments and innovations emerging on the European battlefields. Maj. Spencer Cosby, the military attaché in Paris, was one of the more keen American observers of the combat on the Western Front. Some of the major trends that he uncovered and reported back the American General Staff were the vast rise of casualties caused by artillery and machine gun fire and the challenges presented by trench warfare. He was able to visit French trenches on an active part of the front and gave a pointed account of the devastation left in the wake of the firepower of the new warfare.

23

In November 1914 Crosby reported that a French officer had estimated that 75 percent of soldiers killed on the Western Front were killed by artillery fire. Nine months later, he relayed to Washington the staggering casualties that the French had suffered in the fighting around Arras in May and June 1915. Although he placed French casualties at around 60,000 to 85,000 for the campaign, he admitted that some of the figures ranged to as high as 200,000 men. In one division that had suffered 800 men wounded in the fighting north of Arras, he noted ominously that "only two were wounded by rifle bullets, all the others by shell and grenades."[27] Given the fact that the American Regular Army's strength in 1914 was a mere 92,482 men, and its doctrine underpinned by a faith in small arms fire, Cosby's reports should have carried more portent.

In addition to reports of their own experiences at the front and of discussions with Allied officers, the military attachés also sent back to the American General Staff translated doctrinal manuals and circulars that detailed the evolution of tactics on the Western Front. On 11 November 1915 Capt. J. W. Barker submitted a report containing a copy of a provocative French circular entitled, *Study of the Attack in the Present Period of the War*. Although Barker did not include the name of the author of the study, it was written by Capt. André Laffargue, a veteran company commander. Laffargue's study offered a detailed discussion of the challenges of attacks against fortified positions as well as some suggestions for overcoming them. This penetrating analysis offered a number of gems on minor tactics, combined arms coordination, small-unit training, and combat leadership that were valuable to all company- and battalion-level leaders attempting to understand the realities of the new warfare.

Although Laffargue's study was published in its entirety in the *Infantry Journal* in September 1916, there is no evidence in American doctrinal publications or in the army's training plans of 1917 that the work had any real influence on American military thought.[28]

Although the Frenchman's work was certainly not the be-all and end-all of World War I minor tactics, it at least offered the Americans a model or foundation from which they could have built a tactical doctrine and training plan that was more realistic than those it followed in 1917.

Another French manual, *Instructions on the Offensive Conduct of Small Units,* ultimately received a more official acceptance within the American army. The French originally published the manual in January 1916, and the American attaché forwarded a translation to Washington on 22 March 1916.[29] The manual drew upon Laffargue's earlier manual as well as the combat wisdom of other French officers. Unfortunately, the work languished in the War College Division until after the United States entered the war. The War Department hastily published and issued the work in May 1917, but because the tactics within the manual involved synchronized attacks using artillery, light and heavy machine guns, hand and rifle grenades, one-pound guns, and other implements of modern warfare that the US Army lacked, the publication's utility was limited.

When the United States entered the Great War, the army's expectations of junior officers and NCOs were a jumbled mélange of traditional and modern practices. When it came to leadership, the army's long-standing concepts of gentility, paternalism, and noblesse oblige contended with the burgeoning cult of professionalism and the drive for scientific management and efficiency. On one hand, the Regular Army expected its officers to possess the intellectual know-how derived from civil and professional education; on the other hand, it expected its junior officers to develop their tactical and leadership abilities during a prolonged apprenticeship under the tutelage of their long-service superiors. The NCO corps remained firmly rooted in eighteenth- and nineteenth-century practices. Infantry sergeants and corporals gained their professional knowledge almost exclusively by on-the-job training. Without any formal means for selecting and training NCOs, the sergeants retained their positions based on the whims and desires of their line officers. Although the ad hoc method of training junior officers and NCOs in unit apprenticeships worked relatively well in the peacetime army, it would prove nearly impossible to replicate during a mass mobilization.

The US Army's tactical doctrine and its expectations of junior leaders in combat were also beset by uncertainty and internal contradictions. For over thirty years its officers had striven to understand the dynamic changes in warfare that had been wrought by a host of new and collectively more lethal weapons. The officer corps' realization that modern weapons forced units to thin and spread out to survive on the battlefield led many to believe that the new warfare required junior leaders able to exercise initiative and semi-independent command outside of the traditional direct supervision of their superiors.

Simultaneously, the failure of changes in battlefield communications to keep pace with weapons innovations exponentially increased the difficulty for leaders at all levels to command and control their units effectively. However, these developments were in constant conflict with doctrinal trends and command traditions that sought to retain as much centralized control of small units as possible in the hands of superior officers. The tension between the initiative of junior leaders and control by their superiors was never resolved before the war, but the American tendency, reinforced by its doctrine and training methods, leaned more heavily upon the side of centralized control. All of these factors ultimately meant that the US Army entered World War I without a clear understanding of modern war or a cogent vision of what the new warfare demanded of combat leaders in small units.

3 "We Find Ourselves in Need of a Vast Army of Officers"
The Stateside Selection and Training of Officers

AFTER THE UNITED STATES HAD BEEN AT WAR for more than eight months, the editor of the *Infantry Journal* had grown tired of the constant carping in the officer corps about the training of officer candidates for the wartime army. He rebuked his readers by reminding them, "We find ourselves in need of a vast army of officers. Two alternatives present themselves—to follow our old policy of selecting men for social, personal or political reasons, or of selecting on a basis of individual merit."[1] His remark highlighted the great break from historic practice that the army had engineered in its plans for selecting and training its wartime officer corps.

Unlike previous wars, in which units were officered largely by men selected by state governors, elected by the men of the ranks, or directly commissioned by the federal government, the majority of the company-grade line officers for the AEF would be subjected to a more or less standard system of selection and training, all created and controlled by the Regular Army. The Great War was to be the grand Uptonian moment: the chance for the regulars to prove the superiority of their methods and their "proper military policy" over the appointments of the National Guard or other ad hoc methods of obtaining officers.

The question of how to mobilize a wartime cadre of officers was not a new one. In his 1913 annual report, Chief of Staff of the Army Leonard Wood warned, "I . . . invite attention to the necessity for building up, with as little delay as practicable, a reserve of officers qualified to serve as company officers for reserves or volunteers. If we were called on to mobilize to meet a first-class power, we should require immediately several thousand officers; where are we to get them? This is a matter of vital importance, and one which should be attended to at once and not left to the rush, hurry, and confusion proceeding a war."[2] In theory, Wood's question, "Where are we to get them?" should have been answered by the federalization of National Guard officers and the recruitment of military-trained students from the nation's land-grant colleges. Unfortunately, neither of these sources delivered the quality or quantity of trained officers and officer candidates that the army needed.

In the middle of the Civil War, Congressman Justin Morrill proposed a bill offering federal land grants to colleges and universities "where the leading object shall be . . . scientific and classical studies, and including military tactics." Although the Morrill Act should have provided a corps of potential officers with at least a basic understanding of military discipline and tactics, the splitting of responsibility for the implementation of the act between the War and Interior Departments, and the lackluster support of military instruction by both the army and the college administrations, prevented the land-grant colleges from providing ready officers in any meaningful sense. For example, the *War Department Annual Report* of 1913 noted, "At the last annual inspection it was generally found that sufficient progress had not been made in practical instruction; that too much time was spent in close order [drill] and ceremonies at the expense of good theoretical instruction and practical fieldwork." In 1914 Capt. Richard Stackton argued, "The average ex-military-school student is not . . . suited for a commission. . . . In most institutions, mere *drill* is given, and the youth graduates with the impression that a faultless parade and the ability to form a line of skirmishers and fire a few blanks . . . are the sole requirements of a complete military education."[3]

The utility of using the land-grant colleges as a source of reserve officers was further hindered by the War Department's own inefficiency or indifference. Until 1915 the army made no effort even to maintain a list of those students who had graduated from college military programs. In 1916 the War College noted that during the preceding ten years 287,952 students had received some degree of military training from Regular Army officers at American colleges and universities. The army considered only 44,592 of these students to have completed the entire course of its college military programs and admitted that of those who graduated before 1912, "nearly all have, no doubt, lost touch with things military, and have consequently forgotten what little they learned before their graduation."[4]

When it came to building a reliable reserve of officers, the National Guard also proved a rather brittle and unreliable instrument. In 1903 the Act to Promote the Efficiency of the Militia, generally known as the Dick Act, was the first in a string of legislation designed to transform the National Guard into a reliable military force. However, despite these reform efforts, a wholesale change in the professionalism of National Guard officers and NCOs proved elusive in the years leading to World War I. Due to the lack of effective federal oversight of the professional education of guardsmen, a shortage of qualified Regular Army instructors, and the limited time that guard units could devote to officer and NCO training, the trend toward

27

increased professionalization touched only a limited number of guard officers before 1916. As late as January 1916 the chief of the Militia Bureau, Maj. Gen. Albert L. Mills, admitted to the House Committee on Military Affairs that the guard had failed to meet the goals of the Dick Act "by a considerable measure."[5]

The shortcoming of guard officers was made evident during the National Guard's mobilization during the Mexican border crisis of 1916 and the subsequent call-up in 1917. In the spring and summer of 1917 the War Department called 12,115 guard officers to active duty. Between 5 August 1917 and 10 May 1918, the army had discharged 511 of these officers for physical disability, efficiency boards had removed 352 more, and 648 were encouraged by the army to resign. During the war only 6 percent of the army's officers were to be National Guardsmen. In the final analysis, historian Timothy Nenninger was on the mark when he noted, "The National Guard of 1917 was nearly as unprepared for war as the state militias had been in 1898."[6]

Before 1917 the Regular Army also pinned its hopes for creating reserve officers on civilian military camps of instruction. Between 1913 and 1916 the "Plattsburg" camps of instruction trained over twenty thousand businessmen and college and high school students. The actual military value of these training camps is debatable. The training in the camps stressed close-order drill, route marching, basic marksmanship, and the bare basics of field craft and tactics. Little in the training reflected the realities of modern war as was being waged on the Western Front. Furthermore, the short duration of the camps and the large number of attendees provided little opportunity for the participants to serve as leaders or learn the responsibilities of company-grade officers.

It is interesting to note that although the army had granted nearly eight thousand Plattsburgers reserve commissions before 1917, it was so unsure of the quality of their training that it still required them to attend the three-month-long Officer Training Camps after the war began. In a damning indictment of the quality of screening and training of the Plattsburg officers, the army admitted that many of them were "found entirely unqualified for commissioned grades and . . . were reduced in grade."[7] In many cases the army demoted or removed these officers for incompetence or unsuitability to command.

From the moment the United States entered the war, the crush of time and events overseas influenced the way the nation mobilized its forces and trained its officers. Within weeks of the United States' declaration of war, the Wilson administration made the decision that the nation would send an expeditionary army to fight in France. The bulk of this force would be composed of draftee National Army units, with the remainder made up of existing National Guard and Regular Army units brought up to strength with volunteers and conscripts. Congress passed the Selective Service Act in May to fill the enlisted ranks of the mass army and assumed that the measures it had authorized in the National Defense Act of 1916 provided the means for raising an officer corps. However, authorizations on paper are not the same as warm bodies in uniform, and the administration's plan to build an expeditionary army of two to four million men brought the issue of who would lead the legions to a head. As Leonard Wood had predicted in 1913, the war found the army short of many thousands of officers, and the "rush, hurry, and confusion" that he had feared had become a bewildering reality in 1917.

With the pressing need for officers, the army scrambled to create a system for identifying and training candidates for commission. The regular officer corps fully understood the great limitations of the Plattsburg camps in turning out trained officers. Shortly before the United States entered the war, the army had taken its first hesitant steps towards developing a more coherent and reasoned approach to expanding the officer corps. In the fall of 1916 and the winter of 1917, the army held two officer training camps at Fort Leavenworth to instruct and commission provisional lieutenants for the Regular Army. The courses were three months in duration, and the General Staff intended that "the system of discipline and preliminary training adopted [follow] that of West Point as far as was practicable." The first class, which ran from September through November 1916, commissioned 386 provisional lieutenants, and the second, which was held from January through April 1917, commissioned 338 more. These camps were the model for subsequent wartime officer training.[8]

Few regular officers had any illusion that these first tentative measures were a panacea for the ills of officer training or a solution to the overwhelming issue of scale. It was one thing to train 724 provisional lieutenants in six months, but it was quite another to identify and train 200,000 more to meet the pressing demand for lieutenants, captains, and majors. From the start, few regular officers deluded themselves that ninety days of training would produce a finished product. James McAndrews warned the second class of provisional officers on 17 April 1917, "The time given you is all too short for the ground that must be covered. Officers fitted to command first-class troops cannot be the product of a course of three months' training and instruction, no matter how strenuous it may be. But if you do your part, three months are long enough to give you a good start in your profession, to give you something of an insight into the duties of subalterns and above all to give you a safe foundation upon which to build your future efficiency."[9] The best that could be hoped for was to give the officer aspirants a sound start and then pray that the novices would have the time to hone their leadership and tactical skills within their units before being committed to combat.

Although the training camps for provisional regular lieutenants had allowed the army to sort out some of the issues of officer training, the spring of 1917 was still marked by rush and confusion. This was reflected in the steady stream of messages that flowed from Brig. Gen. Joseph Kuhn, the chief of the War College Division, to Army Chief of Staff Hugh Scott. As the War College did double duty as a school for staff officers and as a planning agency of the General Staff, Kuhn was responsible for developing the mobilization and training plan for officers. Kuhn's first report to Scott on the status of officer training offered a gloomy assessment: "There will probably not be equipment available for the training of any forces except the Regular Army and National Guard before the latter part of next fall. For that reason the establishment of the training camps for reserve officers can well be postponed until the first of June. This will also make it possible to use at these camps officers now detailed to schools and colleges and a large number of officers from the Military Academy." The War College staff recommended that the training camps be established under the provisions of section 54 of the National Defense Act of 1916 and that the period of training was to be for three months, with the course "to be based on that given to provisional second lieutenants at Fort Leavenworth." Although Kuhn's assessment was accurate, the ugly press of events meant that the officer camps would start much earlier than he had either wanted or anticipated.[10]

Only four days after Kuhn's initial report, the adjutant general, Brig. Gen. Henry McCain, sent a warning order to the commanding generals of the army's departments directing them to establish sixteen OTCs at fourteen posts and camps within their areas. The OTCs were to be located in camps whose locations could serve a number of regional division mobilization sites. Each OTC was to train a maximum of twenty-five hundred candidates and, in theory, provide enough officers for at least one division.[11]

On 23 April, McCain sent the commanders additional guidance on establishing the new camps. It mandated that the first OTCs would open on 8 May 1917 and that the camps had to be ready to receive the candidates by 15 May. This gave the camp organizers only a week to get their sites up and running. McCain stated that the intent of the first OTCs would be to provide the required officers for the first sixteen divisions "by the time the necessary machinery can be put in motion for procuring the enlisted men."[12] That meant that the departmental commanders had only fourteen to fifteen weeks to establish a system for screening applicants, receiving and in-processing those selected, building the camp infrastructures, training and evaluating the candidates, and commissioning and posting the graduates before the arrival of the first draftees.

From the beginning the War Department had to fight off those who would bypass the training camp system and return to the former ad-hoc methods of directly commissioning officers. Hugh Scott recalled that despite political pressures to issue direct commissions, Secretary of War Newton Baker's "close adherence to the policy of making applicants for commission earn them in camp, kept our corridors so free from politicians and their constituents that it was as quiet in my office as on a Sunday morning in time of peace." Not all of the departmental commanders were in full agreement with Baker's resolve or Scott's desire to retain a quiet Sundaylike office. The officers responsible for establishing the camps certainly understood the monumental tasks before them, and some believed that the press of events demanded that the army follow an easier path to officer mobilization. The commander of the Southern Department, for example, telegrammed Kuhn that he was "overwhelmed with applications for commissions as lieutenants" from college students and stated that "these would make [the] best officers we could get for immediate commission into regular service and would be efficient for any class of troops."[13]

31

With the backing of Baker and Scott, Kuhn stuck by his recommendation that all such applicants for commission still be required to attend an officer training camp as a precondition for commissioning. Whatever the lure of expediency and easy solutions, few in the higher echelons of the War Department were going to let the overwhelmed local commanders disrupt the army's great Uptonian moment. This commitment was reflected in the statistics. In raw numbers, those commissioned out of the OTCs ultimately accounted for 48 percent of all wartime officers. After factoring out physicians, chaplains, and those civilians given direct commissions on account of their technical skills, the OTC graduates made up 74 percent of the war's officers and the vast majority of the army's company combat leaders.[14]

Although the War Department senior staff was unwavering in its determination to follow through on the training camp system, it was a bit vaguer in its guidance in describing the attributes that local commanders should seek in the applicants to their camps. In selecting candidates, the adjutant general advised, "These should be preferably mature men and the most experienced natural leaders that the country possesses. . . . With the basic experience [of the OTCs] supplemented by natural aptitude for handling men as demonstrated in business or otherwise, a splendid corps of 10,000 reserve officers should be available by the middle of July."[15] The only hard rule was that the candidates had to be American citizens between the ages of twenty years and nine months and forty-four years to attend the camps. However, following long-standing army preferences, McCain did recommend that the commanders seek college students, as they were "especially fitted" for selection to attend the camps.

The department commanders and their divisional subcommanders had to shoulder most of the responsibility for selecting the camp attendees. This task would have been impossible in the short amount of time available for the task if it had not been for the efforts of the Military Training Camp Association (MTCA). The MTCA had been formed by Grenville Clark in 1916 to aid in recruiting and lobbying for the Plattsburg training camps. Shortly after the declaration of war, Clark and the other members of the MTCA's executive committee offered the War Department its files of past and prospective Plattsburg candidates and its administrative assistance in recruiting and communicating with potential officer candidates. As the War Department had been woefully negligent in even maintaining lists of those students who had received military training in college, Secretary Newton Baker jumped at the MTCA's offer.[16]

Despite the short timelines and occasional bureaucratic obstacles, the army generally was able to attract a very high quality of officer candidates for its 1917 OTCs. It is possible to gain an idea of the general education and experience level of the candidates who attended the first two OTCs by examining the backgrounds of a random sample of the graduates of the Fort Sheridan OTC. In 1920 the Fort Sheridan Association, a mutual aid organization formed by the fort's OTC graduates, published *The History and Achievements of the Fort Sheridan Officers' Training Camps*. This publication contained biographic sketches for most of the 267 of the Fort Sheridan OTCs' candidates who had died while in military service between 1917 and early 1920. From these sketches one can gain an idea of the ages, education, professions, and previous military experience of those who sought to become officers in 1917.[17]

Of the graduates of the first two sessions of OTCs at Fort Sheridan who died in the war, 3 were majors, 25 were captains, 103 were first lieutenants, and 129 were second lieutenants. Six candidates washed out of the course but later became NCOs, and one died during the training. In education, two of the majors were college graduates and one was a graduate of public school. Of the captains, 71.4

percent had some college education, with 62.5 percent being college graduates. Nearly 79 percent of the lieutenants had some college education, with 57.5 percent being college graduates. The difference between the number of captains and lieutenants who had graduated from college was a result of the number of younger men who had left school when the United States entered the war to attend officer training. Slightly over 20 percent of the lieutenants and 8 percent of the captains listed their profession as "student" when they entered the OTC.

The selection of this high percentage of men with some degree of college education to attend officer training was in line with long-standing Regular Army assumptions that education conditioned the officer's mind to absorb and process knowledge and thus gave him the faculties to continue to learn his professional skills and overcome any shortcoming in his initial training.

Ralph Perry, the secretary of the War Department Committee on Education and Special Training, reflected this belief when he noted that the OTC candidates were "by education, experience and natural aptitude especially qualified for leadership." He admitted that while the novices "were not trained officers," they were still "picked men who had mastered the rudiments and knew how to profit by the experience and ordeal that awaited them." In his 1917 report to Congress, Secretary Baker echoed these beliefs, praising the fact that the nation's colleges "poured out a stream of young men whose minds had been trained in the classroom and whose bodies had been made supple and virile on the athletic field." He gushed, "They came with intelligence, energy, and enthusiasm and, under a course of intensive training, rapidly took on the added discipline and capacities necessary to equip them for the duties of officers."[18]

The occupations of the officers in the Fort Sheridan sample also reflect the army's existing assumptions about the social class and job experiences it wanted in its reserve officers. The occupations of the first two Fort Sheridan OTC graduates who achieved the rank of captain in the war, and the number in each occupation, are as follows:

33

Businessman, 4	Manufacturer, 1
Engineer, 3	Newspaperman, 1
Farmer/stockman, 2	Pipeline gauger, 1
Student, 2	Regular Army enlisted, 1
College administrator, 1	Real estate, 1
Advertisement, 1	Sales manager, 1
Insurance, 1	Salesman, 1
Lawyer, 1	Unknown, 1
Lumberman, 1	

The occupations of those who were first or second lieutenants during the war, and the number in each occupation, are the following:

Student, 48
Businessman, 24
Lawyer, 22
Manufacturer, 13
Teacher, 10
Salesman, 11
Insurance, 9
Clerk/office work, 8
Bank worker, 7
Farmer/dairyman, 7
Accountant/auditor, 6
Office/sales manager, 6
Stock broker, 6
Chemist, 5
Engineer, 5
Real estate, 5
Secretary, 4
Advertising, 3

Industrial worker, 3
Architect, 2
Newspaperman, 2
Regular Army enlisted, 2
School administrator, 2
Industrial foreman, 1
Lumberman, 1
Patent clerk, 1
Physician, 1
Plumber, 1
Police chief, 1
Policeman, 1
Postal worker, 1
Printer, 1
Purchasing agent, 1
Railroad worker, 1
Unknown, 11

Although it is difficult to pigeonhole the graduates into precise class grouping, after factoring out the students and those candidates whose occupations were unknown, we can classify approximately 84 percent of captains and 93 percent of lieutenants as having been employed in professional or white-collar jobs. Also, only 16 percent of the candidates in the sample had any degree of military training or experience before attending the OTC, and in a number of cases even these experiences had been decidedly limited. All-in-all, the sample of the Fort Sheridan OTC graduates seems to indicate that those drawn to the first two sessions of training camps were the caliber of men that the army had long sought as its officers. Based on evidence from candidates from other OTCs and the comments of later candidate school commanders, on the whole the Fort Sheridan sample is a fair representation of the men who flocked to officer training in 1917.[19]

The available evidence also indicates that the officer candidates of 1917 were some of the most enthusiastic soldiers to serve in the war. William M. Briggs, for example, recalled that he and his classmates from Valparaiso University, like most of the candidates of the first OTCs, were driven by "an impelling desire to get in the service as soon as possible." All of his fellow law school senior classmen signed up to attend the OTC at Fort Benjamin Harrison, Indiana. He was so eager to go to the camp, and so afraid that he would not make the army's minimum weight standard, that he drank two to three quarts of milk daily before his report date to add some bulk to his slight frame. In the fall of 1917, John E. Hull was a medical student at Miami University. He recalled that "everybody in college became interested in getting into the army one way or another" and noted that at least 75 percent of his university's football squad went to the training camps. He later stated that "my ambition was to get a commission and get into the Army. If they'd have sent me a commission as a paperhanger I probably would have accepted it."[20]

The motivations and actions of men like Briggs and Hull were far from unexpected. Since the nation's colleges were hotbeds of pro-Allies sentiments and the preparedness movement, it was no surprise that college men would flock to the training camps. World War I was a period of hyperpatriotism in the nation, and the war struck many Americans as a crusade to protect civilization against barbarous "Prussianism." As one machine gun officer wrote home upon his departure for France, "We are finally on the way to show the Huns that the Americans are not too proud to fight, to make the world safe for democracy, to assure supremacy of Freedom of the Seas and the rights of Smaller Nations." The hyperpatriotic crusading fervor was especially pronounced in the middle- and upper-class men who made up the majority of the wartime officer corps. Most of these men came of age during the presidency of Theodore Roosevelt, and visions of his charge up San Juan Hill shaped their perceptions of war as an exciting, manly, and glorious endeavor. These young Progressive Era men generally shared his passion for the "strenuous life" and his belief that education and social standing brought with them the duty of noblesse oblige. Although these tendencies were most pronounced in the candidates of 1917, much of this spirit continued to motivate officer aspirants throughout the war. In the end, most of the young college and middle-class men who flocked to the OTCs during the war generally were highly motivated to serve and ideologically committed to the cause.[21]

Whatever their backgrounds or motivations, the one thing that the candidates needed was a great deal of training. On 5 May 1917 the adjutant general published Special Regulations Number 49, the definitive guide for the establishment and conduct of the first training camps. This regulation included all of the previous correspondence related to the OTCs that had been sent from the War Department since 17 April 1917, the final regulations concerning who could attend officer training, and the three-month training plan for the May OTC class. The final plan was to give all candidates a one-month common core of training, followed by two months of focused train-

35

ing in the candidate's arm of service. The total course of training for an infantry officer candidate was to be 625½ hours. The branches of service that required more technical training of their candidates ultimately squeezed more training time in the schedule. The first OTC for field artillery candidates, for example, was to consist of 700 total hours of training.[22]

All the wartime OTCs and later Central Officers' Training Schools (COTSs) followed the same overarching goal. As the regulations specified,

> The prescribed courses are designed to teach, as thoroughly as possible in the short time available, the duties of an officer as . . . ,
>
> (a) Instructor: by subjecting our future officers to the same drills and individual training that they in turn must give to their future commands, with the rigid discipline and attention to detail that they must exact when they become officers of an organization that is to be trained.
>
> (b) Manager: by subjecting them to the same mode of life that will obtain with respect of their future commands, supplementing the same with instruction in the proper method of supplying, messing, administering, and disciplining organizations, and caring for [the] health, welfare, comfort, and sanitation [of their soldiers].
>
> (c) Leader: by illustrating the tactical employment of troops and by giving each the opportunity for practice in tactical leadership.[23]

By producing officers who could simultaneously serve as instructors, managers, and combat leaders, in theory, the graduates of the OTCs would serve as the ideal solution to the army's great challenges of mobilization. Given the scarcity of Regular Army officers, these young OTC graduates would be the ones who were to train and deploy the legion of draftees that would descend on the army beginning in the fall of 1917. They would also be the ones who would physically lead the legions into battle. The Regular Army banked on the ability of the OTCs to accomplish this, and, in 1917 it had no other choice.

The first session of OTCs was held from 15 May to 11 August 1917. The sixteen camps trained over thirty thousand civilians and 7,957 officers who had previously been commissioned in the Officers' Reserve Corps after attending a prewar Plattsburg camp or having been recommended by a board of officers. Ultimately, the first OTCs commissioned 27,341 officers, of which only 238 were appointed to ranks above captain. More than half of the new officers (14,484) were commissioned in the infantry.[24]

From the beginning, many regulars doubted the OTCs' ability to train officers in three months, but most also realized that the situation presented them with no other option. Reflecting this wait-and-see attitude, in June 1917 one officer mused, "When the war was declared . . . we needed officers, and we needed them at once. Not everyone will agree that the solution adopted was the best, and undoubtedly there are many men in the regular service who have been hit hard by the methods employed, while time will show whether all the officers who have come from the training camps in fact measure up **37** to their responsibilities. But it must be remembered that those methods were adopted because the Army itself had not provided for such an emergency." A few months later, the editor of the *Infantry Journal* addressed the continued doubts in the regular officer ranks about the purpose of the camps and their final human products. He wrote, "The trained soldier finds difficulty in reconciling himself to the idea that the profession in which he has spent a lifetime of work and study can be mastered by another in the short period of three months. Let such remember that the object of the work is to not make finished soldiers, but to produce practical fighting men. . . . Their training, as far as it goes, has been uniform and thorough." Despite the editor's positive, if somewhat resigned, reassurance, those regulars with doubts about the effectiveness of the training given were justified in their skepticism.[25]

An examination of the training plan for the first two OTCs reveals a number of the shortcuts and missteps that the army made in training its candidates to lead on the modern battlefield. The subjects, and the time the War College devoted to them, demonstrate the army's continued adherence to obsolete tactical ideas as well as its overall lack of preparation to teach a large influx of men. Tables 3–1 and 3–2 illustrate the subjects taught in the first two sessions of OTCs as well as the hours that the War College staff intended to be devoted to the topics.

The hands-on training conducted during the common core phase mostly focused on building the physical stamina of the candidates and on the fundamental skills required of all soldiers: formation drilling, route marching, and basic rifle marksmanship. The core course was also intended to provide the students with the theoretical basis of the tactical employment of units. However, most of the theoretical instruction throughout the three-month course was to be given during scheduled conferences. These conferences were intended to be seminars in which the instructors guided the candidates through a discussion of the tactics, skills, or concepts that the students would practice during the week of training. During the first month of the OTCs, the conferences focused on the *FSR, Infantry Drill Regulations (IDR)*,

Table 3-1. First Month Common Core Course for All Candidates (First and Second OTCs)

Subject/Activity	Hours of Training	% of total Training
In-processing	8	4
Conferences (lectures)	64	29.2
Evening study	46	21
Physical training	11	5
Practice marches	9.5	4.3
School of Soldier and Squad	7.5	3.4
School of the Company (half close order, half extended order)	17.5	8
School of the Battalion	2.5	1.1
Bayonet training	4.5	2
Saber training	1	0.5
Signaling	20.5	9.3
Musketry sighting practice	7.5	3.4
Gallery range practice	9	4.1
Interior guard duty	2.5	1.1
Field craft and patrolling	7.5	3.4
Total training hours	218.5	

Manual of Interior Guard Duty, and *Regulations for the Army.* For example, candidates at the first OTC at Fort Sheridan attended conferences consisting of "lectures by the instructors on American methods of warfare, continuing into the present day methods of foreign armies" to "establish a comprehensive understanding of the subject in the minds of the candidates."[26] Conferences held during the last two months of training delved deeper into the doctrinal material related to the candidates' branch as well as studies of the law of land warfare, field sanitation, the *Manual for Courts Martial,* and other subjects related to officership and administration.

Unfortunately, far too many of these conferences consisted merely of the instructor's reading from the given manual or lecturing instead of a seminar intended to elicit any meaningful degree of student interaction. Despite the fact that the conferences accounted for nearly 30 percent of the training in the core phase and 25 percent of instruction in the branch phase of the course, their overall value was questionable. Gus Dittmar, a veteran of the first OTC at Leon Springs, Texas, recollected that he and his comrades gained little from the lectures and conferences that occupied their evenings. He noted, "The classes were held in the mess halls, which were crowded and hot," and "everyone was tired and full of food and little interested in the dry language of the manuals."[27] This reality did not appreciably change over the course of the war. In its guidance for the upcoming fourth OTCs, in April 1918 the War College warned instructors that "for-

Table 3-2. Second and Third Months of Training for Infantry Candidates (First and Second OTCs)

Subject/Activity	Hours of Training	% of total Training
Conferences (lectures)	102	25
Evening study	64	16
Physical training	10.5	2.5
Company drill	21	5
Battalion drill	10.5	2.5
Pistol training	2.5	0.6
Tent pitching	2	0.5
Bayonet training	5	1.2
Range firing practice	38	9.3
Field tng: patrolling and scouting	10.5	2.5
Field tng: battalion in attack and defense	12	3.1
Field tng: battalion overnight camping	12.5	3.2
Field tng: battalion in trench defense	5	1.2
Field tng: company on outpost, advance and rear guard	5	1.2
Field tng: company in attack and defense	5	1.2
Machine gun drill	4.5	1.1
Platoon combat firing	4	1
Company combat firing	4	1
Battalion combat firing	2	0.5
Trench warfare (included grenades, gas and trench attack and defense)	19	4.6
Three-day maneuvers	60	14.7
Lectures on infantry, cavalry, and artillery	8	1.9
Total training hours	407	

mal set lectures should be resorted to very infrequently, as they become tiresome to the student and nonproductive in results. Better results are obtained when practical work is interspersed with short informal talks or conferences. The most important prerequisite to an interesting talk is that the speaker know his subject thoroughly."[28] Apparently, this admonition did not take, and throughout the war it seemed to be rather rare for a lecturer to retain the attention of his students.

Another reason for the failure of the conferences to be a more effective medium for educating the officer aspirants was the lack of time that the students had to prepare for the instruction. The evening study period was to be the candidates' time to read the materials to be discussed during the conferences or to prepare for the next day's training. Unfortunately, the candidates spent this time preparing their boots, uniform, equipment, and rifle for the next morning's inspection rather than using it for any deep study of the mysteries of the military art. This dilemma of time management was satirized by one late-war candidate:

Police the room and sweep the floor,
Shine each piece and clean the bore,
Scrub your neck, align the cot,
And wash your clothes before they rot,
Each tiny "Fob" must wear a shine,
But do these things in your Spare Time.

Pace o'er the hill and down the side,
Then make a scale to fit your stride,
Make a map of all you see,
And learn the *Field Service* from A to Z.
Estimate distance, do not resign—
But do these things in your Spare Time.[29]

Although cadre members seemed to be aware of this evident quandary with their pedagogy, they remained unable or unwilling to do anything to eliminate the problem. Thus, while conferences and evening study accounted for nearly half the training time allocated for the first two OTCs, what the students actually gained from these theoretical tactical discussions is debatable.

The type and quality of the practical hands-on training that the candidates received during the core phase was also problematic. All too often the training was nearly identical to that given to prewar enlisted recruits, and little time was devoted to leadership development. One attendee noted that the School of the Soldier and the School of the Squad occupied most of the training in his first month in camp. In a letter home, Charles Sorust provided a thumbnail sketch of his daily schedule that highlights the basic-recruit training approach that his cadre used during the first month of the course. On 20 November 1917 he wrote,

We get up at a quarter to six, have physical exercise at 6:15 A M, breakfast at a quarter to seven, drill at 7:30 A M, then comes bayonette [*sic*] exercise at 8:30 A M, then medical hygiene . . . at 9:40 A M, then drill and then locker inspection at 11:30 A M, and dinner at 11:45, then a[t] 1 o'clock we have bomb and hand grenade throwing, we only throw bricks, and at 2:40 P M we dig trenches, and at 3:25 P M we have French class until 4:15 and then supper at 4:45 P M. . . . After supper I am either washing clothes or cleaning my rifle, because at 6 P M it is very dark.

Another candidate at Fort Riley described his training as merely "throwing a gun around and hiking out in the country, taking bayonet exercise jabbing imaginary enemies through imaginary bodies, waving the semaphore signals and drilling in squads."[30]

Although candidates with no prior military experience did need to experience and understand the basics of soldiering, close-order drill and bayonet practice were given far too much emphasis by the cadre. The frequency with which candidates mentioned close-order drill, bayonet practice, and "wigwag" flag signaling in their letters and memoirs suggests that they spent more time in these areas, and less time in field training, than was specified in the War Department's central training curriculum. John Hull, for example, admitted that while his OTC "didn't teach you everything," it at least "gave you a start [so] you didn't feel like a complete stranger when you joined a regiment." Nonetheless, he still characterized much as his training at the camp as "quite a bit of close order drill."[31]

41

These deviations from the prescribed War College curriculum were mainly due to shortages of instructors, equipment, and training areas or other limitations that were specific to individual camps. The biggest problem with these variations in training was that when the candidates became officers responsible for the training of their own soldiers, they tended to fall back upon what they had learned in the OTCs. This explains why much of the unit training of the war's draftees displayed a great penchant for marching and bayonet practice. The differences in the training regimen of the various OTCs also meant that there were often great variances in the experience levels of the graduates when they reported to their units. Since the early OTCs tended to feed officers to specific divisions, this accounted, to some extent, for the differences in training and readiness from division to division.

Some of the practical training in the core phase also highlighted problems with the army's prewar doctrine and the inconsistencies within its emerging tactical thought. One of the great ironies of officer training was that while John J. Pershing proclaimed that the tactical doctrine of the American army would be built upon individual rifle marksmanship, the OTC students' marksmanship training was conducted only to the level that the army had considered to be the bare minimum for a regular army recruit in 1916.[32] Since these novice officers were to be the primary trainers for the National Army, this did not bode well for the future.

Some other aspects of the candidates' training underscored how out of touch the army was with the conditions on the Western Front. The signal training, for example, consisted of the students' learning how to send and receive messages using visual signals, mostly employing wigwag semaphore flags. Prewar doctrine had stressed that these types of signals would be one of the primary means of command and control in battle at the battalion level and below. Given its doctrinal importance, the students spent nearly 10 percent of their time in signal training. Unfortunately, this training had little to no practical value. A soldier or officer waving a signal flag in combat became a ready target for any observant enemy. As one candidate later stated, "Those who got to France were never able to recall having seen a semaphore flag used for anything other than a scarf around the neck of some mademoiselle, or as a pillow top in a French farm house."[33]

The practical training during the last two months of the OTC was to instruct the candidates in the tactical, technical, and leadership skills they would use in their respective branches in combat. This phase would also introduce the candidates to the weapons and tactics of trench warfare. The War College staff intended that the average infantry candidate would spend over 25 percent of the training time in the second phase of the course practicing platoon, company, and battalion tactics. Seven percent of the remaining time would be spent on learning the weapons and tactics of trench warfare (to include trench construction and machine gun drill), and 13 percent of the time would be devoted to rifle practice and unit firing. With over 45 percent of the candidates' training time committed to hands-on tactics and skills of trench and maneuver warfare, on paper the War College training plan was focused and logical. In practice, however, the training fell well short of the War College's goal.

There can be no doubt that the second phase of training was rigorous and physically demanding. Dwight Eisenhower, an instructor at the Fort Oglethorpe OTC, noted, "The training was tough—designed as much for weeding out the weak and inept as to instruct." Candidate Gus Dittmar recalled, "The fact that some men, who had been barely holding on by their finger nails and their teeth, now had to relinquish that hold, brought regret for these unfortunates and considerable concern for themselves, from those that were still in the running." But rigorous and physically demanding training does not always equate to effective training.[34]

Although the tactical training at the OTC was demanding, in many areas it failed at adequately preparing the students for the tactical and leadership challenges that they would face in France. Part of this problem was that the War Department never fully understood exactly what skills and training the candidates needed to be sound combat leaders. Throughout the war it wrangled with the issue of finding the right balance between training for trench warfare and training for open warfare. In forming its training plans for the OTCs, the War College made the logical assumption that trench warfare would be the predominant condition that American forces would face throughout

1917 and 1918. Still, as late as April 1918 the War College training plan for the fourth OTCs maintained that while open warfare had "given place, at least temporally to a trench or positional warfare . . . there has been no change of basic principles—merely a change in the method of employing them." It also stressed that "to prepare for one class, then, to the utter exclusion of the other, would be the height of folly."[35] Despite Pershing's perception that stateside training was too focused on trench warfare, OTC students actually spent more time learning the intricacies of open warfare.

Regardless of the type of warfare the training focused on, the greatest problem with the candidates' tactical instruction was that it failed to replicate the conditions that they would face in actual combat. Not surprisingly, tactical training for infantry candidates in the last two months of the course was based on concepts pulled directly out of the *IDR*. One candidate noted that during his field training and maneuvers his unit concentrated mostly on moving forward as skirmishers under the direction of the platoon commander's whistle to build up firing lines and gain fire superiority in a manner specified in the regulations. In a similar vein, the training at Fort Sheridan consisted of advance by a "forward attacking line" that rushed ahead and then "flopped to the ground and opened up with rapid flashes." Command was exercised by "crouching figures" who "ran haltingly back and forth along the line." These maneuvers were done in the manner prescribed by the *IDR* for building up a firing line in preparation for an attacker's assault upon an entrenched defender. Throughout all of this training, recalled a Fort Sheridan candidate, the instructors emphasized that "bayonet work was an important branch of infantry specialization" and a vital part of the attack.[36]

The problem was that these tactics and assumptions had long been proven invalid on the battlefields of France. The building up of skirmish lines and attempts to gain infantry fire superiority before assaulting had been shown only to stall attacks short of their objectives and thus subject the attacker to higher casualties as units remained for longer durations in areas swept by artillery, machine gun, and rifle fire. This point was not lost on those officers who later commanded in combat. Looking back on his officer training, one combat veteran noted, "Our army had learned no lessons of modern warfare as developed in Europe in the two years the war had been going on. This was again in evidence in the 1st Training Camp for officers. . . . Much time [was] wasted in learning methods . . . which were useless in Europe." The lack of realism at the Camp Root OTC led F. L. Miller to dismiss his training as "three months spent . . . learning wig-wag

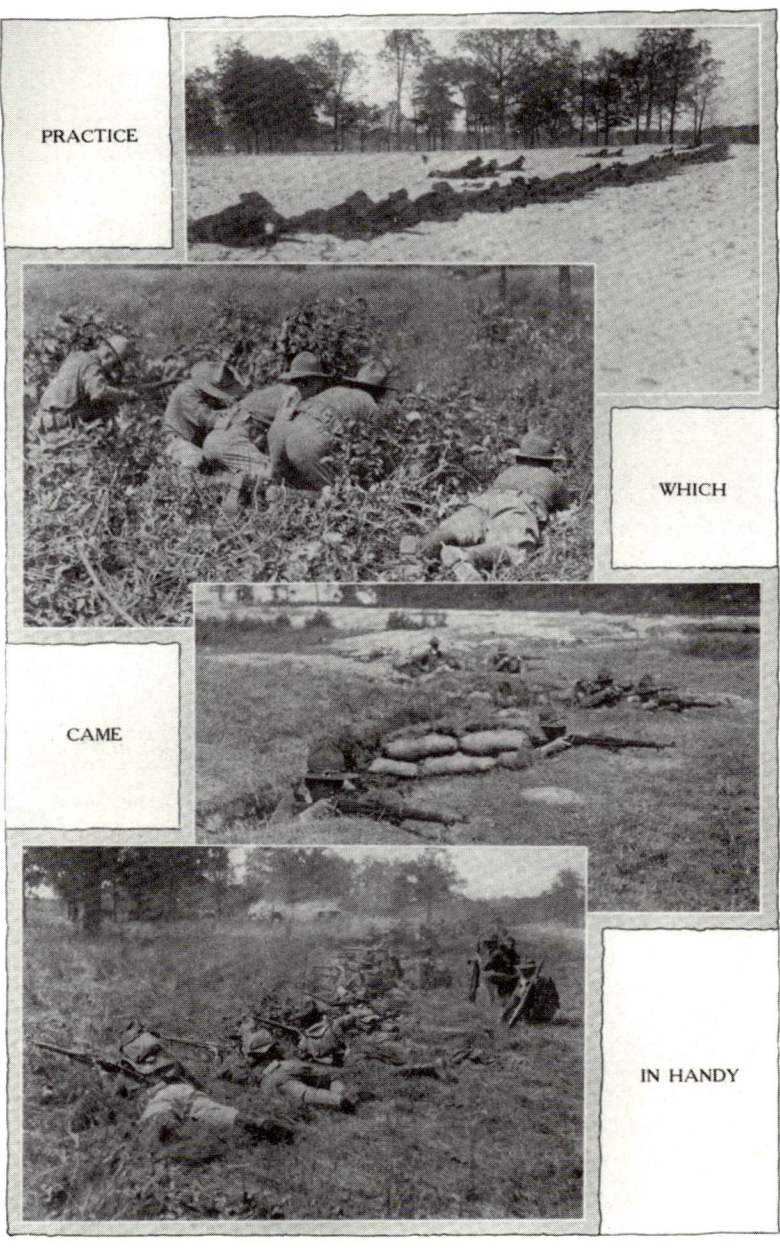

PRACTICE

WHICH

CAME

IN HANDY

Field combat training at the Fort Sheridan Officer Training Camp in 1917. The spacing of the candidates on the firing lines, the formations being employed, and the construction of the trenches used in training illustrate the outdated state of American tactical thought in 1917.
(Photograph from Fort Sheridan Association, The History and Achievements of the Fort Sheridan Officers' Training Camps.*)*

Officer candidates digging trenches in an undisclosed officers' training camp. Although officers had to be familiar with trench construction, having the candidates physically dig the trench works was probably not the best use of valuable training time. The digging of a straight trench line, as pictured here, was also not consistent with practices on the Western Front.
(Photograph from the author's collection.)

and semaphore signaling and reenacting Civil War combat problems through the mosquita [*sic*] filled swamps of Arkansas."[37]

Even the training in trench warfare left a lot to be desired. Much of the time devoted to this subject was actually spent by the candidates physically digging the trenches they would use in training. Although there was some value in giving the future officers understanding of the intricacies of laying out trenches and the time and physical exertion required to dig them, the value of this experience was not in line with the time and effort devoted to it. After being set to dig trenches, one Fort Sheridan aspirant bluntly asked, "What is the use of my learning to do this? I could get a Dago to do it better than I could for a few dollars a day and it will cost me more than that to get fixed up after this mess." Some of the candidates at the Leon Springs OTC in 1917 also wondered at the utility of physically digging trenches. One wit offered that at least, "When you get discharged from this man's army you can always get a job as a grave digger."[38]

Once the candidates completed digging the trenches, the tactics they learned in attacking and defending them were problematic. The methods used for attacking trenches were those listed in the *IDR* and generally the same that the candidates had learned for open warfare: waves of attackers steadily building up a firing line to gain fire superiority over the defenders. As Candidate Dittmar noted, "After the fundamentals of attacking a trench system had been mastered, the companies staged simulated attacks on the enemy positions. These followed to the letter the latest techniques extracted from secret War Department manuals. . . . With wild yells, gritted teeth and much colorful language they poured into the trenches, sticking, knifing, and clubbing the simulated defenders. It was a bad day for the defenders."[39] Although he mentioned "secret War Department manuals," it is clear from his other descriptions of his training that the tactics stated in these unnamed guides must have differed little from those of the *IDR* or had not been followed as intended. Also, his simulated attack was not met with the shell and small arms fire that had tended to make more bad days for the attackers than for the defenders of the Western Front.

This lack of realism was not localized at Leon Springs. In a letter written shortly before his graduation from the Fort Riley OTC, Milton Bernet wrote, "For the past week we have been studying barbed wire entanglements and trench warfare as it is now fought in Europe. . . . We charge from one trench to the next, stabbing the dummies as we go in with our bayonets, occupying and investing the trench and then go on to the next."[40] It is interesting to note that no mention was made by either candidate of any supporting fires by artillery, trench mortars, or machine guns to aid them in their assaults. However, the fact that both wrote of wild bayonet attacks on the enemy speaks volumes of the focus and assumptions of their training.

The War Department's leadership was not unaware of the problems with training at the first OTCs. On 1 August 1917 Brigadier General McCain warned the chief of staff that the candidates "have not received adequate instruction in the methods of modern war by officers familiar with [the] new conditions." As a stop-gap measure, he recommended the establishment of divisional schools for musketry, grenades, trench mortars, trench construction, and gas protection in which at least one officer per company would be trained to be instructor in those subjects. Maj. Gen. William Snow was so disgusted by the training at the OTCs that he acerbically observed, "The only uniformity among them was that each was distinguished for its wholly inadequate course of instruction, its incompetent instructors, and its insufficient equipment."[41]

The sad reality of this lack of realism and effectiveness in training was that this was an area that the army easily could have addressed. As was shown in chapter 2, the army had indications that its tactical doctrine was being called into question by events in Europe even before

America entered the war. Once in the conflict, several sources questioned many of the Americans' tactical assumptions. In early 1917, Harvard University recruited the French lieutenant colonel Paul Azan to assist its American ROTC instructor in teaching military science. While at Harvard, Azan published two books, *The War of Positions* (1917) and *The Warfare of Today* (1918). Both of these works offered a stark repudiation of American doctrine and an honest appraisal of the ugly attritional realities of modern war. He informed his young charges, "Today, the effects of artillery on the earthworks and guns of the enemy is the decisive factor in success; victory goes to that army which has guns in good condition and abundant munitions, as against the enemy whose supplies has [*sic*] given out. The flinging of innumerable infantrymen against batteries that are still intact results in nothing but useless slaughter." Azan was unequivocal in his statement that "infantry is powerless without the aid of artillery" and that it "can make no attempt against a position unless the artillery has destroyed the accessory defenses, smashed the trenches and demoralized their occupants." He concluded that for an attack to succeed, the infantry must "advance with prudence and method, with the constant support of its artillery."[42]

Warnings also came from other foreign officers inspecting the training conducted at the OTCs. The American officer accompanying a general from the Russian military mission passed on the Russian's observations: "The advance to attack by rushes, the attack both at drill and in the maneuvers which he saw, could be, in his opinion, with one machine gun, properly concealed, entirely wiped out in thirty seconds leaving no one except the dead and wounded. Such training is to his mind worse than a loss of time. It inculcates wrong ideas in future officers. . . . He was quite astonished that no apparent effort was ever made for invisibility, [of] either men, trenches or guns." The Russian general further recommended that in addition to resolving these glaring problems, the officer training needed more emphasis on methods of conducting a relief in place, ensuring lateral and rearward communications, proper employment of barb wire, and night patrolling.[43]

One wonders why, if these shortcomings were so apparent, the War Department did not do more to eliminate them? Part of this problem was due to the Regular Army officer corps' faith that their methods and doctrine were innately superior to those of the Europeans. Despite obvious evidence to the contrary, nothing shook this fundamental belief in American exceptionalism throughout the war. Nowhere was that more evident than in the words and actions of Pershing and his senior commanders and staff officers. Pershing insisted that the stalemate in France was an aberration and that the American army's superior drive, morale, and marksmanship would ultimately force the Germans out of their trenches. Once free from the trenches and into open warfare, the Americans' greater skill and ability at maneuver would allow them to corner and destroy the inferior German army.

Furthermore, many Americans, from Pershing to the most junior lieutenant, convinced themselves that years in the trenches had blunted the offensive edge of the Allies and had sapped their aggressiveness, initiative, and will to win. As one senior GHQ staff officer argued, "In many respects, the tactics and techniques of our allies are not suited to American characteristics or the American mission in this war. The French do not like the rifle, do not know how to use it, and the infantry is consequently too entirely dependent upon a powerful artillery support. Their infantry lacks aggressiveness and discipline. The British infantry lacks initiative and leadership."[44] The American could argue with some justification that perhaps the French and British tactical methods were far from proven, given their high casualty rates. However, these American attitudes did make it difficult to create training plans for the OTCs, and the rest of the army, that were grounded in the combat realities the Americans would face in France.

Further evidence of these American attitudes, and their baleful influence on officer training, can be found by examining the debates within the General Staff over the use of Allied instructors in the American OTCs. Despite the abiding faith of the officer corps in the superiority of American methods, some senior officers admitted that perhaps it would be wise to have Allied officers at least conduct some degree of training on the unique aspects of trench warfare. On 16 April 1917 the adjutant general recommended that the War Department press the British and French to provide "a corps of instructors competent to teach our prospective officers ... those military subjects of the first importance in connection with the character of military operations they are most likely to engage in." Although Brig. Gen. Joseph Kuhn, the chief of the War College Division, supported the adjutant general's recommendation, he insisted that the foreign officers be limited to giving lectures and demonstrations on trench weapons and other technical subjects. Kuhn justified this narrow use of Allied officers on the basis that American organizations and tactics were "so different from the European methods that foreign officers would be at a decided disadvantage if compelled to adapt themselves to a new system."[45]

Not all members of the War College Division agreed with the recommendation to use foreign instructors in even this limited manner. The General Staff's Col. William Johnston argued that employing French and British officers would be "a decided reflection upon the ability of officers of the United States Army to teach the duties of company officers to candidates for the Reserve Officers' Corps." Johnston maintained that the candidates "will have ample [subjects] to learn of an elementary nature, without attempting to acquire personal expertness in throwing bombs and other arts peculiar to warfare on the Western Front." He stressed, "To give instruction of this character would be equivalent to teaching chemistry and philosophy by lecture to one not yet able to read." Johnston concluded, "After our forces have had the year's training essential, according to all previous plans, and if there is *then* contemplated sending an expedition to the Continent, the peculiar training now useful in France can be given [to] our forces after landing there." A vote by the fifteen staff members of the War College Division on the proposal to use foreign instructors laid bare the passions that the subject evoked. Johnston and five other officers (40 percent) dissented from Kuhn's recommendation.[46] Both Kuhn's and Johnston's positions offer insights into the attitudes and prejudices of the regular officer corps. Any thought that the American doctrine was flawed or any admittance that foreign officers had something important to offer the US Army seemed to them to be a stain upon the professionalism that they had striven so hard to build since the late 1800s.

These parochial attitudes continued to hobble the training of American leaders, both in the United States and in France, throughout the war. As late as August 1918, Col. Harry Eaton, commander of the Infantry Central Officers' Training School (ICOTS) at Camp Lee, Virginia, complained to the War Plans Division about the assignment of two veteran British officers to his unit to serve as tactics instructors. Eaton reported that "it is not practicable to use these officers to advantage in this kind of instruction." He went on to state, "British officers are not desired for purposes other than bayonet and physical training, gas, and scouting and patrolling. Inasmuch as the policy is now to develop our own instructors, work is progressing along that line and the one British officer on duty to supervise that instruction is sufficient. It is recommended that these two officers be relieved and sent to some place where their services can be utilized to better advantage."[47] It is amazing that in a time when he and other commanders were hard pressed to find qualified American instructors for their candidate schools, Eaton was so quick to dispense with men with a great deal of combat experience.

Although the regular officer corps was reluctant to accept foreign instructors, they showed much less reticence in accepting French and British manuals and doctrinal publications. Ironically, this willingness to use these works contributed to the confusion and lack of focus in the training of American officers and units. In the spring and summer of 1917 both the War College and the AEF embarked on a printing spree of foreign manuals. Unfortunately, there seems to have been little thought given to reconciling discrepancies among the various works. For example, the War College's July 1917 translation of the French *Manual for Commanders of Infantry Platoons* describes and illustrates formations and tactics that were different from the British ones they printed in *Notes on Recent Operations, No. 3* just a month later. Both of these differ from the tactics and formations printed in the War Department's June 1917 *Instructions for the Training of Platoons for Offensive Action* and the AEF's August 1917 translation of the French *Manual of the Chief of Platoon of Infantry*.

To add to the confusion, the War College's *Manual for Commanders of Infantry Platoons* and the AEF's *Manual of the Chief of Platoon of Infantry* offered two different translations and editing of the same French work. The discrepancies in these editions lay with what the two entities chose to omit or emphasize. The War College manual omitted much of the material that the original French work had devoted to "the formation and movement" of both the platoon and company. The little discussion that the War College manual did include in these areas was sketchy and poorly illustrated. All in all, the War College's edition also gave more weight to the manual's discussion of defense than that of offense. To further muddy the waters, in May 1917, the War College had also translated and printed the French 1916 *Instructions on the Offensive Conduct of Small Units*. While this work was similar to the *Manual for Commanders of Infantry Platoons,* it was poorly illustrated and different enough in the details it illustrated of the attack as to sow confusion in the mind of any reader of both manuals. By the time the War Department finally resolved this issue in the early summer of 1918 by reprinting the AEF's latest edition of *Instructions on the Offensive Conduct of Small Units,* with the supplement illustrating the new formations and deployments (issued by the AEF in April 1918), it was too late to be used in the United States by the majority of American junior officers who saw combat during the war.

The overarching purpose of any tactical doctrine is to establish a common understanding in the minds of an army's officers and men of how units are to operate in combat. A doctrine specifies the role of leaders in battle and provides an outline for how units should react to given tactical situations. In the confusion of the battlefield, as units become intermeshed and leaders become casualties, having everyone on "the same sheet of music" is of vital importance. With the A E F "freelancing" its own doctrine and the War College printing a host of foreign materials with no effort to reconcile their discrepancies or illustrate how they meshed with American doctrinal thought, one wonders if it can be said that the US Army never truly had a doctrine during the war.

51

In addition to all of the foreign manuals the army was printing, the War Department never stopped printing its own doctrinal materials. In the spring of 1917 the War Department published *Notes on Infantry, Cavalry, and Field Artillery*. This publication was a compilation of lectures given by American officers to the second class of provisional Regular Army lieutenants at Fort Leavenworth. In mid-1917 the War Department reissued the work as a "special reprint for Officers' Training Camps" and sent it to the various OTCs. When it came to infantry tactics, Maj. Harold Fiske asked the rhetorical question, "Will our drill regulations require radical modification to conform to the experience of the great war?" His answer was a confident, "No," because, "The formations contemplated and the principles taught in our drill regulations have been proven in the main correct."[48] With this profusion of contradictory official sources, and with no official guidance as to their use, it is no wonder that the instructors at the OTC either attempted to reconcile the sources themselves, thus further fragmenting doctrine, or simply fell back to what they knew: the obsolete tactics of the *IDR*. In the tactics described by Dittmar, the instructors seemed merely to have cloaked old ideas with new terminology.

It is also important to discuss how the problem of obtaining qualified instructors contributed to the poor training of the officer candidates in the first two OTCs. From the beginning of mobilization the War Department knew it was going to have a problem finding enough qualified instructors for the camps. On 23 April 1917 General McCain warned the departmental commanders, "The number of instructors is not nearly as great as it should be, but it is hoped and believed that many reserve officers and candidates for appointment as such who will be in attendance at those camps will have special qualifications to teach certain subjects and in this to be of material assistance to the Regular Army officer." The selection of these reserve and student instructors was left up to the company commanders of the OTCs. A number of the instructors ended up being recent Plattsburg graduates or regular army NCOs attending the camps as candidates. The looming teacher problem led General Kuhn to recommend to the chief of staff on 25 April 1917, "With the probable shortage in instructors, and these instructors differing widely in experience and ability, the schedule must follow rather definitely certain prescribed text-books to obtain any sort of uniformity in instruction." Unfortunately, in the same memorandum he noted that the army would also be very short of the exact manuals needed to attain his desired uniformity.[49]

One of the major challenges that the army faced in April 1917 was how best to employ its corps of regular officers. The adjutant general had to juggle the competing demands of providing officers for existing Regular Army units, filling the vacant commissioned ranks of National Guard and National Army units, and assigning instructors to OTCs. Naturally, in the excitement of the early days of mobilization, a number of regular officers themselves muddied the water by actively searching for plum combat assignments and chances for advancement. There were simply not enough regular officers to go around.

The adjutant general did the best he could in reconciling the various calls for regular officers, and the first OTCs did get a large slice of the personnel pie. The number of regular officers assigned to a given OTC varied according to the number of classes held at the post and the general availability of officers for reassignment. At Leon Springs, Texas, for example, the OTC cadre consisted of 101 regular, National Guard, and reserve officers. The camp administrative staff consisted of 28 officers under the camp commander, a Regular Army colonel. The other members of the staff included a camp adjutant, a quartermaster, an ordnance officer, and mess officers (generally from the regulars), each with his own assistant officers; two Regular Army sergeants major; and three medical officers and twelve corpsmen.

The instructional staff consisted of seventy-three officers under the direction of a Regular Army senior instructor in the rank of lieutenant colonel or major of infantry or field artillery. The senior instructor, aided by two Regular Army assistant senior instructors, oversaw the training given by sixteen infantry, five field artillery, one engineer, and two coast artillery instructors, all drawn from the regular officer ranks. These instructors, in turn, were supported by forty assistant instructors from the National Army or reserves.[50]

The first OTC at Fort Sheridan was commanded by a Regular Army colonel and had a regular lieutenant colonel as the senior instructor and two regular majors as the battalion commanders. Each of the twenty-eight companies of candidates was commanded by a Regular Army captain or first lieutenant. The company commander was the senior instructor of his company and was supported by two assistant instructors drawn from the ranks of the recently created reserve officer corps. Most of these reserve officers were graduates of the prewar Plattsburg camps, and few had any real military experience to draw upon in their instruction.[51]

Although the regular officers took pride in their professionalism, being a good officer did not necessarily make for a good instructor. This meant that even in the first two sessions of OTCs, in camps with the largest cadres of regular officers during the war, the quality of instruction could vary wildly from company to company. Some candidates spoke admiringly of their Regular Army cadre, but others were much more critical of their cadre members. In May 1917, future World War II general Lucian K. Truscott was a private serving in a cavalry regiment on the Mexican border when he decided to apply for officer training. Truscott later admitted, "Military education at the training camp had been austere and elementary" and "had been conducted for the most part by instructors who seemed to know little more than the candidates." William M. Briggs's experience was similar to Truscott's. He recalled that while his chief instructor was a West Point graduate and an experienced officer, "he was still going by tactics learned in the Civil War." He also noted that some of his regular officer instructors "were strict to the point of ridiculousness" and that "the old West Pointers and regular army men were more interested in drilling, and were not aware of the newer means of carrying on a war as was going on in Europe."[52]

Poor instruction left a lasting impression on the officers who went through the first OTCs. When the War Plans Division's Morale Branch surveyed officers returning from France in 1919, many remained bitter about the poor training they had received at the OTCs. One first lieutenant noted that "officers acting as instructors at these camps were often poor judges of men and lacking in knowledge of methods and subject matter." An infantry captain later bluntly recalled, "I have never seen such pathetic attempts at instruction as I saw in the First Officers Training Camp."[53]

One candidate, Raymond Phelan, was so concerned about the training he was receiving at the first OTC that he wrote directly to Secretary of War Baker. In August 1917, Phelan informed Baker, "I share the pretty general opinion expressed in many different ways that there is room for improvement in some important directions in conducting our training camps. In the first place, the system of instruction, with the very best intentions . . . on the part of the officers, has so worked as to encourage belief in the American idea that 'getting by' is the proper test of right and wrong." He roundly criticized the "recitation system" used by many instructors and their lack of "pedagogic skill in questioning." The candidate concluded, "It would be better also if all instructors set down, or thought out their questions before coming to class" and that "much would be gained if officers and other leaders planned out action on problems before the company falls in."[54]

Although Baker was aware of the instructor problem, there was little he, or anyone else in the War Department, could do about it. The pace of mobilization prevented any meaningful effort to find or train competent instructors for the OTCs, and the War Department's own personnel management system further hobbled this effort. For example, as soon as the first OTC class at Fort Sheridan graduated in August 1917, the War Department reassigned most of the camp's Regular Army instructors and staff officers, leaving only a handful of experienced men to prepare for the next camp. Shortly after the start of the second OTC, the army reassigned nearly all of the remaining regulars and replaced them with reserve officers who had only recently graduated from the first OTC.[55] This unfortunate process would be repeated throughout the war.

One of the few viable recommendations for coping with the instructor shortage was offered by Col. Chauncey Baker in July 1917. Baker headed a mission sent to France by the General Staff to observe training in the AEF. He recommended that select officers with frontline experience in France be returned to the United States to serve as instructors. He also asked that the War Department send the AEF 208 "extra reserve officers" to aid in the establishment of the AEF's school system and to serve as instructors in France.[56] Not surprisingly, Pershing was unwilling to strip the AEF of experienced officers, and the War Department had no extra reserve officers readily available to send to France. Ultimately, only a small number of officers were ever reassigned from the AEF to teach at stateside schools during the war.

In addition to problems with doctrine and instructors, candidates in the OTCs found that their training was hindered by chronic shortages of equipment and the other growing pains of mass mobilization. The army entered the war with no mortars, tanks, grenades, or gas masks, and many of its machine guns and artillery pieces were obsolete. The OTCs had to scrape by using whatever they could borrow from local posts or requisition from the army's overburdened supply system. The shortages of machine guns and artillery were particularly acute. One OTC graduate recalled that meaningful artillery training was stifled in his camp because "guns and horses were sadly lacking." He added that after being bombarded with "such weird sounding terms as angle of site [*sic*], mil and corrector" without the actual equipment, "at the end of the lecture, heads whirled like a rotating band blown through a rifled tube.[57] These shortages contributed to the lack of realism in the first OTCs and certainly hindered the candidates' ability to understand the challenges they faced leading small units in modern war.

The crucial problem with the lack of realistic training in the first OTCs was the long-term impact that this shortfall had on the overall effectiveness of the American army. As the OTC graduates were to become the primary instructors of the new army, any flaws with their training were redoubled in the instruction of the larger force. There was also a psychological price to be paid for failing to present in the OTCs' training an accurate view of the war in Europe and the harsh challenges that the officers would face as combat leaders. The graduating officers of the OTCs often left the camps with an overly unrealistic and romantic view of war in general and warfare on the Western Front in particular. Photographs from the various OTCs show men training in immaculate trenches and making mock attacks in formations that would have made Frederick the Great smile. Shortly after graduating from the Plattsburg OTC, one officer gushed that America's entry into the war would return the Western Front to "the warfare of the old days, the warfare of our own West and South, when sabers flashed to the beats of galloping horses, and men went miles over the top instead of yards." In a similar vein one OTC student wrote home, "Just think of it! The future may hold in store for me the chance to charge with thousands of other horsemen the retreating Germans being hurled back to Berlin."[58]

These unrealistic and romantic views even appeared in the semi-official handbooks popular among reserve officers. The *Plattsburg Manual,* which along with Moss's *Officers' Manual* was one of the most popular semiofficial works purchased by wartime officers, assured young officers that trenches and machine guns could be overcome by a vigorous attack pushed to the point of "bayonet against bayonet, man against man, and nerve against nerve." In the attack the soldier's "clear eye and steady nerves, his soul's blood and iron, constitute a better defense than steel and concrete."[59] The officers' own self-delusions, their sketchy training at the OTCs, and the army's visions of an American *offensive á outrance* all combined to overshadow the need for tactical know-how. Unfortunately, this war would not be like the Civil War, in which a junior officer's leadership was judged by his personal bravery and ability to keep a dressed line moving for-

55

ward. It would instead be a war that required a deep understanding of how to combine the effects of a host of weapons and one in which the junior officers often operated in the loneliness and isolation of the "empty" yet deadly battlefield.

The army held the second session of OTCs at sixteen army posts from 27 August to 27 November 1917. There were very few changes to the training programs in these camps, although the issue of instructors was more acute than in the first session. The War Department estimated that it received approximately seventy thousand applications to attend the second camps, and it ultimately selected twenty-thousand to attend. The second OTCs commissioned 17,237 new officers, with infantrymen again accounting for over half of the total number. With the exception of fifty-nine men commissioned as majors, the rest were all company-grade officers.[60] The quality of the candidates in terms of education and prior military experience was close to that of the first OTCs. It should be noted that the War Department also established a special OTC for African American candidates at Fort Des Moines, Iowa, on 18 June 1917. This class commissioned 639 officers on 18 October 1917.

One improvement in the second OTCs was that the War Department was a bit clearer in defining the characteristics that it was seeking in its officers. In the 4 June 1917 memorandum that established the second OTCs, the adjutant general required all applicants to list any previous military experience they had before entering the camp. The candidates were also to attest to any "mental training" they had gained from "study in educational institutions, or systematic and extensive study at home, or from constant dealing with difficult, complex, or technical problems in business or professional life." Furthermore, they were to document past evidence of "Executive Experience and Opportunities for Leadership" from their business, professional, civic, or other practices that indicated the "character of responsibility assumed and success attained." The army made clear that "while it is desired to give full opportunity for all eligible citizens to apply, no man need make application whose record is not in all respects above reproach and who does not possess the fundamental characteristics necessary to inspire respect and confidence." The only major difference between the first and second camps was a slight change to the ages that the War Department was seeking in its applicants. Although the age range remained twenty years and nine months to forty-four years, the memorandum noted that "in order to obtain the experienced class of men desired preference will be given to men over 30 years of age [with] other things being equal."[61]

With two sessions of OTCs under their belts and 1917 drawing to a close, the army took few steps to change the direction of officer training as it approached its third session of camps.

There were certain indicators that the training of the candidates was not as thorough as needed or intended. Hugh Scott later explained, "While these camps did not by any means provide a finished military education to fit an officer for war in the short time possible, they were the utmost that could be provided under the circumstances. Their establishment enabled the sorting out of the unfit, and an enrollment and organization. It gave an opportunity to place men in the positions which they were best fitted to fill and gave them an inkling of discipline and the life of a soldier."[62] Although there was much truth in Scott's admission, the army could have done a better job of focusing the tactical training of its candidates to prepare them for the challenges ahead. Unquestionably, the pace of mobilization and the systemic problems with obtaining weapons and instructors were major obstacles to officer training throughout the war. But the training of combat leaders at the company level was not just about properly employing weapons; the most important part of their education was conditioning the mind of the leader to react quickly to unexpected situations occurring in the confusion of battle.

Much of combat by small units is the execution of battle drills designed to meet the common challenges presented by a general tactical situation. For example, there are general steps that the small-unit leader should take to destroy a machine gun position without taking undue casualties in his unit. The key to training small-unit leaders was thus to ensure that the officer or NCO understood how to identify the proper battle drill, or combinations of drills, to meet the given situation. The leaders also had to be able to think on their feet so as to adapt a drill designed to meet a general situation to the specific realities confronting them in combat. This required that the leader make a rapid assessment of the situation, taking into consideration the terrain, the strength of the enemy, the strength of his troops, and what he must accomplish to achieve his overall mission, and then issue the proper orders to deal with the problem confronting him.

It is clear that the army understood this process in World War I. The "estimation of the situation" had been a basic concept of the Leavenworth schools for over a decade. In fact, the schools' 1908 *Studies in Minor Tactics* was written in a manner to lead the student through the process of tactical problem solving. Although the work focused mainly on the operations of battalion or larger units, the thought process for the leader was the same for all echelons. Harold Fiske even

57

went so far as to argue that once the leader understood how to properly estimate the situation, any given tactical problem could be solved "without the conscious use of reason." Sadly, little in the OTCs encouraged the candidates to achieve this nearly intuitive level of tactical decision making.[63]

Although the press of time in the spring and summer of 1917 may have curtailed the army's range of options for officer training in the first six months of the war, there was no need for it to have continued pursuing a path that so many within its ranks knew was problematic. Although ninety days was an awfully short time to train officers, the problem was not so much of a lack of time itself as of how the OTCs used the time they had. This can be seen by comparing the differences between the American and French methods of training officers.

At the request of the War Department, the American Military Mission in France studied the French method of training officer aspirants. In July 1917 Capt. Dawson Warrington submitted a report that detailed not only what the French trained at the Saint-Cyr Aspirants' School, but also their philosophy of officer training and their expectations of the skills and attributes that their army needed in their junior commanders.

What impressed Warrington the most was that the French approached officer training "as if it were a branch of business." He noted that the French put little stock in detailed training in the new weapons of war and were content to give their candidates only enough instruction with these tools to be "all-round competent men." What the French did emphasize, however, was training that forced the candidates to think on their feet in times of stress and to "develop such steadiness and resourcefulness, such independent reasoning powers, that the most unexpected circumstances cannot take them by surprise." The French accomplished this by constantly placing the aspirants in a series of tactical leadership problems that forced them to make quick estimates of the situation they faced and then give their units clear orders to react to the dilemma. Under the Saint-Cyr method,

> The future aspirant must be a resourceful man, not a machine acting by rule. Therefore the leader of the company or of a platoon is not told, 'Do this,' but is informed, 'You are in such and such a position,' and is asked, 'What are you going to do?' This keeps his mind constantly on the alert. . . . Then the commander must go on to explain how he will utilize the ground and fight effectually. Afterwards, he is questioned as to the lessons to be re-

tained for similar circumstances if they should reoccur. . . . Here, then, there is a double lesson: How to get out of present trouble, and how to avoid trouble another time.

In other words, the French were training their candidates in a form of "estimating the situation" that would have been readily identifiable to any prewar Leavenworth graduate.[64]

Warrington also noted that the French cadre went to great lengths to have their maneuver training reproduce, as closely as possible, the conditions and confusion of the battlefield that their aspirants would face in combat. They continually updated their curriculum to take advantage of the new tactical, organizational, and technological changes that had developed at the front. However, the key for the French was not only to "manage under conditions as they may develop next week or month, but furthermore to face conditions as they may develop two years hence: because the cadet's mind has been accustomed to reasoning and to adaptability." In combat training, "The actual conditions of modern warfare are minutely reproduced, down to the explosions of shells of firing calibers, barrage fire, bombs, machineguns, machine-rifles, rifles, and mines. Even the cartridges used are not blank cartridges, having cardboard bullets which can hurt at 20 meters." Under these conditions, he observed, "some of them struggle against the nerve-shattering effects of the smoke, the noise, the vibrations, the flying bits of earth, and I have seen them conquer fear as they would have to conquer it at the front if the lesson had not been taught them at St. Cyr."[65]

Warrington observed that the terrain and trenches within the French maneuver area, unlike the pristine trenches that most American candidates trained in, were pock-marked with shell holes and laced with barbed wire and other impediments to tactical movement and weapon employment. The sensory inputs that the cadets were subjected to were designed to replicate those combat realities at the front that hindered effective small-unit command and control. As with the cadets' earlier training, throughout the maneuvers the cadre grilled them on their understanding of the situation and their response to the crisis. The senior instructor demanded of the cadets "a clear and simple statement of what was being done, a statement which none of the men could have failed to understand, and which at the same time would have sufficed for the information of any officer who had just come up."[66]

Since the French officer training focused more on the mental rather than the material aspects of warfare, nearly all the elements of their training could have been replicated by the Americans. So, why wasn't more made of Warrington's report? Like an alcoholic, the army first had to admit that it had a problem. However, to admit that perhaps the French had a more effective system for officer training would also have entailed calling into question the professional competence of the Regular Army officer corps. The more intractable problem remained one of instructors. The French placed much more emphasis than the Americans on selecting, training, and assigning the best possible instructors to its officer schools. With the rare exception of officers such as Dwight Eisenhower, the US Army never displayed a willingness to take its best and brightest officers and assign them the role of developing its officer candidates.

In addition to sorting out the issue of training officer candidates, the War Department also had to devise a system for rating and selecting the best students for commissioning. In June 1917 Dr. Walter Dill Scott of the Bureau of Salesmanship Research of the Carnegie Institution of Technology approached the War Department with a method for "scientifically" rating the abilities of officers and candidates. In true progressive form, the army jumped at the opportunity to replace its ostensibly subjective methods with an equally ostensibly objective method for sorting its aspirants. On 11 July 1917 the adjutant general informed the commander of the Fort Meyer OTC of the army's desire "to work out a more careful and scientific system" for ranking candidates for the second OTCs and that his post had been selected to test the new method.[67]

The ultimate result of the Fort Meyer experiment was the adoption of an officer qualification card developed by Walter Scott in November 1917. With some slight modification, the card would be used throughout the war as both a means of selecting candidates for commission and for commanding officers to evaluate the performance of their subordinate officers. The card listed the soldier's age, education, military experience, and previous occupations. More important, the card contained the senior officer's evaluation of the soldier using the five categories of the "Scott Scale": physical qualities, intelligence, leadership, personal qualities, and general value to the service. Scott used the following criteria for each of the categories:

> **Physical Qualities:** Physique, bearing, neatness, voice, energy, endurance. Consider how he impresses his command in these respects.
> **Intelligence:** Accuracy, ease in learning; ability to grasp quickly the point of view of the commanding officer, to issue

clear and intelligent orders, to estimate a new situation, and to arrive at a sensible decision in a crisis.

Leadership: Initiative, force, self reliance, decisiveness, tact, ability to inspire men and to command their obedience, loyalty, and cooperation.

Personal Qualities: Industry, dependability, loyalty; readiness to shoulder responsibility for his own acts; freedom from conceit and selfishness; readiness and ability to cooperate.

General Value to the Service: Professional knowledge, skill and experience; success as administrator and instructor; ability to get results.

61

While the first four categories used a scale of three to fifteen (the higher the better), the "General Value to the Service" category was more heavily weighted and used a scale of eight to forty. Although the Scott Scale was little more objective than the ad hoc methods used by instructors in the first OTCs, at least it allowed the army to believe, in the best manner of the Progressive Era, that it had used science to achieve efficiency in its selection of officers.[68]

Every OTC candidate was constantly under the observation of his commander and instructors. These cadre officers were on the lookout for any flaws in the candidate's character and abilities as well as any violation of orders, regulations, or post policies. If they discovered one of these infractions, the cadre member noted it in the candidate's file and meted out punishment for these "skins." One candidate recalled, "'Skins' could be incurred for a great variety of errors, misdemeanors, miscalculations, mental lapses, acts of fate, and ill luck. Regardless of the cause, when a man got 'skinned' he was confined to camp for the ensuing weekend; and a mark went against his name in the company commander's 'little black book.' Too many marks and he received an order to appear before the Benzine Board."[69] Many candidates considered this system to be more harsh and arbitrary than it needed to be and marveled at the pettiness that the army system could produce. After one candidate had not been able to go to the firing range because he was in the post hospital, his comrades were astonished when, during the next inspection, the inspecting officer cited the man for having powder residue in his rifle even though he had never fired it.[70] Often the candidates were put off by the fact that they were charged with a violation of rules that they did not even know existed. This experience also left the candidates with a skewed idea of "what right looked like" when it later came to enforcing discipline within their own units.

At some point in the candidate's schooling he appeared before the dreaded "Benzine Board" (named for the post–Civil War board convened to remove superfluous or incompetent officers from the army). If a candidate had demonstrated a pattern of ineptitude, indiscipline, or flaws in character during the course, these boards of officers were convened to judge his fitness to continue officer training. Gus Dittmar recalled, "No inquisition chamber in the Dark Ages ever controlled the destiny of people more completely than did the Benzine Board, as it judged the frailties and capabilities of these aspirants for commission." He did note, however, that it generally was not a candidate's failure to do well in command or poor tactical knowledge that resulted in his being called before the Benzine Board; rather, it was "repeated failure to react properly, physical weakness, lack of confidence, poor cooperation and slack interest, and an inability to adjust to the strange and often harsh environment" that caused a student to be boarded.[71]

At the end of the course, all candidates also went through a Benzine Board. The graduation board was the last check of a candidate's suitability for commission and an opportunity for the cadre to test his overall military knowledge. Like much of the OTCs, the quality of these boards varied widely from post to post, cadre to cadre, and OTC session to session. Candidate Fred Wheeler remembered that when he appeared before the final examination board two of the star questions were, "What kind of whiskey do you prefer?" and "How many buttons are on your coat?" After those preliminary questions Wheeler described what happened next:

> Finally some officer thinks of some question; the answer to which he desires to know, states it in some new language and the candidate must reply. So he makes an attempt, usually getting by as the officer [who gave the question] chooses to conceal his ignorance. By this time every one of the other officers have a question not in Moss's Manual and they open "Rapid Fire" at once. . . . Some officer requests a description as to how some simple movement should be executed. It is then the duty of the candidate to give, verbatim, the paragraph from the 1925 paragraphs of the I.D.R.[72]

The candidates make clear that the graduation Benzine Board was one of the most stressful events in their short military careers, and a percentage of those who made it through the course had their hopes dashed for a commission in the last days of the OTC.

Notwithstanding the boards' apparent arbitrariness or harshness,

THE NIGHTMARE OF A CANDIDATE

This cartoon, "The Nightmare of a Candidate," by a soldier attending the fourth Officer Training Camp at Camp Sherman, Ohio, reflects the anxiety that some candidates felt in officer training. Always lurking in the background was the specter of failure. *(From Camp Sherman Officer Training Camp,* The Gold Bar.*)*

they served two vital roles: to weed out the incompetent and to encourage the students to do their best. Given the pace of mobilization, the first function was all-important. A wartime battalion commander hailed the fact that "careful selection at the training camps" had "undoubtedly served to weed out the more defective material which presented itself for commissions." The second function of the boards was also important. Candidate Largron recalled that when the going got tough, "a little voice whispers to you 'Get busy, boy' . . . , if you flunk those two exams, they'll have you up before the 'Benzine Board' and then it'll be good-bye shoulder straps and then what will the folks at home say?" In a similar vein, Candidate Russell Frazier noted that the fear of being labeled quitters drove him and many other candidates to persevere in the training camps just at the moment "when you were sick of it all." The army undoubtedly understood the power of ego and peer pressure. The harsh whip of pride and the fear of the Benzine Boards proved a powerful motivator for the candidates during the war.[73]

One of the challenges of the training camps was to determine not only who would receive a commission, but also what rank the new officer would hold upon graduation. This was a particularly difficult decision in the first two OTCs because the camps also had to select those few candidates to be commissioned as majors and captains and thus deemed suitable to command battalions and companies. Only a small handful of commissions for majors and captains were allotted to each graduating class. At Camp Sheridan, for example, only one captain and one first lieutenant slot were allocated to each training company. In making this determination, the instructors weighed "age, previous experience and training camp record." When it came to these qualifications, "age was considered of greatest importance" in the selection of men for these higher ranks.[74]

When the selection system was properly working, the training camps seemed to have done a fair job in choosing the best candidates for commission. The whole process was described in an August 1918 report by Maj. William Gunn, the commander of the Camp Sherman fourth OTC. Gunn noted that his camp selected its final slate of candidates for commissioning after evaluating each soldier's scores on the Yerkes psychological test, his scores on written examinations, the results of his personal interview with the camp's senior instructor, and the rating he received from his platoon instructor on his overall military and leadership abilities. The last was based on daily observations of the candidate by his chain of command using the Scott Scale. Gunn maintained, "Each instructor carried at all times a book in which he made notes, from time to time, as he gained impressions of the men under his observations."[75]

Gunn's mention of the use of the Yerkes psychological test to evaluate candidates highlights one of the fascinating aspects of the war. For the first time in history, the US Army used psychological testing on a massive scale to help classify and employ men by their supposed mental capacity. On the very day that the United States declared war on Germany, a group of experimental psychologists, led by Robert Yerkes, met to gauge the practicality of using psychological testing as an aid in properly utilizing military manpower. After months of debate within the military and the continued refinement of the examination plan, on 24 August 1917 Secretary Baker gave permission to hire or commission psychologists to finalize the test and to conduct experimental evaluations of officers and enlisted men.[76]

Despite initial skepticism, the Regular Army officers involved in the experiment warmly embraced the psychological tests and sent glowing reports to the War Department on the success of the program. In many ways it is no surprise that Yerkes's psychological testing gained such an enthusiastic response in what was generally a conservative regular officer corps. The new science was a reflection of the Progressive Era spirit. It allowed the army to use science to help solve some of the pressing problems of society, in this case the mobilization of wartime manpower. The tests promised to replace subjectivity with objectivity in the quest for the great "holy grail" of Progressivism: efficiency. In an army with precious little time and beset with the pressing need to rapidly pigeonhole its legion of new soldiers into the most efficient positions that they could fill, the mental tests seemed to be a godsend. With Yerkes, Taylorism came to the army.

The rapidity with which the army accepted psychological testing was also due to the fact that the tests seemed to confirm the correctness of all of the officer corps' social Darwinist prejudices and assumptions about race, character, education, and "native intelligence." Time and time again, officers scored much higher than enlisted men on the tests. Of the first 1,116 officers tested at Camp Lee in the fall of 1917, 44 percent scored in the A category, most intelligent; 32 percent in the B, or superior, category; and only 24 percent in the C, average, category. The 76 percent of officers scoring in the A–B range was much higher than the percentage scoring in that range from the ranks of the NCOs and privates. The results seemed incontrovertible; the army had done a fine job of selecting the right men to serve as its officers from the first OTCs. If this were true, then the officer corps' assumptions about the proper education, class, and character needed for leadership must also be valid. As one officer noted, "The results of the psychological tests are fully borne out by actual observations of the abilities and capacity of various officers in the performance of duties assigned them."[77]

The Regular Army officers of units mobilizing throughout the country were quick to grasp the utility of using the mental tests as a means of selecting men for officer training. One went so far as to maintain, "It is doubtful whether applicants should be admitted to school who have not, according to the psychological examinations, made a score equivalent to 'high average' intelligence (C+)." Another camp commander revealed that in addition to an evaluation of their personal bearing, military knowledge, and physical abilities, applicants for officer training in his unit were also given mental tests. He noted, "If they do not rate A or B in this examination they are rejected."[78]

This practice of using the psychological examinations as a deciding factor in the selection of officer candidates was so widespread that on 14 August 1918 the War Department issued General Order 74 mandating that the tests were only "assisting guides" for selection boards and that "no particular psychological rating shall be declared as a minimum to be attained by any such candidates." However, this

65

stricture did not keep unit commanders from continuing to place great weight on the mental tests as a tool for weeding out officer candidates. As late as the fall of 1918 the commander of Camp Lewis, Washington, reported, "One of the most important services [of the exams] has been to assist in selecting candidates for the officers' training schools. It was demonstrated that a certain minimum of intelligence was essential to success in the training school, and that candidates failing to reach a given psychological rating failed to receive commissions. Approximately 17 per cent of candidates of the fourth officers' training school were thus eliminated by purely objective standards with considerable saving to the army."[79]

The tests also followed a pattern in which test scores all fell out according to rank. In other words, majors scored higher than captains, who in turn scored higher than lieutenants, and so on. Yerkes claimed that this progressive stepping in scores based on rank had nothing to do with education. To prove his point he compared officers with the equivalent of an eighth-grade education to "native born white recruits of high school and college education." The end result was that the undereducated officers still did slightly better on their intelligence tests than did their more highly educated enlisted soldiers. Yerkes made no real effort to explain this phenomenon other than to assert, "It is evident that the examination is measuring other qualities, in which officers stand above recruits, to a greater extent than it is measuring education."[80] Although this might invalidate the army's preference for college-educated officers, the army could still use these figures as proof that it was a democratic institution where natural talent allowed everyone to rise to their fullest extent.

Of course the greatest problems with Yerkes's psychological tests were that they were culturally biased and, contrary to his assertions, dependent on a degree of general education to correctly answer the questions. Despite its inherent flaws, the psychological tests served a valuable purpose. Knowingly or not, the tests rewarded white, educated, middle- and upper-class men — men who mirrored the characteristics and perhaps the social outlooks of the West Point graduate.

Looking back on 1917, the War Department could claim some notable accomplishments in its creation of an officer cadre for its rapidly expanding divisions. The first two sessions of OTCs had commissioned 44,578 new officers. Given the fact that the army began the war with only 17,000 Regular Army, National Guard and Army Reserve officers (and many of these had less than a year in service), the OTCs had accomplished a prodigious feat. The army's desire for college men and business leaders for its candidates meant that overall the graduates of the training camps were some of the best educated and motivated officers that the nation had ever been able to commission at the beginning of its wars. Moreover, it can be stated that the army's insistence that all new officers for the National Army have (in theory) at least three months of standardized instruction before commissioning meant that the new officers of 1917 were also better trained than the wartime volunteer officers of any previous American conflict. It is clear that the army endeavored to establish a rigorous system for selecting and evaluating its officer candidates, to the point of turning to the new science of psychology to evaluate their intelligence and mental capacities.

Despite all these accomplishments in the training of officers, however, the devil was in the details. Because of shortages of instructors, equipment, and facilities, the training at the OTCs never reached the degree of standardization that the War Department expected. The training itself was too mired in obsolete prewar doctrine and unrealistic views and assumptions about modern war to accurately and adequately prepare the candidates for the tactical trials that lay before them. As these new officers were to be the primary teachers of the wartime legions, this fundamental flaw in their training was redoubled in the tactical instruction of their later units. Given the fact that OTC graduates made up the bulk of the junior officers in the wartime Regular Army, National Guard, and National Army divisions, few units escaped this elemental problem. Unfortunately, the officer training situation only grew worse in 1918.

4 "By Improvised and Uncoordinated Means"
Officer Selection and Training in 1918

IN HIS ANNUAL REPORT TO CONGRESS FOR 1918, Chief of Staff Peyton March accurately noted that the army's efforts to build a wartime cadre of officers was accomplished only "by improvised and uncoordinated means."[1] Creating the first OTCs had been a monumental undertaking, and despite their great inadequacies in training, they generally succeeded in bringing to the army a high quality of human material to serve as its new officers. But the question remains: why did officer training in 1918 continue to operate under the "improvised and uncoordinated means" that March so lamented?

Although some within the Regular Army expressed concern over the training given at the OTCs, most of the regular officer corps seemed to have been either content with the system or resigned to the fact that there were no other viable options open to the War Department. With the army satisfied with, reconciled to, or perhaps blissfully ignorant of the outcomes of the first OTCs, it is no surprise that the third session of OTCs (5 January–19 April 1918) ushered in only a few minor changes to officer training. With the continued demand for more officers, the War Department expanded the camps from sixteen to twenty-four schools. The army also opened small schools in Hawaii, the Philippines, and the Panama Canal Zone. Other changes to the camps included giving preference for selection to attend the camps to enlisted soldiers and commissioning new officers to company-grade ranks only. The only change to the camps' training regimen was to extend the courses by two additional weeks to give the candidates instruction in "Army paper work and company administration."[2] The army made no material change to the existing OTC plan for tactical training.

The army's desire to admit more enlisted men to OTCs ultimately led to 90 percent of the camps' candidates coming from the ranks. There were also further changes to the requirements and obligations of the candidates entering the camps directly from civilian life. These men could be no older than thirty-two (as opposed to forty for the enlisted candidates) and were now required to enlist in the army as a precondition for admission. The latter change had been pushed by

Adjutant General McCain as a way of reducing costs. McCain argued that while the first training camps were successful, they were also expensive. In December 1917 he recommended forcing all candidates to first enlist in the army, because "camps composed of enlisted men bring practically no expenses additional to those of the normal expenses of training enlisted men." While the candidates from the ranks continued to be paid based on the enlisted rank they held before entering the camp (that is, a candidate who was a sergeant continued to be paid as a sergeant), the candidates from civilian life were now paid only as privates first class (approximately thirty-three dollars a month) instead of the stipend of one hundred dollars a month enjoyed by previous civilian candidates. Also, if the civilian candidate failed to be commissioned, he was required to serve the rest of his term as an enlisted man.[3]

The third OTCs eventually graduated 11,659 candidates, commissioning 8,165 in the infantry, 3,347 in field artillery, and 147 in the cavalry. The War Department originally intended that only a small number of the graduates from the third OTCs would be commissioned immediately. The rest would constitute a reserve pool of trained personnel who would be commissioned as needed when officer vacancies opened as a result of casualties or the continued expansion of the army. Until such time as these men were needed, they would return to their units to serve as NCOs. Unfortunately, unforeseen officer losses in the spring of 1918 forced the War Department to commission all of the third OTCs' graduates on 29 May 1918. This left the army with no ready reserve of soldiers qualified for immediate commissioning if the need arose.

As the third OTCs witnessed the first large influx of enlisted men into officer training, there were differences in the levels of education, military experience, and occupation between these candidates and those of the first two camps. The available evidence from the third OTC at Camp Devens, Massachusetts, illustrates some of these changes. The Camp Devens class began with 716 candidates. This number dropped to 625 by the end of March 1918. Of the 91 losses, 31 had accepted commissions in the Engineer and Tank Corps, 2 had transferred to aviation and ordnance schools, and 56 had resigned or failed out of the course.

Due to the preference given to enlisted men for the third OTCs, over 78 percent of the candidates at Camp Devens had prior service in wartime units. One should not make too much of this military experience. Few of these soldiers had been in the army more than a few months, and much of their training had been limited to the School of the Soldier. The remaining 21.3 percent of the candidates were students drawn from local New England colleges and universities. The average age of the candidates was twenty-five, at least two or more years younger than those of the 1917 Fort Sheridan sample.

Unfortunately, the education and occupations that the Camp Devens candidates listed in their graduation book were not nearly as detailed as those given at Fort Sheridan. Ninety-eight of them (13.6 percent) did not give an occupation. Those candidates who did list their occupation were classified as follows:

Business, not specified as to type, 319 (44.5 percent)
Student, 152 (21.3 percent)
Engineer, 48 (6.7 percent)
Professional, 31 (4.3 percent)
Teacher, 31 (4.3 percent)
Farmer, 28 (3.9 percent)
Lawyer, 7 (1 percent)
Regular Army enlisted, 2 (0.02 percent)

If the available Camp Devens statistics are representative of other of the third OTCs, the statistics do suggest a steep decline in the education levels of the candidates. Only one of the four companies in the graduate book listed the education levels of its members. The first company stated that 69 (46 percent) of its 150 candidates had some level of college education; 65 (43.3 percent) had graduated from high school or preparatory school, and 16 (10.7 percent) had only a grade school education. If one combines the number of college students, teachers, engineers, lawyers, and professionals for all of the OTC's companies, it is a safe inference that at least 38 percent of the all of the candidates had some degree of higher education.[4] Since the "business" category was so vague, it is impossible to make even the barest guess as to the social class of the Camp Devens students.

The changing demographics of the OTC candidates illustrate the personnel challenges that the army faced during the war. These statistics in no way suggest that the army was repudiating its long-standing preference for college-educated officers. In fact, soon after the graduation of the second OTCs, Col. P. D. Lochridge expressed his belief to the army chief of staff that "in the present crisis the best material for officers should be sought from all colleges and representation not limited to those colleges in which military training has been given."[5] However, Lochridge also admitted that college graduates were a finite resource, and the army would have to be content with the stopgap measure of sending the best qualified enlisted men to the OTCs.

The commissioning of enlisted men proved to be a double-edged sword. The army had to balance the perceived military experience of the enlisted candidate against its desire to have men with the mental flexibility that it believed came from a college education. The army, as an institution, also had to match itself to the expectations of the larger American society. The elitism of education had to compete with the national expectations of equality of opportunity.

Throughout the war, many regular officers claimed that entry into the officer corps was based on merit and natural talent. Dwight Eisenhower claimed that "in our Army, it was thought that every private had at least a second lieutenant's gold bars somewhere in him and he was helped and encouraged to earn them." Lochridge's replacement as the chief of the War College Division, Brig. Gen. Lytle Brown, went so far as to argue in July 1918: "It is true that a college education is an advantage, but many men have made good officers who have not had college education, and it is believed that in the democratic army we are building up under our draft system every man should at least be given the chance." The desire for college men had to coexist with the idea that every white American citizen in the ranks carried a field marshal's baton in his knapsack, or at least a pair of lieutenant's bars in his pocket.[6]

The belief that the wartime force was a vast democratic army was also held by some of the candidates themselves. Russell Frazier, a student at the Camp Lee COTS in the fall of 1918, recalled,

> As we help that Pennsylvania coal miner with the slope scale, he forgets his labor union and his animosity for the Plute [plutocrat]. Over in the other corner the son of a United States Senator and a former carpenter are stretched out on one bunk telling of their future hopes and fears. All are dressed alike, the snobbery of the first days are gone and all is harmony. We are a little puzzled as we look over this peaceful scene and then the great light breaks; "Gosh!" we fairly shout, "this is what is meant by Democracy, this is what we are going to fight for."[7]

In Frazier's vision was the stated hope of the prewar progressive UMT advocates: that military service would "Americanize" the nation and bank the fires of class warfare. It is clear that a democratization of the officer corps did occur to some extent during the war, but this was due much more to the ever-gnawing need for junior officers than to any high-minded commitment to equality of opportunity. By the end of the war the army would prove itself more than willing to change its age standards to get men with some college education than to continue down the path of the democratization of the officer corps.

Although the army wanted college-educated officers, the decision to commission NCOs was not unexpected. The army had long in-

71

tended to commission a number of regular NCOs upon mobilization. On 1 May 1917 the adjutant general placed these plans into effect by notifying the departmental commanders to activate the commissions of all enlisted men who had previously passed the examination for becoming reserve officers. Adjutant General McCain reminded the department commanders that since the "need for expansion of military forces is urgent," they must ensure that the commissioned enlisted men must "take over legitimate duties of an officer."[8] The commissioning of regular NCOs quickly expanded beyond just those who had taken the officer examination.

In the spring of 1917, regular sergeant Sam Woodfill was still serving with his unit on the Mexican border. When the war broke out, Woodfill recalled that "all of us old rookies had visions of being rushed to the firing line" and that "they'd shoot us across on the next boat."

Woodfill and his comrades soon found out that "they needed officers and, although we were typical old-time non-coms of the regulation buck private breed, they shot us over to San Antonio for a couple of months' training and then gave us temporary commissions."[9] It is interesting to note that despite Woodfill's having nearly twenty years of service in the regulars, the army still sent him through an OTC before granting him a commission.

Although Woodfill would be touted by Pershing as the AEF's finest soldier, not all of the enlisted men admitted to the officer corps did as well as he. Comments by some OTC students and cadre members suggest that there were some tensions between the old regulars and the other candidates. Of the regular soldiers attending the Leon Springs OTC with him, Gus Dittmar noted that the old-timers brought both experience and humor. Unfortunately, "the humor that they contributed was unintentional, resulting mainly from their volatile language, lack of imagination, and close conformity to the 'old army' thinking and habits."[10] In 1918 another candidate penned this ditty poking fun at the supposed superiority of the long-service regular NCO going through officer training:

> He's been in the service for sixteen years,
> And he's spent all his days shooting blanks.
> At blowing he really has not any peer,
> And he's spent all his life in the ranks.
>
> We marched one day to the P.T. field,
> And the C.O. called his name.
> The shock of the thing made the old boy reel,
> But he tottered out just the same.

He tried his best to lead that bunch,
But he lost his poise and grace.
He knew how to act in ranks and at lunch,
But in front he was out of place.[11]

Perhaps the traditional jaundiced American view of the regular enlisted soldier was simply too great for many of the college men to
overcome. But if these sentiments were common, then Frazier's barracks room utopia was far from a success.

The college men were not the only ones with patronizing views
of the regular soldiers. Regular officers sometimes showed their own
prejudices, especially when enlisted men failed to make it through
officer training. One officer training school commandant disapprov-
ingly noted, "Among the men from the old Regular service there was
a large proportion that seemed to have studied but little, outside their
military subjects; these men had a hard time with the course and at
times seemed to feel they were at a disadvantage, and they were,—the
fault was their own; they had never studied when they were young
and when called upon for some elementary knowledge requiring fig-
uring, elementary geometry, etc., [they] floundered hopelessly."[12] Officer training was perhaps less of a democratic meritocracy and more
of a Darwinian jungle in which the regular officers believed that the
jumped-up enlisted men were out of their depths and destined to
fail. This attitude can also be seen in what became of many of the
NCOs commissioned during the war after the Armistice. While Lucian Truscott, an officer with a bit of education and no overseas service, retained his commission after the war, the uneducated Medal of
Honor recipient Sam Woodfill reverted back to being an NCO so as
not to lose his pension.

The inclusion of enlisted men in officer training presented as many
problems as it solved. In selecting enlisted men for officer training, the
army had to balance the need for retaining good NCOs in its units
with the equally pressing need for building a corps of competent junior officers. Throughout the war the army chose to err on the side
of producing officers over retaining good NCOs. On 21 November
1917, Lochridge airily dismissed complaints from field commanders
that the War Department's evolving OTC attendance policies were
denuding their units of good soldiers: "The withdrawal of several of
the best noncommissioned officers from the companies at the end of
the training period will, for the time being, weaken them, but companies must expect to make preparation in advance for the loss of their
best men. Part of the duty of the company commander must be the

development of officer material in the company."[13] This cavalier attitude toward the NCO corps would later have a major impact on the cohesion and effectiveness of the AEF's small units.

Although the third OTCs witnessed the first large influx of enlisted men into officer training, other aspects of the camps were far less novel. The problem of finding qualified instructors, for example, remained intractable. Following a trend that had begun with the second OTCs, as the regular officer instructors were posted to new assignments, their billets for the third camps were being filled by newly commissioned officers with less experience. Of the twenty-one instructors (out of twenty-two) of the third OTC at Camp Devens who listed their commissioning sources, two were reserve officers commissioned from the 1916 Plattsburg camp, fourteen were commissioned by the first OTC, three were commissioned by the second OTC, one was a National Guard officer, and one was a regular NCO commissioned for war service. Sixteen of the instructors had some degree of college education. Half of the instructors also had some level of prewar military experience. Four of the instructors had served in the prewar regulars, with one, John Schweitzer, having served over eighteen years in the ranks as an infantry NCO. Seven more had served as enlisted men in the National Guard, but all for less than six years.[14] Camp Devens was actually fortunate to have even this amount of military experience in its instructors. The instructors in many other camps were far less seasoned. However, the fact remained that none of the Camp Devens instructors had been officers for more than fifteen to eighteen months when they began training the camp's candidates.

In addition to continuing problems with instructors, the third OTCs also faced other challenges. Shortages of equipment, especially in machine guns and the new weapons of trench warfare, continued to limit hands-on training. Even the weather seemed to be against the army. The winter of 1917–18 was exceptionally cold. Training camps in the deep South saw accumulations of snow and frigid temperatures that played havoc with the OTCs' training schedules. In the end, weather, the continued adherence to obsolete doctrine, and the shortages of qualified instructors and equipment all combined to make the third OTCs as ineffective at preparing their graduates to face the realities of combat as had been the first two sessions.

Despite the thousands of new officers commissioned by the first three OTCs, the expanding army's thirst for leaders was insatiable. In April 1918 the War Department directed the commanders of the twenty-four divisions then training in the United States each to establish an OTC at their post to provide officers for their units and to serve as a general reserve for all other officer vacancies in the army as a whole. This fourth session of OTCs was scheduled to run from 15 May to 1 September 1918. To fill the camps, the War Department required that all divisions and nondivisional units or organizations (excluding the Coast Artillery) stationed in the United States provide a quota of 2 percent of their total enlisted strength to attend officer training. The total enrollment at the schools was 13,114, almost all of whom came from the enlisted ranks.[15]

The War Department hoped to address some of the previous shortcoming in officer training by adding new classes or adjusting the time devoted to certain subjects in the fourth OTCs. For example, the training time for infantry officer candidates expanded from 625½ hours of instruction in the third OTCs to 718 hours of training in the fourth OTCs. Unfortunately, much of these additional hours came from expanding the candidates' study time and from adding classes on "practical administration": paperwork, company supply, and mess hall management. Although there was some effort to make the tactical training more focused and realistic, the overall amount of time devoted to minor tactics remained largely unchanged from that of the previous OTCs. The changes made to tactical training were little more than the reallocation of time from one subject to another. Thus, while the first two OTCs devoted 84 hours to trench warfare (to include 60 hours of maneuvers), by the fourth OTCs the candidates were spending 106 hours studying trench construction, weapons, and tactics.[16]

The tactical training that the candidates of the fourth-session OTCs received remained problematic. Notwithstanding the War Department's best efforts to improve instruction within the camps, local commanders continued to deviate from the standard training plan because of local conditions and shortages or their own tactical proclivities. Ironically, the War Department had actually exacerbated the problem of standardization by making officer training the responsibility of the division commanders. For better or worse, the individual division commander's personal interest in his OTC often determined the quality of officer training in his camp. Given the systemic problems of raising and training the larger division, many commanders seemed to have followed the path of least resistance in their OTCs. This often meant that close-order drill and bayonet practice continued to occupy far too much of the candidates' time. In a parody of the British trench song "Drunk Last Night," one Camp Dodge candidate penned,

We drilled last week and we drilled the week before,
We're going to drill next week like we never drilled before;
For when we drill we're as awkward as can be,
I don't see how the hell they'll make an officer out of me.[17]

The War Department and some division commanders did attempt to make the training of the candidates more realistic by using the latest Allied tactics. A few of the third OTCs had also experimented with the new tactics in April 1918. Regrettably, these efforts do not seem to have met with much success. With the advent of the new two-line formations that were to replace the old skirmish lines of the *IDR,* one of the candidates acidly observed, "If you don't understand the new formation, ask your officers about it; they don't either." He also noted that under the latest system, "We have two lines of skirmishers in the new formation. This is to prevent one-half of the platoon from finding out what the other half does not comprehend." Another candidate recalled the confusion that resulted from a French officer's efforts to teach his company the army's latest tactical formations for the platoon in the attack. In executing the drills, the candidate described his tactical training as wading "promiscuously through the woods" following commands that he did not understand while assuring himself that he was "now ready to advance on the enemy and should be able to inflict fifty thousand casualties without sustaining a single one" himself. Although these observations were made for amusement, it is clear that the instructors were unsuccessful at explaining the underlying concepts upon which these tactics were based, and the candidates merely went through the motions of the maneuvers rather than truly understanding them. Despite its obsolescence, the *IDR* remained the paramount doctrine at the OTCs if for no other reason than it was relatively simple in concept and execution and in its expectations of junior leaders.[18]

By the time the fourth OTCs began training, the United States had been in the war for a year. Some regular officers believed that the army's policy of filling the camps with enlisted men had caused a decrease in the quality of its officer candidates. After an inspection of the fourth Engineer OTC at Camp Lee, Virginia, Gen. Lytle Brown concluded that "the qualifications of the students at this camp are far below the standard previously maintained at other Reserve Officers' Training Camps." He placed the blame for the inclusion of "undesirable men" in the OTC at the feet of division commanders and their subordinates for failing to screen applicants for commission properly.[19]

An examination of the candidates of the Camp Sherman fourth-session OTC provides a glimpse into the changes and continuities in demographics of the officer candidate population and a way to test the validity of Brown's assertion. The average age of the candidates at Camp Sherman in the summer of 1918 was twenty-five, the same average age as that of the students in the Camp Devens third OTC. Of the 362 Camp Sherman candidates who listed their educational background, 249 (68.8 percent) were college graduates or had some degree of college education, 41 (11.4 percent) were high school graduates, 64 (17.6 percent) had some high school education, and 8 (2.2 percent) had attended only grade or common school. If these levels of education where indicative of the other fourth OTCs, then these camps actually witnessed an increase in education of their candidates over that of the third OTCs.[20]

Other evidence also points to the possibility of a rebound in the quality of candidates between the third and fourth OTCs. Robert Yerkes noted that the psychological testing of over nine thousand candidates attending the third OTC "seemed to indicate that the student officers in the third series of training schools are inferior to the first and second schools in intelligence." This trend seemed to reverse itself as test scores for candidates in the fourth OTCs returned to levels similar to those of the 1917 classes. However, scores for infantry officer candidates going through training in the first COTS class in the summer and fall of 1918 were more than ten percentage points lower than the average score for infantry officers the previous year.[21]

In addition to increases in education, more than three-quarters of the Camp Sherman candidates had served at least three months as enlisted men before attending the OTC (with nearly all of these being wartime enlistments). This was less than the percentage of enlisted men attending the third OTCs. Of these enlisted men, 11 (3 percent) had been regimental or battalion sergeants major, 12 (3.2 percent) had been first sergeants, 106 (29.2 percent) had been sergeants, 33 (9 percent) had been corporals, and 120 (33 percent) had been privates. The high percentage of NCOs attending the course (over 44 percent of the total class) represented a major drain of junior leadership from the 84th Division and the other units that filled the Camp Sherman OTC. It is interesting to note that 91 of the 120 privates who attended the course either were college graduates or had spent at least a year in college. The same could be said of the 82 men who were admitted to the camp without any prior military service. Only 14 of these men (17 percent) lacked any college education.[22]

The detailed records of the Camp Sherman Fourth OTC also

allow us to draw some tentative conclusion about the social class of those seeking commissions after a year of war. The occupations the Camp Sherman candidates listed upon entry into the course, and the number of men in each, were:

Student, 44	Mechanic, 3
Salesman, 33	Advertising, 1
Clerk/office work, 31	Baseball player, 1
Teacher, 28	Buyer, 1
Farmer/rancher, 21	Contractor, 1
Businessman, 19	Draftsman, 1
Accountant/auditor, 16	Editor, 1
Industrial worker, 8	Farm manager, 1
Office/sales manager, 7	Lineman, 1
Banker, 5	Plumber, 1
Actor, 3	Real estate, 1
Druggist, 3	Watchmaker, 1
Insurance, 3	Unknown, 89

Using the standard applied to the Fort Sheridan OTC for determining social class, and factoring out the students and those candidates who listed no profession, 86.6 percent of the candidates could be classified as having been employed in professional or white-collar jobs. It appears that in many ways the quality of the candidates at the fourth OTCs actually increased or at least remained constant with that of the third OTCs.

Whatever the quality of the candidates, training in the fourth OTCs was still undercut by the intractable problem of securing qualified instructors. A year into the war, only the camp commandants and perhaps a handful of staff officers were regulars. This meant that short-service officers were responsible for training and nurturing the army's leadership seed corn. The OTCs at Camps Dodge and Sherman illustrate this ongoing problem. While Camp Dodge's OTC at least retained a Regular Army officer as its camp commander, the senior instructor, Capt. Harold Schaub, was a graduate of the first OTC at Camp Snelling, Minnesota. Of the remaining thirteen instructors and staff officers, nine had received their commissions from the first OTCs and four had been commissioned only since November 1917. Only two had had any military experience before the war. First Lt. E. G. Kelsey had served three years as an enlisted man in the Idaho National Guard, and Capt. Marion Drake had served briefly with the Indiana Naval Reserve. Although it was bad enough that all of the instructors had less than a year of commissioned service, the military experience that they had gained since their commissioning was also suspect. Of the thirteen instructors, nine had served only in depot brigades or other OTCs before joining the Camp Dodge camp. This meant that their actual leadership experience and depth of military knowledge were limited.[23]

The instructor situation of Camp Sherman was equally grim, and actually made worse by the War Department's eleventh-hour personnel decisions. After the Camp Sherman class had been in training for only a month, the War Department abruptly reassigned the camp commander and all of his instructors. The replacement commander, Maj. William Gunn, had been commissioned after the first OTC and had less than ten months in uniform when he took over the camp. His senior instructor, Capt. Jesse Marshall, had held a commission only since 21 April 1917. Of the eight assistant instructors, one was commissioned in June 1917, five were graduates of the first OTCs, and two were graduates of the second OTCs. All told, this meant that all of the assistant instructors, those officers who were directly responsible for most of the training of the candidates, had only seven to twelve months of military training themselves when they began teaching the course in June of 1918. This lack of experience was not lost on the OTC students. One candidate only half jokingly warned future candidates, "If on a deep tactical question, do not embarrass the instructor by asking him to explain [it] before the entire command." Given the dearth of knowledge that the Camp Sherman and Dodge instructors brought to their teaching, it was truly a case of the "blind leading the blind" when it came to officer training.[24]

79

If all these woes were not enough, last-minute decisions and policy changes from the War Department also hobbled training in the fourth OTCs. On 15 May 1918 the army opened an OTC at Camp Zachary Taylor, Kentucky, to provide officers for the 84th Division, then organizing at that location. After only three weeks of training, the OTC accompanied the division on its move to Camp Sherman, Ohio. Due to this major disruption in training, the War Department at first granted the OTC commander permission to extend the course by one month, but later reneged on the agreement. In August 1918 Maj. William Gunn, the OTC commander, reported that his students' instruction on the automatic rifle had been limited to a theoretical discussion of the tactical employment of that weapon because the school had been "unable to procure the rifles" for hands-on training. He also noted that the candidates had not received rifle range practice or instruction on company administration because of the move of the OTC from Camp Taylor to Camp Sherman and the War Department's early closure of the training camp.[25]

These instructors from the fourth OTC at Camp Sherman, Ohio, illustrate the challenges that the army faced in finding qualified instructors. The camp's senior instructor, Maj. William Gunn (top photo) was commissioned in the first OTC in August 1917. He had been in the service for just one year when his fourth OTC class graduated. Of the instructors in the lower photo, one had served in the Regular Army for three years prior to the war, five had been commissioned in August 1917, and four had been commissioned only since November 1917.
(From Camp Sherman Officer Training Camp, The Gold Bar.)

Sadly, these disruptions were far from rare. Like Camp Sherman, the fourth OTC at Camp Dodge was also hindered by last-minute changes imposed by the War Department. After a month of training at Camp Dodge, the instruction of candidates ground to a halt as the department filled the camp with a large levy of new aspirants. The department took this step in an attempt to overcome the ongoing shortage of officers. At the same time, the department reassigned nearly all of the instructors back to their original divisions for deployment to France. Soon after training resumed with a new cadre of instructors in mid-June 1918, the War Department again disrupted the camp by sending candidates slated to be artillery and machine gun officers to the new Central Officers' Training Schools (COTSs) at Camps Taylor and Hancock.[26]

As with the third OTCs, the War Department originally intended to commission only a fraction of the candidates as officers upon their graduation from the fourth OTCs. This was again an effort to build a pool of personnel trained to be officers. Unfortunately, no one had thought to consult the Germans on the American plan. In March 1918 the German army began a series of attacks along the Western Front designed to end the war before the Americans could bring their weight of numbers to bear. The German offensives forced the commitment of the AEF to battle much earlier than Pershing or the War Department had anticipated. As American officer casualties mounted in the spring and summer of 1918, the War Department tried to meet the crisis by commissioning all the third OTC candidates

With the deepening officer crisis, on 2 August 1918 Adjutant General McCain recommended that the army immediately commission those fourth OTC and COTS students who had received at least three and one-half months of their four-month officer training course. He argued that this would help to alleviate the shortage of infantry officers by filling the junior commissioned ranks of those divisions departing for France in August and September 1918. The director of the War Plans Division concurred with McCain's recommendations and admitted that since the newly formed COTSs were still short of both instructors and facilities, the early commissioning would gain him some time to sort out the growing pains of the central schools. During this crisis the adjutant general even directed camp commanders immediately to commission some candidates who were still far short of the three and one-half months of training he had earlier proposed. In early August 1918, with more than a month of instruction left in the course, the War Department ordered Camp Dodge's OTC commander to commission sixty-eight candidates early so they could deploy with the 88th Division.[27]

As the War Department had not expected to need replacement officers in any number until 1919, it was caught flat-footed by the demand for junior officers in mid-1918. To a great extent, poor planning by the General Staff had contributed to this problem by not linking the attendance at the training camps to accurate projections of long-term officer needs. It was not until 18 May 1918 that the adjutant general warned Chief of Staff Peyton March that despite the recent commissioning of an additional 11,657 officers from the third OTCs, the army was in danger of draining its remaining pool of surplus officers. He estimated that "only 2,000 Infantry and 750 Field

Artillery [officers] are available to be absorbed." He further noted, "It is believed to be a conservative estimate that the replacement needs of officers in the several arms will approximate not less than ten percent per month of those actually engaged in front line service. Beginning with the month of August and assuming four army corps to be actually engaged, replacement needs alone will approximate 2,000 officers per month, therefore, for the remaining five months of the present year, provisions should be made for at least 10,000 officers, for replacement purposes alone."[28]

Although the War Department waited until the crisis was upon them to act, it did take steps in the spring of 1918 to try to solve its glaring shortages of officers. Shortly before the commencement of the fourth OTCs, some members of the General Staff were already questioning the wisdom of placing the OTCs under the control of division commanders. On 14 March 1918 the acting chief of the War Plans Division, Col. D. W. Ketcham, recommended to March that the army replace its system of OTCs with four centralized camps for training officer candidates. Ketcham argued that the war was likely to be a long one and that "there is a possibility of necessity for very large officer replacements." In a telling indictment of the current OTC training, he stated that recent inspections had revealed that "the present schools in divisions are, with few exceptions, failing to produce the desired results" and that "some of the schools have been described as so ineffective as to be farces."[29]

Ketcham was far from being alone in his desire to reform officer training. Leonard Wood had long been dissatisfied with the officers he was receiving from the OTCs. He unequivocally stated in April 1918, "The three months' course does not turn out sufficiently well trained officers and it has been necessary to give them comparatively long periods of training in France." He recommended that the OTCs be expanded from three to six months. McCain agreed with these sentiments and pleaded with March immediately to establish COTSs. He estimated that such centralized officer schools could turn out at least two thousand new officers per month and would overcome the problem of training replacement officers after divisions in training left for France and closed the OTCs.[30]

Unfortunately, change was slow in coming. It was not until early June 1918 that the War Department finally issued orders closing the OTCs in the divisional training camps and establishing three infantry COTSs at Camps Gordon, Lee, and Pike; a machine gun COTS at Camp Hancock, Georgia; and a field artillery COTS at Camp Zachary Taylor, Kentucky. As already noted, this late decision disrupted training in the fourth OTCs and complicated the establishment of the new COTSs.

Regardless of branch, all of the COTS classes were to be four months in duration. In designing the course, the War Plans Division sought to continue to balance training in both trench and open warfare and directed that in the interest of time "the course of training must not be interrupted by unnecessary parades, reviews, practice marches," or other distractions.[31] During the first month of the new course for infantry officers the instruction would focus on the theoretical basis of the *IDR,* the *Manual of Interior Guard Duty,* field sanitation and first aid, and the *Small Arms Firing Regulations.* Training in the second month continued to emphasize the *IDR* but put more time into the theory and practice of minor tactics, topographic sketching, administration, paperwork, and the "interior economy of company messing and property." The third month of training continued to emphasize the *IDR* and topographic sketching, but now focused mostly on musketry training and practice (sixty-two hours) and on minor tactics. The final month of training introduced map reading, field fortifications, the *Manual for Courts Martial,* the tactical employment of machine guns (three hours), and "trench warfare, grenades, trench mortars, gas, etc." for twenty-six hours. Physical training, bayonet training, and signaling were emphasized throughout and given the same amount of time during each month of training. This meant that the candidate still spent forty-nine hours in bayonet training over the course of the school.

The continued reliance on the *IDR* also demonstrates the persistent use of obsolete tactical doctrine in officer training. It is amazing how little the tactical instruction of officer candidates had changed over the course of the war. In the fall of 1918 candidates were still being subjected to training that in no way reflected the realities of combat in France. The subjects taught in the infantry COTSs, and the time devoted to them, are listed in Table 4–1.[32]

The ill focus of much of the training at the COTSs was reflected in the questions the candidates were given in their written examinations. An examination given to Camp Gordon candidates during their first week of training, for example, asked them, "What is the present status of Bulgaria in this war?" "What are the 'Eagle Boats' and who makes them?" and "Who wrote the following: *Evangeline, The Last of the Mohicans, The House of Seven Gables, Ben Hur, The Call of the Wild?*" On an examination in the eleventh week of training the aspirants were required to "Name the eight rules that govern the carrying of the piece." An examination given COTS students at Camp

Table 4-1. Four-Month Training Plan for Infantry COTS Candidates

Subject	Hours of Training	% of Total Training
In and out processing, and misc.	45	4.6
Evening study	176	18.3
Physical training	49	5.1
Bayonet training	49	5.1
Infantry Drill Regulations (33 hours theoretic and 121 hours hands-on)	154	16
Small Arms Firing Regulations and fundamental musketry	42	4.3
Practice march	23	2.3
Manual of Interior Guard Duty	23	2.3
First aid, hygiene, and field sanitation	16	1.6
Military discipline and courtesy	8	0.9
Signaling	24	2.4
Field firing	8	0.9
Range practice	48	5
Military topography and sketching	43	4.4
Manual for Courts Martial	13	1.3
Field engineering and fortification	6	0.7
Sand table and map exercises	15	1.5
Company administration, paperwork, and interior economy	24	2.4
Field Service Regulations	6	0.7
Minor tactics	17	1.7
Advance guard and outposts	14	1.4
Combat offensive and defensive	8	0.9
Map problems	12	1.2
Tactical walks and terrain exercises	19	2
Employment of machine guns and automatic rifle	6	0.7
Trench warfare (grenades, trench mortars, gas)	26	2.7
Tests and examinations	36	3.7
Inspections and ceremonies	42	4.3
Lectures	10	1
Total training hours	961	

MacArthur on 21 September 1918 included some weighty questions drawn from materials in the *IDR,* such as, "Why is close order drill essential?" "Explain the command, By the Right Flank, March," and "Describe the Hand Salute." The essay portion of the test asked that the students, "In not more than one hundred words state the value of bayonet work from the following standpoints: physical benefits to be derived; need for alertness of mind as well as quick and decisive movements of the body; discipline; developing the individual fighting spirit of each man." As had been the case with the previous O T C s, the problem with the training in the camps had more to do with what was taught than how much time the course had to teach its students.[33]

The poor quality of the training was not lost on either the students or certain senior American officers. On 4 September 1918 Leonard Wood expressed his concern to the adjutant general that the COTSs were failing to give "proper weight to the quality of leadership, judgment, character, and initiative" of the candidates. He went on to note, "There are plenty of men who can squeeze through a training camp, memorize commands and make fair recitations; who are absolutely unfit to be officers. . . . It is a dangerous policy to continue turning out partially instructed men. This war demands thoroughly trained officers, and no class is more important than the platoon commander, who is generally a young officer just from training school. They are not coming to us at the present time, so trained as to be really competent to perform this duty."[34]

85

Wood's sentiments were shared by the Camp Gordon COTS candidate Henry P. Frey, who in a July 1918 letter to an acquaintance in the War Department wrote that "there are a lot of defects in the system" of officer training. Fry was a graduate of VMI, a former commissioned officer in the National Guard, and a recent sergeant in the National Army. He maintained that "the theoretical instruction given in the school is exceeding weak" and singled out instruction in musketry, bayonet work, topography, and minor tactics as being particularly flawed. He also believed that not enough time was being devoted to subjects that a young officer needed "to make him think." Candidate Robert O'Hair simply believed that his training did not entail "enough attention to detail or preparation for actual battle."[35]

Wood and the candidates were not alone in their anxiety over the training at the COTSs. Capt. V. S. Hebbert, a British officer assigned to Camp Gordon, reported on 25 July 1918 that the training plan for the COTSs was in disarray because there was little uniformity in training. He also noted that the candidates had arrived at Camp Gordon from several defunct OTCs, each of which differed in the content and methods of instruction. An inspector from the General Staff's Training and Instruction Branch discovered similar problems upon visiting Camps MacArthur, Pike, and Gordon in the same month. Maj. Elvid Hunt noted that the COTSs he visited all lacked adequate training in "leadership and drillmastership," and overall the schools suffered from a lack of high-quality instructors.[36]

Despite the War Department's best efforts to "fix" officer training with the COTSs, recurring problems with instructors and equipment and the AEF's voracious demand for replacement officers combined to short-circuit its goals. Ultimately, only the classes that began on 15 September 1918 received the full four-month course, and that was only because the War Department allowed them to continue for over two months after the Armistice. Due to the crush of events, nearly all of the COTS students who were commissioned before the Armistice received only two to three months' training.

It is amazing that in the last four months of the war the army was still wrestling with unresolved problems in officer training that dated back to mid-1917. The British Captain Hebbert viewed the issue of instructors as the Americans' greatest obstacle. He noted, "Great hindrance is also caused by the shortage of instructors. . . . The company commanders, platoon commanders, and students are all employed as instructors. . . . In many cases the students chosen as instructors are really insufficiently trained to act in that capacity, and owing to the instructor being of the same category as the students subconscious slackness results in the students."[37] When presented with Hebbert's criticism, an American General Staff officer simply retorted that it reflected nothing more than "the English desire that their fixed system be used in Central Officers' Training Schools." This officer expressed his horror that another Allied officer was going to "recommend that fewer hours be given to bayonet work, etc., and more to Offensive and Defensive Warfare." The American concluded that it was best to allow the American commander on the scene to "work out his own system, selecting from British and French criticism what is best and applicable."[38] The staff officer seemed oblivious to the fact that allowing local commanders to work out their own system of training was partly to blame for the Americans' predicament.

Throughout the late summer and fall of 1918 the COTS commandants wrote a steady stream of reports to the Training and Instruction Branch bemoaning their problems with obtaining suitable instructors. On 26 August the Camp Pike commander informed the War Department that "the general character of the instructor personnel which has to be employed leaves much to be desired in the point of efficiency." A few weeks later, he stated, "The greatest need is that of qualified instructors in charge of the various battalions to supervise the work of the company instructors, most of whom require a great deal of instruction . . . to qualify them for the work." In late September 1918, the engineer COTS at Camp Humphreys, Virginia, was 150 officers short of its goal of graduating seven hundred officers per month and had been forced to decrease the length of its course from four months to only three, with some students receiving only five weeks of instruction. The camp's commander attributed this decline to a serious shortage of qualified instructors.[39]

As with the earlier training camps, at times the War Department added to the commanders' litany of woes by reassigning COTS in-

British bayonet training NCO, Camp Logan, Texas, 1917. The American army tended to limit Allied trainers to teaching only technical tasks rather than tactical instruction in how weapons and formations would be used in combat.
(Photograph from the author's collection.)

structors with little or no notice. The commander of the Camp Lee COTS, Col. Henry Eaton, sent a blistering letter to the chief of the Training and Instruction Branch in September after one such occurrence. He complained that just as his cadre was adequately trained to serve as trainers, the War Department reassigned his senior Instructor and twenty junior instructors. He angrily replied, "If the War Department will let my officers alone and give me a chance to develop them, I shall be able to turn out satisfactory Lieutenants."[40]

At other times the camp commanders themselves helped to exacerbate their instructor problems. Only a month before launching his tirade against the War Department, Eaton himself had reported his dissatisfaction with having recently received fourteen captains from France to serve as cadre at the school. Eaton complained that since these officers were due for promotion to major, he had no use for them. They would be too senior to serve as company commanders, and he did not want them as senior instructors or battalion commanders because they would not be "thoroughly conversant with training school systems." In October 1918 Col. Edan Frey, the Camp

Gordon COTS commander, recommended that for the sake of their morale, some of his instructors should be given the chance to be reassigned to combat duty in France. While he admitted that "this will result in the loss of trained instructors," such a move would help to overcome the "slump" he believed his officers were stuck in by holding out the carrot of a sought-after assignment. Although both commanders believed that they were doing right by their instructors, neither seemed to consider the detrimental effect that these actions would have on their students' training.[41]

The General Staff and the War Plans Division sympathized with the camp commanders, but by late 1918 they were powerless to remedy the situation. The chief of the Training and Instruction Branch admitted to the commanders that "in most cases the instructors will be officers of limited experience." His only solution was for the commanders to ensure that their instructors made "careful preparation of the problems given" and were provided "a logical solution of the problem which will bring out the tactical principle involved."[42] By this Pontius Pilate–like move, the War Department washed their hands of the mess and placed the burden on the schools' senior instructors to prepare all of the tactical problems and to ensure that their assistant instructors understood how to teach the class properly.

The available evidence suggests that the camps faced worsening problems with instructors in late 1918. By examining the background of instructors at Camps Lee and Hancock, it appears that the experience level in the cadre continued to wane as the war went on. The company commander and senior instructor for the class of August through October 1918 for the 8th Company, Camp Lee Infantry COTS, Maj. Gordon Hunter, was commissioned at the first OTC at Plattsburg but since then had served only as an instructor in other OTCs or depot brigades before arriving at Camp Lee. Hunter's five assistant instructors had followed a similar path. Two of them were graduates of the first OTCs, two others were graduates of the second OTCs, and one had a reserve commission that predated the war. Only one of the instructors had any wartime experience leading soldiers outside of depot brigades or OTCs before his Camp Lee assignment.[43]

Of the ten instructors that taught the 6th Battalion, Camp Lee Infantry COTS, from mid-October 1918 to January 1919, five were commissioned from the first OTCs, two from the second OTCs, and one from the third OTC, and two others had been students in the Camp Lee COTS class that had just graduated in early October 1918. While seven of the instructors had some level of college education, only three had any military experience before April 1917. First Lt. William Rodenberger's military experience consisted of only one year's service during the Philippine Insurrection. First Lt. George Stevens had served five years as an artilleryman in the British army before immigrating to the United States. Second Lt. John Teter was a first-generation Polish immigrant who had served as a coast artilleryman in the Regular Army from 1914 to 1917.[44]

The assignment of officers who had just graduated from the course to the cadre was not limited to Camp Lee. Of the fifteen company commanders and senior instructors at the Camp Hancock Central Machine Gun Officers' Training School, four had graduated from the first OTCs, five from the third OTCs, and six from the fourth OTCs. Only three of the twenty-one assistant instructors had been commissioned for at least a year. The remaining officers had all been recent graduates of the machine gun COTS itself. Five of these instructors had graduated in September 1918, and thirteen had graduated in early October 1918. This meant that when the last machine gun COTS class began its training in mid-October 1918, half of its instructors had been officers themselves for only mere days and weeks.[45] This trend of using recently graduated officers is evidence of both a steady decline in overall instructor quality and a continued adherence to the unintended policy of the blind leading the blind.

Surprisingly, the COTSs continued to face shortages of equipment and instructional material to the end of the war. On 15 September 1918 the commander of the Camp MacArthur COTS, Col. John Boniface, informed the War Plans Division's Training and Instruction Branch that he had concerns about his instructors and the glaring shortage of textbooks. He admitted that his new instructors were "eager and willing" but hoped that their "efficiency will develop as time passes." The shortage of textbooks, however, was of greater import, and the colonel noted that until the army could provide the required texts, he would make do by "borrowing books from the post library and other sources." Over a month later he was still lamenting, "The shortage of text books is terribly trying; I have again purchased them locally . . . and I hope that this meets with your approval in this." He noted, "We *must* have text books and I am sure you will not disapprove my buying when I cannot get them any other way." A General Staff inspection of COTS in October 1918 revealed that the Camp Gordon school was still short two thousand copies of the *IDR* and other publications.[46]

Textbooks were not the only instructional materials in short supply at the COTSs. As late as November 1918, camp commanders were still reporting shortages of key equipment. The Camp Grant C O T S commander, Col. C. E. Reese, reported on 4 November that "no rifles, bayonets and scabbards are available for issue to October and November classes." He noted that while the camp had been able to supply the September class with rifles, a great number of them "are unserviceable as no spare parts are obtainable in camp." The same day, the chief instructor of the Camp Gordon Infantry Replacement and Training Camp, Col. Robert Getty, reported that he was so short on gas masks that they had to be shared among his different organizations, thus making the supervision and scheduling of "practical gas instruction" very difficult.[47]

The greatest single complaint or concern broached by the COTSs in the last four months of the war was not about instructors, textbooks, or equipment but rather the perceived decline in the quality of candidates attending their camps. Nearly all of the camp commanders raised this issue in the fall of 1918. As early as August, the Camp Lee and Camp Pike COTS commanders were reporting what they viewed as the precipitous decline in the standards of their students. The Camp Lee commander, Colonel Eaton, lamented that the decline was not only evident in the enlisted men being sent to the school, but also in the attendees that he was being sent from the civilian sector. He maintained that the local draft boards had "not been as strict as required for the infantry course and candidates sent here have had to be rejected after being here but a short time." He later noted that a number of the enlisted men "are being sent who are entirely lacking in the instincts of a gentlemen" and "were uncouth and their language was far from being refined." He concluded that it was "quite impossible to make gentlemen out of men of this type in four months." The Camp Pike commander, Col. Charles Miller, echoed Eaton's unease and reported that "many of the enlisted candidates who reported in July and August were lacking in educational qualifications and a considerable number were lacking in physical qualifications, and many had to be eliminated at once."[48]

The situation only worsened in September and October. On 26 September 1918 the Camp MacArthur COTS commander complained that the quality of candidates he was receiving was not what it should be. He was angered that a recent levy of enlisted candidates had included two men with venereal disease and others who clearly lacked the qualities needed for officership. In October he furiously reported, "We have discovered two or three men that have the most abominable records and how they were ever allowed to come here is a mystery and shows that company commanders still have to realize the grave error of sending men here with bad records. . . . Today I found one candidate's record that showed something like 21 company punishments and 13 courts-martial, yet he was sent here to become an officer." The problem had become so glaring that even the assistant secretary of war had to admit that the army had experienced great "difficultly in securing the proper number of qualified candidates for the Officers' Training Schools, and within the greatly enlarged military program this difficulty has threatened to become acute."[49]

The camp commanders all blamed the candidates' company commanders for sending unqualified men to officer training. The Camp MacArthur COTS commander accurately noted that

> some of the company commanders failed to select suitable men as candidates; such mistakes are bound to occur where the selection is made by inexperienced officers back in their companies. . . . My observations of the candidates, as a whole, impresses me with the fact that previous classes have drained the service and we must expect many poor candidates to arrive. . . . Please urge company and regimental officers to appreciate the grave importance to the service of selecting their VERY BEST men, regardless of whether that means they must send their first sergeants, sergeants majors and troop clerks.[50]

In establishing the blame for the problem, however, the colonel had also stumbled onto the reason for the decline. Not only had many of the best soldiers already gone through the training, but the unit commanders had also grown tired of seeing their companies decimated by the War Department's ceaseless levies of personnel.

The demands of officer replacement drove the War Department both to cut standards for candidates and to increase its quotas for attendance to officer training in the last months of 1918.

In August 1918 the severe shortage of infantry officers led the War Department to depart from its long-standing policy of having all line officers graduate from established officer training courses. The adjutant general directed that commanders of depot brigades, replacement centers, and infantry and cavalry regiments still in the United States hold boards to "examine enlisted men for direct appointment as 2nd lieutenants of infantry and cavalry." Those enlisted men already attending an officer training school, but not having finished the course of instruction, could also apply to their school commander for an appointment. By this commissioning sleight of hand the army added a further 539 infantry and 309 cavalry second lieutenants to its ranks. On 27 September 1918 Secretary of War Baker directed that the commandants of the three infantry COTSs graduate and commission the students who had entered the course in August 1918 even though they had received only two and one-half months of training.[51]

Although it intended that the COTSs' candidates would come mainly from the army's enlisted ranks, the War Department found that the supply of high-quality enlisted men was running dry. This was particularly apparent in candidates for field artillery and other technical branches. The field artillery, for example, had to accept over half of its candidates from civilian life in order to find men with the requisite mathematical skills that the branch required. This meant that while it was physically capable of processing over a thousand students per week, in October 1918 the field artillery COTS at Camp Taylor, Kentucky, was only reaching a weekly intake of five hundred new students. When it came to worthy applicants, Colonel Cabell reported, "the Army had already been thoroughly combed for good material and the normal draft calls did not supply the quality of material required."[52]

The War Department tried to address this problem by first expanding the quota of soldiers that units had to provide to the COTSs. In the fall of 1918 the adjutant general raised the quota for infantry units from 2 percent to 5 percent of their total enlisted strength. It set the quota for artillery and machine gun units even higher. The shortage of infantry officers became so acute that in September and October 1918 the army even added three more infantry COTSs at Camp Grant, Illinois; Camp MacArthur, Texas; and Camp Fremont, California. While few of their graduates ever left the United States, the COTSs ultimately commissioned 20,563 officers between September 1918 and February 1919.[53]

In many ways it was this increase in quotas for the training camps that led to some of the worst abuses of the system. All too often, harried company commanders used the levies as a means of ridding themselves of the troublemakers or shedding the nonessential men from their units. The Camp Lee commander noted that one sergeant from the 29th Infantry Regiment was sent to the course against his will because his company commander had a quota to fill for the school and the sergeant was the most expendable. In another case, a first sergeant from a unit in Panama entered COTS "solely to get back to the United States with hope that he could secure a furlough to visit his home."[54]

Nearly all the COTS commanders also reported great spikes in the number of candidates who reported that they had been sent to officer training against their will or without any previous consultation by their commanders. In his final report the Camp MacArthur COTS commander noted, "There were candidates stating, quite frequently, that they had been sent here to school without their wishes being consulted, in fact against their expressed wishes and that they desired to be relieved from the course. . . . [This] impeded the work of the school, caused additional work of elimination boards at the schools, and the elimination of such men created vacancies in the candidate body at a time when the Government needed every officer that could be obtained from such schools."[55] Given the frequency with which these comments appear in the record, this seems to have been a rather common occurrence.

In August 1918 Congress and the army had begun to explore ways to better utilize the manpower that was being held in the nation's colleges and universities. This came to fruition with the Man Power Bill of 12 September 1918. The bill authorized the enlistment of all of the nation's able-bodied college students into the Students' Army Training Corps (SATC). Beginning 1 October 1918, the SATC militarized the nation's educational institutions by essentially turning them into mills for the production of junior officers.[56] In fact, soon after the establishment of the SATC units in the colleges, the assistant secretary of war made clear in a memo that the units were to be "a recruiting ground for the central training schools for line officers" and that the army had already asked the SATC to provide eight thousand students to the COTSs per month.[57]

Faced with a deepening officer manpower crisis, the War Department also used the Man Power Bill to justify raising the maximum age for admission of candidates from the ranks from forty to forty-five years. Two days after Congress passed the bill, Secretary Baker further authorized the commissioning of SATC candidates who were at least eighteen years old and had been recommended by a local board of officers. By essentially lowering the minimum age for commissioning from twenty years and nine months to eighteen, the army squeezed out another 3,264 second lieutenants before the end of the war. The changes also meant that by mid-October the COTSs were commissioning approximately one thousand officers every four weeks. Of these candidates, 60 percent were enlisted men, 20 percent were SATC students, and 20 percent were coming directly from civilian life.[58]

On 21 October 1918 the adjutant general further lowered the standard for admission into officer training by allowing 5 percent of COTS candidates to be drawn from "limited service men whose physical defects [such as poor eyesight or lack of a full range of motion in the arms and legs] were not glaring, but were of a minor nature." The idea was that upon graduation and commissioning, these limited-service officers would be sent to fill positions in depot brigades and other postings not requiring field service. This would release physically qualified men then occupying those jobs for active service. In the week before the Armistice, the adjutant general was also successful in lowering the acceptable physical standards for officers to those of the minimal physical standards for draftees. This allowed enlisted men to apply for attendance to COTS who had previously been disqualified because of poor eyesight, color blindness, or other minor defects. Despite this steady lowering of the standards, Colonel Cabell still argued that had the Armistice not been signed, "it would have been difficult to have gotten the November classes to the Central Schools to the authorized strength."[59]

The lowering of the standards for entry into the COTSs was also met with alterations to the training conducted in the camps. On 6 November 1918 the chief of the War Plans Division's Training and Instruction Branch issued new guidance to the infantry COTS commanders on the length of the course and subjects that were to be taught. It directed that the course be reduced from four to three months. The new three-month course drastically reduced training time devoted to musketry training, minor tactics, and the new weapons of warfare. Signal training was limited to the use of wigwag signal flags. It is interesting to note that under the new course, thirty-three hours were devoted to bayonet training and eighteen hours to wigwag signaling, while the candidates received only fourteen hours in small arms firing, eight hours of gas instruction, and three hours of map reading.[60] Even in the face of severe reductions in training, at the eleventh hour the army was still devoting an inordinate amount of precious instruction time to subjects of questionable importance.

Were the men attending the COTSs in late 1918 truly as poor as the Regular Army officers who trained them claimed? A study of candidates from Camps Lee, Hancock, and Taylor does allow for some tentative answers to this question. The statistics for the candidates of the 4th Battalion, Machine Gun Central Officers' Training School, at Camp Hancock, Georgia, who began their training in August 1918 actually show a slight increase in levels of education over the sample taken from fourth-session OTCs. Over 70 percent of the candidates in the battalion's 21st Company had some college experience. Of the 111 candidates of the 22nd Company, 75 (67.5 percent) had some college education; 31 (30 percent) had only a high school education; 3 (2.7 percent) had attended vocational school, and 2 (1.8 percent) had less than a high school education. There also seemed to have been no major change in the candidates who had held professional or white-collar jobs before the war. The occupations given by the 121 candidates of the 20th Company, and the number in each occupation, are as follows:

Clerk/office work, 19	Engineer, 3
Student, 15	Office/sales manager, 3
Lawyer, 14	Banker, 2
Businessman, 11	Mechanic, 2
Farmer, 10	Stock broker, 2
Industrial foreman, 8	Architect, 1
Salesman, 6	Insurance, 1
Teacher, 6	Dentist, 1
Industrial worker, 5	Musical director, 1
Newspaperman, 4	RA enlisted, 1
Secretary, 4	Unknown, 1
Accountant, 3	

95

Using the same criteria applied to the Fort Sheridan sample, after factoring out the students and those candidates whose occupations are unknown, approximately 85 percent of the candidates could be classified as having been employed in professional or white-collar jobs.[61]

Statistics from two different classes at the Camp Lee Infantry COTS illustrate the shifts in demographics among those attending officer training after the COTSs replaced the OTCs. They also highlight the influence of the SATC on entrants to the COTSs in the last months of the Great War. The statistics for the candidates of the 8th Company, Camp Lee Infantry COTS (who went through the course from July through October 1918), indicate the continued shift toward selecting candidates from the enlisted ranks of the army. Due to the requirement that all those seeking commissions had to first enlist in the army, every candidate listed his military rank at time of entry into the course. What is interesting is that a majority of the candidates (78.7 percent) had been NCOs before attending officer training. Of the 8th Company's 173 candidates, 38 (22 percent) had been corporals; 68 (39.4 percent) had been sergeants; 3 (1.7 percent) had been sergeants first class; 19 (11 percent) had been first sergeants, and eight (4.6 percent) had been regimental or battalion sergeants major. The high number of NCOs attending officer training illustrates the army's continued policy of robbing Peter to pay Paul when it came to company-level leadership. It also explains the irritation of many company commanders in being forced to give up their NCOs to officer training. The constant drain of solid NCOs from line units, as will be seen in the next chapter, played havoc in the training and cohesion building of companies, platoons, and squads.[62]

The average age of the 8th Company's candidates was twenty-six, and 35 (20 percent) had military service in the National Guard, Regular Army, Navy, Marine Corps, or Coast Guard or with a foreign army before America's entering the war. The company also contained four soldiers who had attended, but failed out of, previous officer training camps. One of these men, William Pierson, had washed out of both the first OTC at Fort Benjamin Harrison and the third OTC at Camp Sherman. The occupations given by the 8th Company candidates before joining the army are as follows:

Accountant, 21	Artist/designer, 2
Office/store manager, 21	Mechanic, 2
Clerk/office work, 20	Purchase agent, 2
Salesman, 17	RA enlisted, 2
Businessman, 14	School admin., 2
Lawyer, 9	Secretary, 2
Banker, 4	Advertising, 1
Farmer, 4	Contractor, 1
Industrial foreman, 4	Police sergeant, 1
Insurance, 4	Tree surgeon, 1
Industrial worker, 3	Writer, 1
Railroad worker, 3	Unknown, 4

By the same criteria as those applied to the Fort Sheridan sample, the data show that approximately 85 percent of the candidates could be classified as having been employed in professional or white-collar jobs.

The statistics for the candidates from the 6th Battalion, Camp Lee Infantry COTS, who went through the course from October 1918 to January 1919 illustrate the dramatic change to the demographics of officer training schools after SATC students began to flood the camps in the fall of 1918. Of the 167 candidates in the 6th Battalion, nearly 71 percent had been born between 1898 and 1900, thus making them eighteen to twenty years old. This influx of younger candidates lowered the average age of the course to twenty-one. The youngest candidate, Robert G. Hunt, was born on 24 June 1900, making him eighteen years and four months old at the time of the Armistice.[63]

The vast majority of these younger candidates were entering the schools directly from college and university SATC programs. This raised the overall education level of the men in the course. Of the 167 candidates, 153 (91.6 percent) were college graduates or had some level of college education. Of the remainder, three (1.7 percent) had graduated from preparatory school; five (3 percent) had graduated from high school, and six (3.5 percent) had less than a high school education. The influx of college students also witnessed a precipitous decline in those candidates with prior military service. The only candidate with any real military experience was one who had served with the French army for seven months as an ambulance driver.

Based on the samples from these camps, the regular officers may have been overhasty in their characterizations of the COTS candidates. With the addition of the SATC cadets, one could even argue that the overall education level of the candidates, and thus their perceived quality, actually improved at the end of the war. This being said, it is still clear that the army was facing an uphill battle to fill the ranks of the training schools in late 1918 and had correspondingly lowered standards to fill the officer schools. However, the number of obviously unqualified candidates sent by irate company commanders to fill the War Department's endless quotas may have magnified the issue of poor overall candidate quality in the eyes of the COTS commanders.

At no time in the war did the Regular Army officer corps ever stray too far from their preference for college-educated officers. In his final report, the Camp MacArthur COTS commander still maintained, "It was found that the brightest men were those from the universities and colleges; also that it was these men who gave the greatest enthusiasm for the work."[64] In the end, the SATC cadets seemed to offer a panacea for the War Department's officer manpower woes by filling the ranks while still being college-educated men. Unfortunately, for all its allure, the SATC path came with serious consequences. First of all, given the ages that these cadets were when entering the COTSs, it is obvious that they had spent very little time in college. Thus, whatever intellectual broadening or conditioning the army hoped to gain by selecting these men was largely moot. More important, one should not disregard the inherent danger that the army was willing to accept in placing immature youths in command of combat units. Of these callow youngsters, one COTS commander warned, "The incoming candidates from the Students' Training Corps look like boys; quite a few of them seem to be about eighteen, underdeveloped, weak, needing the training I had as a boy in prep school. I doubt if your office contemplated such material being sent here to become officers after four months' training. Some will not be nineteen when their four months are up."[65] It is generally difficult enough to turn young eighteen- and nineteen-year-olds into privates; it is quite another thing to place them in command of soldiers who are dependent on their decision making abilities and knowledge. It was perhaps fortunate that the war ended before these adolescent officers made it to the front.

During the war a French officer was reputed to have told his

American counterpart that while raising a four-million-man army was no great feat, the Americans' ability to create an officer corps from virtually nothing was quite miraculous. In all fairness to the US Army of the Great War, the sheer effort it shouldered in identifying, training, and selecting its two-hundred-thousand-man-strong officer corps was a prodigious accomplishment. Shortly after the end of the war, Peyton March admitted that while OTCs were not perfect, the camps ultimately selected "those who showed that they were capable of becoming instructors and leaders of soldiers" and that the method used "was fully justified by the results."[66] Although March's statement is certainly valid, one should not forget that this accomplishment was only achieved by cutting a number of corners.

In the final analysis, the operation of the OTCs and the SATC continued to demonstrate the ad hoc and stopgap nature of the American mobilization for the war. Some of the corner cutting was an inevitable result of the systemic problems of mass mobilization. Others problems, however, such as the ineffectiveness and lack of realism in the candidates' training, were the War Department's self-inflicted wounds. While the army's makeshift officer training system just barely managed to fill the officer ranks, the improvised character of the training produced leaders of wildly uneven abilities, questionable technical and tactical competencies, and unrealistic concepts of warfare. This fundamental flaw would dog the US Army until the end of the war.

5 "Ninety-Day Wonders" and "Jumped-up Sergeants"
Stateside Mobilization and the Challenges of Small-Unit Leadership

UNTIL THE CREATION OF the COTSs, most of the graduates from the officer training camps went directly from school into leadership or staff positions in units in the process of mobilizing or training. As these officers reported to their new units, most Regular Army officers realized that the OTCs were but a bare beginning in the graduates' overall military education. Following the traditions of the Old Army, the novice officer's finishing school would be his on-the-job training within his unit. As a battalion commander later noted of his new officers, "Careful selection at the training camps has undoubtedly served to weed out the more defective material which presented itself for commissions. Three months of intensive exercise and the most superficial training in the theory of leadership have naturally failed to impress this human material, though it is of the finest quality, with the true character of officers. . . . Their intelligence, enthusiasm, energy, and potential capacity for leadership are in no sense satisfactory substitutes for the knowledge and experience which in the main they lack."[1] This fact was not lost on the new officers. Lt. Milton Bernet recalled, "Every candidate realized that if he were fortunate enough to receive a commission, he would have to supplement his actual work with a great deal of further study."[2] Unfortunately, as Bernet and the other officers filling the new divisions in the fall and winter of 1917 and 1918 discovered, the pace and problems of mobilization left little time for completing the "graduate study" of their new trade or even filling in the gaps in their training left over from the OTCs.

With the OTCs often located on the same posts as the National Army cantonments, for the young ninety-day wonders it was a short walk from the OTC graduation field to their lives as company-grade officers. The letters, diaries, and memoirs of those officers reveal that many of them were apprehensive and unsure about assuming their new roles. As one young officer recorded, after attending OTC he and his peers still found it "hard to assume the mental attitude of honest-to-goodness officers." Some realized that it was not going to be an easy task to instill discipline in their independent-minded citizen-soldiers.

An Oklahoma National Guard officer recalled, "Men who had lived all their life in the open and managed their own affairs found it difficult to obey someone else in nearly everything they did, especially as it was not always explained why the thing was to be done."[3]

Even the soothsayers of the new science of psychology gave dire warnings of the fate awaiting the new officers. Noting the effects of individualism and egalitarianism on American society, Harvard professor and army consultant William Hocking wrote,

> Those who say it is hard for an American to take orders may not realize that it is equally hard for the average American to give them. . . . The raw commander is conscious of his individual self, and consequently realizes that the words falling out of his mouth have hardly the weight that should make men obey them. . . . He knows he has to face, not so much the surly criticism as the more searching humor of his men. . . . He needs the manner which only experience can justify, the manner of confidence, authority, prestige.[4]

Thus damned by science and uncertain of his own "confidence, authority, [and] prestige," the young officer went forward to meet the men he would lead into battle.

The beginning of the young officer's acculturation process into the army's expectations of leadership often began with his reading of many of the same semiofficial publications for officers that had circulated before the war. They became key media for inculcating the novices with the army's values and codes of conduct for officers. James Moss's *Officer's Manual* was perhaps the most widely used guide for young officers and officer candidates. Moss reprinted the manual in May 1917, and 135,000 copies of the work were printed during the war. Although the book provided a handy guide to the organization and workings of the army and provided a few hints or general maxims to aid young officers in their daily work, it was still much too concerned with the niceties of the peacetime army's "customs of the service." Neither it nor any of the other reprints of this type of prewar semiofficial publication truly offered any real insights into combat leadership or any help in training a unit for war or preparing it to deploy overseas. What these manuals did do, however, was to serve as a means to inculcate the novices with the service's paternalistic culture and the belief that the officer was first and foremost a gentleman.

Whatever the value of refining the manners of its young officers, the army's expectations of gentility echoed throughout much of their training. Lt. Col. James McAndrews warned the second class of the provisional second lieutenants graduating from Fort Leavenworth to "remember at all times that you are gentlemen worthy of confidence." Throughout their time at the third OTC at Camp Devens, the camp's senior instructor, Lieutenant Colonel Massee, constantly reminded the candidates that "war is a gentleman's game and you will play it as gentlemen." Maj. Christian Bach informed the new officers that these attributes were key to their ability to command because their moral ascendancy over their men came from their being morally straight, physically strong, and having the strength to do what is ethically right.[5]

The seriousness with which the regulars viewed the need for all officers to act as gentlemen was reflected in its wartime use of military law to police its commissioned ranks. The records of courts-martial reveal that not all the army's officers were up to the moral expectations of their rank and position. As the officer ranks swelled, so too did the number of officers brought before courts-martial. From 6 April 1917 to 30 June 1918 the army tried 642 officers, but from 1 July 1918 to 30 June 1919 the number more than tripled to 1,948. Most charges during that period were for offenses that actually occurred before or shortly after the Armistice. Trials of lieutenants accounted for over 75 percent of all officer courts-martial.[6]

Nearly 37 percent of all of the trials of officers in the latter half of 1918 were related to violations of just three Articles of War: drunkenness, absence without leave, and conduct unbecoming an officer and a gentleman. As the case of Capt. John M. Andrews demonstrates, the army took its concepts of "officers and gentlemen" very seriously. Andrews went on something of a binge between 30 August and 9 September 1917 in Des Moines, Iowa. During that time he was seen drunk in uniform on several days, consorted with and "occupied a bedroom with a woman not his wife," and, "while in uniform, consort[ed] with prostitutes and did become drunk with them in a public place." To add insult to injury, the officer went absent without leave to conduct his debauchery. The court found Andrews guilty of all the charges except being AWOL and recommended his discharge from the service. It was clear that the regulars were going to make damned sure that the new "emergency" officers lived by their code and expectations of officership.[7]

The immutability of the officer code of conduct was also displayed in the actions of Lt. Col. Dwight Eisenhower. While commanding the Tank Center at Camp Colt, Pennsylvania, Eisenhower had to deal with a new officer who had been caught cheating at cards by his fellow officers. Ike gave the officer the choice of resigning his commission or

facing court-martial. The officer opted to resign. Shortly after the resignation, Eisenhower was visited by the cashiered officer's congressman
and father, who tried to convince him to tear up the resignation and
transfer the officer to another post and made veiled threats to harm
Ike's career if he did not comply with their wishes. Eisenhower refused
to change his order and argued that if he failed to act, he would merely
pass his problem on to another commander and undermine the foun-
dations of the officer corps.[8]

Beyond the desire to instill a veneer of gentility on the new officers, when it came to the subject of leadership the regulars could offer
little. Col. Charles Miller admitted that "the management of men is
a vast unbound sea upon which the young officer sets sail without
pilot and without chart" and that "there is no textbook nor treatise
to guide him in this, the most important feature of his profession."[9]
Like Moss and the other Regular Army pundits, Miller could only
offer broad rules of thumb and hoary chestnuts for a young officer to
follow when it came to leading soldiers. A good bit of their counsel
sounded much like the advice that Polonius gave Laertes in *Hamlet:*
live within your means, don't drink, don't gamble, don't grumble, and
don't gossip. Unfortunately, when it came to some of the most important issues of combat leadership, dealing with battlefield fear and
stress and motivating men to kill and risk death, Miller, Moss, and the
other "authorities" revealingly were silent.

In addition to the reissued manuals and guides, the war saw an explosion of new War Department and semiofficial privately published
handbooks. Myron Adams, the morale officer for the Fort Sheridan
second OTC, worked with some of the camp's instructors to publish
The Officer's Responsibility for His Men. This small tome not only encapsulated the army's paternalistic outlook on leadership, but it also
melded it with the crusading spirit of civilian progressivism. Adams
admonished his readers, "The officer entrusted with a company in
the National Army has a responsibility in caring for his men, equal
to that in commanding them." Although one of the work's four chapters was devoted to the practical care of the soldier from the aspect of
field sanitation, food, clothing, and shelter, the bulk of the book provided the reader a guide for dealing with issues of recreation, religion,
"moral problems," and the "mental training of the enlisted man."

In true progressive fashion, the paternalistic care of the soldier also
extended to moral uplift and social engineering. Adams informed the
candidates, "The leisure hours of the men can be made valuable or
dangerous," and warned, "An unwise use of leisure hours results in
destroying the good spirit of the company, multiplying delinquen-

cies and discrediting the character of the army among civilians." Thus, a good officer was the one who took "social measures to diminish sexual temptations" and worked to eliminate such vices as gambling and the drinking of alcohol.[10] As civilian progressives sought to uplift the uneducated immigrant street urchin, "fallen women," or men addicted to drink, the social working officer sought to steer the child-like soldier toward cleanliness in body, mind, and soul for the good of the army, society, and nation. Again, while these pleas for uplift and paternalism served a military purpose, they still failed to address the salient issues of combat leadership.

The army did not completely wash its hands of the issue of leading men in battle, but it consistently equated the issue with the concept of discipline throughout the war. In other words, if the young officer inculcated iron discipline in his soldiers, the demands of combat leadership became relatively easy and manageable. The officer simply ordered and the soldier simply obeyed. Iron discipline was to be the remedy for battlefield fear, confusion, and friction. Miller argued that discipline had to be the focus of all training if a unit was to be successful in combat. He defined discipline as an "ever present respect for superiors, and an instant, cheerful obedience" that produced "a spirit of loyalty to leader and organization which will result in *unity and promptness of action, in instant response to the will of the leader.*"[11]

As previously discussed, the debate over the proper mix of obedience and initiative had dogged the regulars long before the Great War. However, the influx of masses of new officers in 1917 and 1918 raised this debate to an ever more strident level. Given their subordinates' lack of experience and training, no colonel or general was willing to relinquish the reins of responsibility for the conduct of battle into the hands of the wet-nosed junior officers. In their tactical training, in their lectures, and in their personal reading the young officers were bombarded by the steady message that the unquestioning obedience to the orders of their senior commanders was much more important than exercising independent action or initiative. McAndrews warned his new officers that when dealing with their superiors, "Do not set up your own judgment against theirs. If they seem to you to be at fault, have patience, as time will probably show you that they were right. And always remember that implicit obedience to orders and wishes of your superiors is the foundation of discipline." Col. C. H. Hitchcock pointedly instructed young officers that "the conduct and orders of your superiors must not be criticized, questioned, or even commented upon. . . . An order is an order and is to be carried out without cavil or evasion." Both McAndrews and Hitchcock made one thing perfectly clear: the young officer was as bound by the rules of unquestioned obedience as the NCO and private.[12]

The regulars wanted the best of both worlds: obedient officers able to exercise the right initiative at the right time. This desire was evident in the wartime revision of the *IDR*. The regulation stated, "Subordinates must . . . be given great latitude in the execution of their tasks," and, "A subordinate who is reasonably sure that his intended action is such as would be ordered by the commander, were the latter present and in possession of the facts, has enough encouragement to go ahead confidently." The manual also warned, however, that "independence must not become license" and that there was still "one supreme will to which all must conform." Ultimately, the young officer was warned that any responsibility for an exercise of initiative that went wrong "rests on the subordinate." This was perhaps less than reassuring to neophyte officers already unsure of their own authority, training, and knowledge.[13]

The army's inability to reconcile the demands of strict control and obedience with the needs of junior officer independence and initiative began a lively and cogent debate in the pages of the *Cavalry Journal* in the winter and spring of 1918. The journal's editor sparked the debate by recounting his experiences of sitting on an examination board for young officers. The editor noted that the "results of the practical examination in Minor Tactics came as a bewildering disappointment." He recounted, "The officers being examined showed no training in leadership; their estimates of the situation were insufficient and inaccurate. . . . Instead of giving orders to their platoon leaders when a situation requiring action was announced [to] them, [they] assumed a far-a-way look, as if they were searching their memory for precedents, and answered as if they were reciting a lesson of abstract theories: 'I would do so-and-so;' 'I would give such and such orders;' 'I would send such and such a message.'" After evaluating the machine gun companies of several regiments, he was astounded by the officers' unwillingness to take the initiative and act without direct orders from their superiors. When he asked the officers why this was the case, "One of the officers . . . replied: 'why, Colonel, all my service I've been jumped on so hard if anything that I attempted to do without authority went wrong, that I have learned the safest way is to do only what I knew beforehand will be approved.' And the other chimed in: 'Same here!'" The editor also maintained that the Regular Army had so long labored under this same problem that it contained far too many officers "whose initiative was, by their training in those regiments, completely destroyed." Overall he worried that the officer corps was largely "untrained to leadership and without initiative."[14]

The response to the editor's indictment of junior officer leadership and initiative was not long in coming. In the journal's next issue one regular officer blamed the problems on the army's own culture and the narrow mental horizons of many of its senior officers. He denounced instances in which "commanders have forbidden their subordinates from conducting tactical exercises, merely because such particular exercises were not specifically described in detail in the drill books." The writer laid the blame for "any observed lack of initiative and spirit of leadership in juniors" on

> 1. A long extant deep rooted struggle hold grip of bureaucratic administration. Papers must be kept straight whether the battle be won or lost.
> 2. Centralization of control, command and instruction.
> 3. Insufficient basic training.
> 4. A failure on the part of a majority of those charged particularly with instruction to realize and visualize the necessities and to take advantage of such latitude as given them to decentralize and to force subordinates to assume responsibility and the initiative.[15]

If the army was going to solve its problems with junior leaders, its senior commanders were going to have to divorce themselves of the desire to maintain absolute control and change the way they trained their young officers.

One of the young officers in question also offered his critique of the army's culture of absolute command, control, and obedience. He argued that the *Cavalry Journal*'s editor was off the mark in his disparagement of the new officers. He blamed much of the problem of junior officer initiative on the army's school system. As he pointedly observed, his instructors were too tied to teaching by "the book," that it became the "all and all of their instruction." The officer recounted,

> I have been red-inked . . . for a departure from the approved solution. I was not informed that I violated any principle but that a second lieutenant could not improve a solution . . . , therefore a departure was necessarily wrong. Thereafter, knowing what my instructor wanted, which was quite a game at these schools, I memorized the solution and gave them verbatim. . . . Immediately I received excellent marks and saved myself the labor of original thought. . . . Thus instead of mental development, the young officer received a training in mental gymnastics. . . . Is it any wonder that the edge of our initiative and keenness to assume responsibility are somewhat blunted?[16]

As he saw it, the greatest failure in his training was that the army, through its schools and instructors, had beaten any desire or ability to exercise initiative out of its junior leaders.

Although the exchange in the *Cavalry Journal* revealed that some officers were aware of the problems of reconciling obedience with initiative within the army's leadership training, regular officers tended to place the burden of repairing these deficiencies back on the junior officers themselves. Time and time again the young officers were browbeaten by their Regular Army superiors to take responsibility for completing the professional education that the army had failed so miserably to provide. Col. C. H. Hitchcock chided, "By law and regulations you are the superior of the men under you. You must make this a fact and not a mere theory. This is the first thing. You must not allow any of the men under you to know more of your profession than you do, and you must not allow them to surpass you in any of the true soldierly quality." Maj. Christian Bach warned his novices that "men will not have confidence in an officer unless he knows his business" and sharply pronounced, "If you have a rotten company, it will be because you are a rotten captain."[17] It does seem that the majority of the new officers took this charge seriously and studiously applied themselves to expanding their knowledge. Regrettably, those officers who tried this course of self-improvement still found themselves stymied by the continued prevalence of obsolete doctrine and their inability to sort the best tactical practices from the flood of information coming from the host of official and semiofficial publications.

The War Department attempted to resolve its monumental problem of preparing its tactical units for war by instituting a standardized training plan for all levels of command within a division. On 27 August 1917 the War Department issued the pamphlet *Infantry Training* to serve as the mandatory guide for readying divisions to fight in France. The pamphlet delineated the responsibility of commanders at all echelons for the training of their soldiers, mandated the establishment of thirteen divisional schools for specialists, and provided a weekly training plan for the instruction of infantry and machine gun companies. The total course of instruction was to take sixteen weeks, and its intent was to make it possible for the division's units "to take their places on the line" with a "minimum of training in France."[18] On 20 December 1917 the War Department extended the standard training plan from sixteen to eighteen weeks. This move was made partially to appease Pershing by adding more emphasis on maneuver warfare. At all times during this training the unit's officers and NCOs were supposed to continue to hone their skills through lectures, unit schools, and hands-on practice. Once the basic training of the companies was complete, the divisions were to receive additional months of training to focus on regimental, brigade, and division operations.

Although the standardized training plan seemed very complete on paper, it proved to be very difficult to execute in reality. The systemic problems of mass mobilization that had so plagued officer training were redoubled in the wartime training of units. Across the country the order to have the divisional cantonments ready to receive the first draftees by August 1917 led to a frenzy of activity and quickly showed the army's lack of preparation for such a massive mobilization. The War Department had to locate suitable land for the training posts and build a complete road, water, and barracks infrastructure on the new sites. Much of this work was still being done when National Guard and National Army units arrived in camp.

Despite the War Department's standard training, these other demands continually drained men from training. One officer recalled, "In their efforts to supply details for all of these objectives, Company Commanders found schedules impossible." On 19 November 1917 alone, the 353rd Infantry had to divert two and one-half companies from training to do engineer work on the post. Furthermore, "when the guard and school details were added to this list, few were left on the drill grounds." One exasperated company commander in the 78th Division recalled that his higher headquarters demanded such an endless stream of reports that "life was a veritable nightmare of typewritten figures" and "drill was carried on in the intervals of lining up for another check or inspection." The young officer concluded that as a result of this frenetic and often pointless activity, "the men, quite naturally, looked upon the officers as a set of lunatics who didn't know their own minds for 10 minutes at a time." This last comment highlighted the ultimate effect of this muddle and confusion. Leadership rests upon a foundation of mutual trust between the leaders and the led. When the leader seems to act in a manner that indicates to his men that he has no idea what is going on, or seems unable to provide those articles that his soldiers need, they lose their trust and confidence in him and in the army.[19]

One of the gravest shortages that the army faced in 1917 was of arms and equipment. The experience of the 82nd Division demonstrates the effects that the equipment shortages had on training and leadership development. The acute shortage of weapons in the 82nd Division forced officers to contract with local sawmills for the production of dummy rifles. The "Camp Gordon 1917 Model Rifle," as the doughboys derisively called the wooden weapons, allowed units to conduct limited instruction in close-order marching and bayonet training but had few other useful purposes. Though the 82nd Division was formed in August 1917, some of the unit's infantry regiments were not completely armed with rifles until the first week of February 1918. Rifles were but one of the shortages that hamstrung the division's training. The division chief of staff, Col. G. Edward Buxton, recalled,

> The training of specialists in the United States was necessarily of a theoretical character. The Divisional Automatic Rifle School possessed about a dozen Chauchat rifles; the regiments had none. Colt machine guns were issued to machine gun companies, although this weapon was never to be used in battle. The Stokes Mortar platoon never saw a 3-inch Stokes Mortar while in the United States, and the 37-mm gun platoons possessed collectively one of these weapons during the last two or three weeks of their stay at Camp Gordon. A limited number of offensive and defensive hand grenades were obtained and thrown by selected officers and non-commissioned officers at the Division Grenade School. The men of one regiment witnessed a demonstration where four rifle grenades were fired.

These shortages not only hindered the training of the unit's weapons specialists, but also prevented the junior officers from understanding the employment and potential of the new military technologies.[20]

Equipment problems were not limited to National Army units such as the 82nd Division. The National Guard 36th Division had to rotate its limited stock of rifles around its infantry units to accomplish the bare minimum of marksmanship training. As late as 18 December 1917 most of the division's machine gun battalions had not been issued weapons and had received only a modicum of training on the obsolete Colt and Benet-Mercie machine guns. Even the Regular Army 4th Division suffered shortages of rifles and machine guns. Its soldiers were dispirited when their unit's arms and equipment were stripped to fit out other units who were to deploy before them. In fact, it was not until after these units shipped out that the 4th Division was able to begin its rifle marksmanship training. Unfortunately, the time available for the training was so short that the 4th Division's 39th Infantry Regiment and a battalion from the 58th Infantry were not able to complete even the basic firing course before the division itself deployed to France.[21]

Although the 14 January 1918 Division Table of Organization stated that divisions would have 768 automatic rifles, in February

Improvised training cannon made from boxes, carriage wheels, and a log at Camp Pike, Arkansas, in late 1917. This photograph gives some indication of the equipment shortages that plagued the American mobilization. *(Photograph from the author's collection.)*

1918 the French military mission reported that of the eighteen divisions that they inspected, seventeen had 32 or fewer of the rifles on hand. They found similar shortages in trench mortars, signal equipment, and artillery. In fact, the only artillery found at Camp Beauregard, Louisiana, was four Parrot guns and three Napoleon guns from the Civil War. In nearly all the divisions, artillery units were most affected by equipment shortages. Through much of 1917 most of the new artillerymen had to content themselves with practicing their trade on wooden guns made from "logs mounted on the running gear of escort wagons" or other improvisations. In fact, artillerymen in the 90th Division did not receive and fire their guns until March 1918, only weeks before the unit departed for France.[22]

The shortage of guns prevented artillery officers from adequately learning how to control and adjust fires and stymied efforts to conduct combined arms training in most divisions. This later contributed to the AEF's problems in mastering infantry-artillery coordination. Regardless of their branch, the young officers' and NCOs' lack of experience with modern weapons limited their professional development and further hobbled their efforts to realistically prepare themselves and their units for combat.

Adding to these problems was a shortage of training aids and publications. The 89th Division's George English noted that his division's artillery units had no drill manuals to assist them in their training until an officer translated the French artillery drill regulations and provided mimeographed copies to the units. Lucian Truscott recalled that when he attended an ad hoc regimental machine gun school while posted at Fort Huachuca, Arizona, "There were no textbooks of any kind. No manuals. And no charts." During the class "the instructor read his own notes, compiled during his course at the Machine Gun School at Fort Bliss . . . and the class copied furiously in their own notebooks." Reflecting back on his World War I experiences, John E. Hull recalled, "We were short . . . of all the training aids that you need to conduct training adequately. But we didn't realize we were short because we didn't know any better."[23]

Ironically, Maj. Gen. John F. O'Ryan faced the opposite challenge. O'Ryan bemoaned the fact that when he began training his 27th Division, there was no general agreement among army officers on which direction the American training should take. Despite the existence of a standard training plan, there continued to be a great debate over what should be emphasized in instructing the troops. O'Ryan recalled that some regular officers argued for the primacy of training the bayonet and rifle, and others for concentrating primarily on machine guns, grenades, or other specialist training, and some argued that since trench warfare had so completely changed tactical principles, "time expended in maneuvers was time wasted." While he was working through these issues, he found that "the training problem at the outbreak of the war was compounded by a veritable avalanche of books, booklets, pamphlets, and bulletins covering every phase and aspect of the conduct of war, which were delivered at the training camps almost daily. . . . Some of them were illuminating and valuable. Many were repetitions of other pamphlets, while a considerable percentage were obsolete."[24] The War Department's orgy of publishing further muddied the water as officers at all levels attempted to sort through the weighty questions of what should be taught to the flood of new recruits filling the divisions in the fall of 1917.

In addition to these systemic problems of mass mobilization, officers also faced other challenges in completing their training that were completely out of anyone's control. The winter of 1917–18 was particularly inclement and wreaked havoc with the War Department's intricate training plans. In February 1918, for example, the 83rd Division's intelligence officer reported that the "unduly severe winter" at Camp Sherman, Ohio, had so retarded the progress of the division's training that they were still working on the eighteenth week of the War Department's training plan when they were supposed to be working on the program for week twenty-three. He also noted that "trench work, of necessity, has been cut down to almost nil up until the present time."[25]

The situation was even worse for the 4th Division at Camp Greene, North Carolina. The rain and snow in the winter of 1918 turned the post's red clay soil into such a viscous mess that "the troops simply could not work out of doors" for weeks at a time. Between 10 December 1917 and 4 March 1918, one officer recorded, the division's units experienced only sixteen days on which any meaningful outdoor training was even possible. Under the constant wet weather, the trench system that the division constructed for training quickly turned into an unusable sea of muck. The only bright side to this natural disaster, one division staff officer quipped, was that it gave the soldiers an unintentionally realistic view of the conditions that they would later face in Flanders and the Argonne.[26]

Adding to the misery of weather-induced inactivity and further hindering training were deadly outbreaks of Spanish influenza, measles, and other diseases. Disease hindered or halted unit training as whole companies were quarantined for weeks at a time to prevent the spread of the sickness. In a letter home to his wife in March 1918, Benson Oakley wrote that a mutual friend in another company had "just got out of his two week quarantine last week when a fellow in his tent came down with the mumps. He together with the others in the tent were moved out into the woods away from everyone else for 21 days more of quarantine."[27] That meant that his friend lost at least five weeks of valuable training time.

By far the deadliest disease that the doughboys encountered was influenza. The army estimated that between 25 and 40 percent of its soldiers suffered from influenza at some time during the war. Influenza eventually killed forty-five thousand American soldiers, almost as many as were killed in action. The army also lost 8,743,102 days of work and training from enlisted men laid low by the epidemic.[28] Forced indoors by the weather or sickness, junior leaders found it difficult to complete their own training or that of their soldiers and units.

The systemic problems of mass mobilization, such as shortages of equipment, lack of weapons, inadequate infrastructure, and uncertainty in the subjects to be trained, along with weather problems and epidemics all had the cumulative effect of hindering the instruction of American units and served as an obstacle to the on-the-job training of junior officers and NCOs. This also meant that the War Department's training plan for the divisions was never as standardized as the army had envisioned. Local conditions often determined the quantity and quality of unit training and produced divisions, and unit leaders, with widely varying levels of ability.

Perhaps the gravest systemic problem that confronted the army and its junior leaders was the issue of how to select and train NCOs. As noted in chapter 2, the Regular Army system for selecting NCOs rested upon the ability of long-service officers to identify and promote potential NCO talent from within their units. Since the army had given little to no thought to raising a wartime cadre of NCOs, the Old Army method became the de facto system for the selection and training of the Great War's NCOs. The keys to the success of the

Old Army system were experience and time: the experience of the officer making the selection and the time the man in the ranks had to learn his trade and demonstrate the qualities the officer sought in his NCOs. Time and experience were, of course, the two things sadly lacking in 1917 and 1918.

The regulars were well aware of the problem of raising NCOs and sought to provide a small cadre of regular NCOs to serve as the basic noncommissioned skeleton for the newly formed units. This last idea proved to be a mixed blessing. Some of these regular NCOs did quite well. The 353rd Infantry received a contingent of thirty-four Regular Army NCOs to train the unit's recruits and serve as the basis of the NCO cadre. Capt. Charles Dienst admired the fact that these men were "soldiers by profession and played the game in a manner worthy of the best traditions of the old army."[29] Unfortunately, Dienst's experience with the regular NCOs seems to have been the exception rather than the rule.

At the outbreak of the war, the Regular Army was already short-handed and overextended. The loss of a number of NCOs to officer training resulted in the rapid promotion of a number of short-service privates to the noncommissioned ranks. Lt. Col. George English noted that the 89th Division received enough Regular Army NCOs to assign two to each infantry and artillery company. However, the officers soon discovered that while "many of these men were of inestimable value in drilling the new recruits," far too many of them had only themselves been recently promoted to sergeant. English believed that the best-qualified regular enlisted men had already been commissioned, "leaving only the less qualified men available for noncommissioned officers." He went on to state, "As a class these noncommissioned officers did not accommodate themselves well to the new conditions, and [were] not so valuable as the better educated and more highly skilled men from civil life, of which there were a number in every company or battery."[30]

The 89th Division's officers were not alone in being somewhat disappointed in the quality of the regular NCOs sent to their organizations. When the Regular Army soldiers reported to the 77th Division's 305th Field Artillery, it quickly became clear that their previous company commanders had used the requirement to transfer their NCOs as a means to rid themselves of their problem soldiers. The men that the 305th Artillery received had a string of disciplinary infractions on their records. One of the officers sarcastically noted, "We pitied those distant, unknown commanders. If these were their best we shrank from picturing their days and nights with the worst." Capt.

Kerr Rainsford, of the 307th Infantry, best summed up the practice of assigning regular NCOs to the new units. He recalled that the regular NCOs "did excellent service as drill sergeants; but on the whole the experiment was not successful, and the greater number returned to the regiments whence they came."[31]

Given the uneven quality of the regular NCOs, most company commanders had to rely on their own instincts when selecting their sergeants and corporals from the anonymous mass of recruits arriving daily in their units. Their Regular Army superiors were quick with advice. Just as the draftees were arriving at the camps, Maj. J. C. Wise warned, "The company officers will find within the course of a few days about 150 recruits committed to their charge—recruits representing all degrees of education and intelligence and every social caste, from professional gentleman with a college education to illiterate city loafer." Wise advised that the new commander must plan his training only after considering the individual soldier's "social caste" and "character, intelligence, and experience." He recommended that junior officers should divide their soldiers into three categories: superior, "business and professional men, tradesmen, [and] skilled mechanics"; ordinary, "uneducated but intelligent laborers and miscellaneous recruits"; and inferior, "recruits of the lowest order of intelligence and character."

Wise suggested that officers select their NCOs primarily from the "superior" caste of their units and that these men be made "temporary acting noncommissioned officers" and be "given a chance to demonstrate [their] fitness for a chevron." He stressed that if the acting NCOs are "intelligent and ambitious they will rapidly acquire military knowledge superior to that of other recruits in order to measure up to their responsibilities." Along the same line, Maj. Charles Tipps noted that during the war he found that "men who have successfully handled six to eight men in civil life as a boss of a group of farm hands, or as the foreman of a small department in some factory, will almost invariably make good corporals, and men with relatively more experience will, in most cases, qualify to fill relatively higher positions." Although Wise and Tips offered sound rules of thumb for selecting NCOs, their advice still required that the commander have the time and experience to identify those with the attributes they described.[32]

Although some officers used the Yerkes mental tests as a way of identifying potential NCOs, most muddled through the best they could with little help from their superiors. On 4 September 1917 the 89th Division's commander, Leonard Wood, tried to cut the gordian knot by simply directing that "the training of the first five percent of the new National Army men will be undertaken immediately upon their arrival, with the purpose of developing among them non-commissioned officers and instructors for the National Army men who will arrive later."[33] Wood's grandiose plan, which was followed by other divisions, still provided only two weeks to train these jumped-up NCOs before the arrival of the second contingent of draftees.

Faced with the pressing need for NCOs, some officers simply opted to assign men to these ranks based on seniority. In these units, NCOs owed their positions to the fact that they arrived days or hours before their comrades. As none of the men who arrived at Camp Funston with Pvt. John Nell had any previous military service, his new company commander merely went down the line selecting every fourth man to serve as a corporal.[34] Given the need to maintain some semblance of order and discipline in the ranks, some NCOs gained their stripes solely on their ability to overawe or bully their fellow recruits into line.

It is no surprise that these ad hoc measures for selecting NCOs produced wildly uneven results. Soon after the war, one officer lamented the lack of any uniform system for picking NCOs. He pointed out, "Candidates for the commissioned ranks as a rule pass through a rigid process of selection, are then sent to appropriate schools for further observation and training, and weeding out at the end of the training period. In the case of sergeants and corporals, the pillars of the army, there are no such schools. Rough observation, personal idiosyncrasies, subjective factors, all enter more or less into the selection of these men." This confusion was not lost on the privates, and some even felt pity for their officers' plight. Pvt. D. B. Gallagher wrote, "Our officers were all men who had respect for those serving under them, and the mistakes that were made in selecting the 'non-coms' were not of their making, but due to the utter lack of any system to be governed by placing men in positions which they were qualified to fill." Nor were these problems lost on the jumped-up NCOs themselves. As one sergeant recalled, "The confusion was unbelievable—it seemed as if nobody knew anything for sure. I was a corporal within three months and knew very little about the army."[35]

Regardless of their method of selection, the vast majority of the newly minted sergeants and corporals had no real knowledge or practical experience to merit their promotion or assure their authority. Capt. John Stringfellow observed that the best method for giving the NCOs the experience they needed was to follow the old Regular Army method of throwing them into the deep water of responsibility to see if they would sink or swim.[36] Although Stringfellow and his peers often had no other options readily available to them to train their NCOs, these unsystematic and idiosyncratic procedures did not produce the strong backbone of NCOs that the army needed for its combat units.

The army made some efforts late in the war to provide a degree of training to its novice NCOs, but these changes came too late to significantly change the course set in 1917. As late as 31 July 1918, a War Department inspector reported of the 84th Division: "The non-commissioned officers are as a rule not thoroughly instructed. Many of them are noncommissioned officers simply because there were no others to make. Many corporals have only a few weeks service and many organizations have not made all their noncommissioned officers for lack of trained personnel."[37] The lack of a strong corps of NCOs to help to train and lead the squads, platoons, and companies of the new divisions had immediate and profound influence on unit training and a lasting effect on how the American army later fought in France.

The most significant consequence of the weak NCO corps was that the company-level officers not only had to serve as the unit commanders, but also took on the training and administrative roles usually accomplished by first sergeants, company clerks, and supply

THE SERGEANT.

Who is that man of haughty mien,
With ample chest and peanut bean
And movement like a Ford machine?
Why, Sonny, that's the Sergeant!

Who yells, "Right Dress" and "Right by Fours,"
And gets as mad as all outdoors—
And sends you out to do the chores?
You're right, that is the Sergeant!

Who cries "Fall in," and when you do,
Says, "As you were, you rough-neck crew;"
"Fours right about!" "I'll put you through?"
Why sure, that's like the Sergeant!

Who's busy as a bumblebee,
To get you up at reveille,
And shouts your name in strident key?
Why, bless you! that's the Sergeant!

Who carries all the world's disgrace
Writ in furrows on his face,
And looks for trouble every place?
Why! That must be the Sergeant!

Why does the poor boy act this way?
Will he be General some day?
No, sonny, quite the other way,
For Hell is full of Sergeants.

A comic, but rather jaundiced, view of American NCOs.
(*From* The Wadsworth Gas Attack and Rio Grande Rattler, *27 November 1917.*)

sergeants. As one officer observed, "Lack of experience on the part of non-commissioned officers at the beginning of training centered full responsibility on the officers." Another infantry officer recalled that the "training of non-commissioned officers [was] slighted almost to the point of neglect. Officers, from the Company Commander down, [were] obliged to spend fifty percent of their time and energy in doing the work of non-commissioned officers." As the officers became the font of all military knowledge and the leader directly responsible for the training, discipline, and care of the soldiers, the NCOs' roles and prestige within the units declined.[38]

Knowingly or not, the Regular Army senior officers and commanders encouraged this process. They merely wanted certain tasks accomplished to standard in their subordinate units and held the officers in those units responsible for these results. One regimental commander expressed no alarm in the fact that "the junior officers really performed the duties of non-coms, in each company one being in charge of quarters, one with an assistant in charge of mess, one in charge of the company office, and so on."[39] With the spotlight on the junior officers to produce results, and a lack of training and readiness on the part of their nascent NCOs to aid them, the officers tended simply to do their jobs as well as those of their sergeants.

The tendency of officers to become de facto NCOs led to a culture of micromanagement and dependency in the small units that made it difficult for NCOs to find their authority and break their unit's absolute reliance on their commissioned leaders. Soon after the war, Maj. Thomas Swann decried the fact that "it was often the practice in the formation of the National Army to have a sergeant always supervised by an officer." He maintained that while the "officers and men had to be trained simultaneously . . . too much supervision [by the officers] was destructive of initiative in the noncommissioned officers, and rather made them dodge than accept responsibility." The 305th Infantry's Capt. Frank Tiebout also ruefully acknowledged that "the officers never ceased to regret the theory of the Division Commander who forbade the placing of any real responsibility upon the shoulders of our non-coms. Far better it would have been at camp and throughout our subsequent experience, if it had not always been required that an officer be present, whether at the fairly simple task of filling a bedsack, or at an inconsequential gathering of any sort." Although Tiebout was quick to denounce the micromanagement of his senior commander, the company-grade officers also perpetuated this officer dependency and undercut their sergeants and corporals. One private observed that when NCOs began to drill their squads, inevitably their lieutenant would step in and correct the NCO in front of his men, all the while lecturing him on the chapter and verse of the drill regulations. At every turn the officer's position was thus strengthened to the detriment of his sergeants and corporals.[40]

While their superiors advised the young officers to follow the old army tradition of trying out acting NCOs before actually promoting them, some officers took this as a license to establish a revolving-door policy for selecting their sergeants and corporals. On 29 July 1918 Capt. Clarence J. Minick, of the 361st Infantry, 91st Division, wrote that he had reduced thirty-one of his NCOs to privates, leaving him with only one remaining NCO in the company. He noted that "these are the ones I tried [out] on the way over" and found they did not work out in their new ranks.[41] Although Minick's case was extreme, it does highlight the company commander's latitude to shape the composition of his unit NCOs to his pleasure. While battalion commanders were, by regulation, expected to be the approving authority for the demotion of any NCOs in their units, few seemed willing to second-guess their captains in these matters.

To add to this litany of woe, neither the new officers nor their proto-NCOs seemed to understand exactly what the roles and responsibilities of the sergeants and corporals were to be.

Some officers, however, were quick to denounce the failings of their noncoms. An infantry battalion commander observed, "Among the so-called noncommissioned officers, who are but the more apt enlisted personnel with chevrons, no high sense of individual obligation to their ill-defined and imperfectly understood responsibilities exists, and being, like those over whom they have been set, but novices at the game, they are lacking utterly in the confidence which is necessary to force them to the front."[42] This officer failed to grasp that the American NCOs generally received no special training and little incentive for their assignment and were often poorly guided and supported by their officers. The position of NCO carried few privileges in terms of pay and status and even fewer responsibilities. That the new NCOs' responsibilities remained "ill-defined and imperfectly understood" was the fault of the army and of officers such as the battalion commander himself.

The NCOs had few guides to help them to understand their duties and responsibilities. In 1914 the War Department issued the *Manual for Noncommissioned Officers and Privates* for each branch of the service. These manuals were updated in 1917 to serve as the recruit's handbook for basic military knowledge. As the title implied, these manuals were also to instruct NCOs on their duties and responsibilities. The information in the books was far too broad and general to be of any practical use to the NCOs. The books covered how to give close-order drill, but nowhere was the new sergeant instructed on the principles of leadership or on his role in combat. The manual stated that the NCO's overarching duty was to enforce discipline and "obey strictly and execute promptly the lawful orders of your superiors."[43] Beyond those admonitions, the inexperienced NCO was given no helpful hints on how to turn a group of civilians into soldiers or get those same men to move forward in an attack. Also, if one followed the manual, there was little to no room for an NCO to exercise initiative or any independent combat action outside of very limited patrolling.

Throughout the war the army made little effort to distinguish NCOs from privates. Even the title *Manual for Noncommissioned Officers and Privates* is suggestive of this attitude. Given the fact that the majority of NCOs had no more experience than the privates they led, this attitude may be understandable. However, this outlook seems to have sidetracked any serious efforts to systematically educate and develop NCOs until late in the war. Throughout 1917 and much of 1918 the only effort to train NCOs was at ad hoc schools established within individual units. Unfortunately, these schools were usually taught by OTC graduate officers who often lacked the knowledge and experience to train their student NCOs adequately. Also, when the units began to face large transfers of soldiers in the winter and spring of 1918, and new drafts of men had to undergo basic instruction, the NCO schools were often the first thing dropped from overcrowded training schedules.

These missteps in raising a corps of NCOs were not lost on the Allied officers assigned to train the Americans. What troubled a number of these officers was that the Americans seemed to overlook the emerging role of NCOs as true combat leaders who would be required to use their initiative and make weighty decisions on the battlefield, rather than being merely an executor of their officers' explicit orders. In a 10 January 1918 report to the American General Staff, an officer of the French military mission tried to broach this subject with the Americans. After noting how the lethality of the battlefield had led to a greater dispersion of units, thus with more responsibility devolving upon NCOs, the Frenchman observed, "The American N.C.O.'s have no authority at the present time and consequently no influence over their men. . . . Under such conditions, they can neither second the officers efficiently nor replace them." He recommended that the ad hoc company or battalion NCO schools be replaced by central divisional schools that could give the NCOs "the power, confidence, and prestige which only instruction can bestow." Two months later, after inspecting training at Camp Oglethorpe, Georgia, the French general Claudon reported, "The [American] N.C.O. is non-existing. . . . At present time they have no authority and they have no right to punish. They are mixed with their men; they fight with them to get a piece of food, etc."[44]

The French were not alone in their concern over the weak American NCO corps. On 24 May 1918 the commander of the British **119** military mission, Brigadier-General Trotter, warned the American Chief of Staff, "In the three divisions that I visited [the 83rd, 84th, and 89th] during my tour, the British Officers commented on the status of the N.C.O. in the US Army. The opinion I formed was that for both training and disciplinary purposes his status as compared to the British N.C.O. is not sufficiently recognized."[45] Trotter stated that he had warned the American division commanders that if the American NCOs were to become effective combat leaders, the officers had to grant them more privileges and do more to give the sergeants increased status and standing within their units.

The War Department's failure to adequately plan for training N C O s meant that most of the divisions that deployed to France before the fall of 1918 did so with sergeants and corporals who had learned their roles mostly through on-the-job training under the constant supervision of their officers. Immediately following the war, the Morale Branch of the War College Plans Division submitted a questionnaire to officers leaving the army to gauge their opinions and attitudes toward their service. Looking back on their service, nearly all the officers polled agreed that they and the army had not done enough to give their NCOs the respect, prestige, or authority to accomplish their tasks or to encourage their privates to follow them.[46] Unfortunately, this hindsight wisdom came too late to do much good.

It was not until the bulk of the American divisions began deploying in the spring and fall of 1918 that the War Department began to take seriously its responsibility for training NCOs. As the divisions departed, the War Department established a series of replacement training centers and depot brigades. In addition to providing basic training for new draftees, the War Department also directed these units to establish schools for training NCOs. However, as the War Department neglected to provide a standardized training plan for these schools, the training regimen and the selection criteria for the students for these courses varied greatly from post to post. For example, Brigadier General Sage, the commander of the Camp Gordon

Infantry Replacement Camp, directed that ten percent of the privates going through instruction at the camp would be selected for "special training in the duties of non-commissioned officers." The NCO school lasted for thirty days and provided the candidates with their basic infantry training focusing on "close and extended order drill, musketry, elements of field service, guard duty, [and] signaling."[47]

On 30 July 1918 the officers at Camp Devens, Massachusetts, recommended that the War Department establish a two-month-long NCO school. After removing the time dedicated to conferences, inspections, and in- and out-processing, the school curriculum consisted of 262 hands-on training hours:

Close-order drill, 55 hours (21% of total hours)

Physical training, 40 hours (15.2%)

Bayonet training, 20 hours (7.6%)

Interior guard duty, 10 hours (4%)

Small arms training and firing, 32 hours (12.2%)

Extended-order drill, 15 hours (5.7%)

Field fortifications, 12 hours (4.5 %)

Minor tactics and field problems, 23 hours (8.7%)

Voice culture (giving oral commands), 6 hours (2.2%)

Map reading, 5 hours (2%)

Practice marches, 12 hours (4.5%)

Overnight bivouac, 10 hours (4%)

Camp and trench experience, 6 hours (2.2)

Modern weapons (machine guns, automatic rifle, one-pound gun, mortars, grenades, and gas), 16 hours (6.1%)

Over half of the practical instruction was devoted to close-order drill, interior guard duty, and bayonet and physical training.[48] Although these matters were important in building discipline and aggressiveness in the nascent NCOs, the relatively short amount of time given to subjects such as map reading and minor tactics reveals the continued problems that the army faced in understanding the realities of modern combat and the skills that its junior leaders needed to face them.

NCO schools later established at Camps Lee and Grant had the same problem in focusing their subject matter as had the Camp Devens school. The Camp Lee school devoted 29 percent of its training time to close-order drill and less than 15 percent of the curriculum to minor tactics. Had it not been for the specter of combat that hovered over the training, some of the material taught in the schools would have been laughable. Despite all the lapses in the training of the army's NCOs, the Camp Gordon morale officer gushed with pride at his ability to have a "special school for singing" established in the camp for selected NCOs to train them to serve as "song leaders in their companies." The officer noted that this was a coup, for "singing is becoming an essential part of training."[49]

It is interesting to note that all of the NCO schools mentioned here were of differing durations. The Camp Gordon NCO course ran for thirty days, while the Camp Grant course was for seven weeks and the Camp Lee course was for two months. This lack of standardization was not the only problem with the training. Just as with the OTCs, the NCO schools faced grave challenges in obtaining qualified candidates in the late summer and fall of 1918. Part of this resulted from the army's own priorities. Maj. Charles Tips reported that at the infantry replacement depot at Camp Gordon, "the best men from the noncommissioned officers' training school are selected to attend the central officers' training school." To overcome this problem, the chief of the War Plans Division suggested that the army use the 35 percent of candidates who failed out of COTS as NCOs, but even he admitted that "this source alone will not furnish the necessary number of noncommissioned officers required."[50]

A postwar board of officers studying the combat effectiveness of the AEF lamented that some of the tactical sluggishness of the US Army was a result of "poorly trained and rather dull non-commissioned officers."[51] Given all of the problems the army faced with selecting and training its NCOs, was it any wonder that its corporals and sergeants often failed to live up to their ill-defined responsibilities and expectations? As the 307th Infantry departed for France, the best that one officer could say of the unit's enlisted leadership was, "[In] every company one or two N.C.O.'s had shown that absolute reliance could be placed upon them as leaders of their men; for a much larger number it was confidently hoped that under war-time conditions their power to command would develop; but the great mass of men still constituted an ununified, unknown, and very insufficiently trained quantity, who had never learned to take themselves serious as soldiers."[52] This rather bleak, though accurate, assessment could have been applied to most of the divisions at the time. The reality of combat in France later meant that many of this "ununified, unknown, and very insufficiently trained quantity" of NCOs would be forced to take command of units because of officer casualties and the arbitrary dictates of the fog and friction of war. While some NCOs rose to these occasions and others did not, in neither case did much in their training adequately prepared them for the eventualities that thrust them into the center stage of combat leadership.

Given the systemic problems of mobilization and the level of leadership and knowledge of the new divisions' officers and NCOs, it should be no surprise that stateside unit training was often ill-focused and incomplete. This was not for want of effort. The vast majority of the officers and NCOs involved with the training poured their hearts, souls, and intellect into preparing for combat. Still, enthusiasm and effort are no substitute for skills and know-how. In a telling incident, Capt. Robert Gill recalled that in August 1917 his commanding officer ordered him to form a trench mortar battery. After accepting the assignment, Gill's only question was, "May I ask, sir, what a trench mortar is?" His commander's only response was, "Damned if I know, but you will soon find out."[53]

With their own experience and knowledge barely above the level of a prewar private's, the junior officers found themselves suddenly responsible for the basic instruction of their soldiers. This left little time for the officers to concentrate on developing their own tactical competence. The undertrained lieutenants and captains frantically scrambled to learn the basics that they were expected to impart to their subordinates. For example, W. A. Sirmon, a lieutenant in the 82nd Division's 325th Infantry, recalled spending many of his mornings in hurried classes so he could give the same lessons to his soldiers later in the day. Soon after being assigned as the commander of the 157th Infantry's machine gun company, Lt. Maury Maverick realized that he faced a grave problem in training his soldiers. He recalled, "I could, with great show, take a machine gun apart, but putting it back together again was another matter. A few pieces would always be left over, or I simply couldn't get it together." Lt. Charles Bolte noted that as he fumbled to teach his soldiers such basic tasks as operating the .45-caliber semiautomatic pistol, "It was a case of the blind leading the blind." The phrase "the blind leading the blind" peppers the writings of the war's veterans and was perhaps the best description of the tragicomic training environment in the mobilization camps.[54]

It is an age-old military truism that leaders can't fool the troops for long. In a case of motivated self-interest, enlisted men historically have observed their leaders closely to determine their strengths, weaknesses, and overall competence. As their daily life and very lives depend upon the personalities and abilities of their officers and NCOs, the men in the ranks have tended to be harsh and honest judges of their leaders. This was certainly the case with Pvt. John Oechsner. He described his officers as "90 day wonders" and "boys just out of school." During his time in training, it was clear to this private that his officers "didn't know what the hell it was all about . . . it was all Greek to them." He noted that even when it came to basic drill, "Our commanding officer didn't know a damn thing." It was even clear to the rookie recruit Benson Oakley that there were large gaps in the knowledge of his officers. Writing home from Camp Hancock in

April 1918, Oakley described his officers as "ignorant." After a week of "quite a little drilling," he determined that his leaders "ought to study up [on] the drill regulations a bit." William Clarke remembered being drilled by an officer fresh out of OTC while he was training at Camp Wadsworth, South Carolina. The officer "got so mixed up we were scattered all over the field and he was unable to get us back into platoon or company line." None of these officers' faults could have done much to inspire trust and confidence in their soldiers.[55]

Many senior commanders tried to aid their new leaders by establishing after-hours unit officer and NCO schools. Unfortunately, the hectic conditions under which the leaders operated left little time for continued professional development and self-study. Although well-intentioned, the unit schools often lacked qualified instructors and hands-on application. Lt. Milton Bernet complained, "This school was valueless and uninteresting" and that "we all tried to duck it as it was so useless."[56]

Lucian Truscott experienced similar problems with continuing his professional education after reaching his unit. The first thing that struck him was that "the instructional content and methods of training in the regiment differed little from that in the officers' training camp." His formal leader training consisted of one of the regiment's veteran officers gathering the new lieutenants under an open-air dance pavilion, where "he would read from the manual the lesson assignment for the day. There were no charts, no diagrams, no photographs, no illustrations, no training aids of any sort. No practical work for the students; no questions period. He read. We listened. Then, the day's reading done, he would regale us with tales and anecdotes of colorful cavalry personalities and past cavalry history." Nor were his senior officers helpful in filling in their new lieutenants' knowledge. Truscott's regimental commander "amazed us young officers.... He insisted that the repeating rifles and machine guns wasted ammunition and encouraged soldiers in careless habits.... Considering that the machine gun was dominating the battlefields of Europe at the time, Colonel Morgan's views provided us junior officers with a great amount of conversational material and did little toward increasing our confidence in some of our superiors." True to the old army tradition, his regimental commander believed that the best way to instruct young officers "was to assign them a task and then let them work out their own solution."[57] Without a strong cadre of competent NCOs to aid them, the junior officers such as Truscott and Bernet were left to sink or swim in the training of their soldiers and units.

In the fall of 1918, Chief of Staff Peyton March reported, "The

The feelings of Lieut. Ivan New Goldbar, U. S. R., on "taking the company" for the first time.

This cartoon from a soldier of the 27th Division is a good characterization of the angst that many green OTC officers felt upon first taking command of soldiers. Those officers who could not overcome this angst were prime candidates for reclassification.
(*From* The Wadsworth Gas Attack and Rio Grande Rattler, *9 March 1918.)*

quality of troops and their value as an effective force depends to a very large extent upon the character and sufficiency of their training, which in turn is dependent upon the officers who are designated to instruct them in camp and lead them in battle."[58] For the US Army in the Great War, no truer words could be spoken. Despite the War Department's well-laid plans for stateside training, the war's participants suggest that far too much of the instruction time in the United States often centered on subjects that the novice officers understood and could easily teach, such as close-order and bayonet drill.

As was the case in the OTCs, far too much of the training in the new divisions was also centered on mastering obsolete battle formations and inculcating unrealistic views of warfare. It should come as no surprise that the young officers passed on to their soldiers the tactics and assumptions that they had so recently learned themselves. Looking back on the training he learned and imparted to his soldiers, one lieutenant confessed, "Too much stress [was] put on form, ceremonies, close order drill and other West Point relics of the Roman phalanx age. . . . Too much valuable time [was] spent teaching 'squads right' and not enough making every man able to use any type of machine gun." This point was also not lost on the 82nd Division's W. A. Sirmon. He recorded that the division's soon-to-be-deployed infantry units were still focusing on "one hour in general infantry training, close order drill, bayonet work, and bombing."[59]

Without their own base of experience to draw upon, the junior officers often found it difficult to instruct their men in the more complex tasks of soldiering. Despite the time and effort that his unit devoted to officer-supervised marksmanship training, the 82nd Division's Alvin York remembered that his comrades remained the worst "shots that ever shut eyes and pulled a trigger." Likewise, an inspection of the 84th Division revealed that soldiers who had been in training for several months were still "found to have a very poor knowledge of subjects as care and preservation of arms and equipment . . . individual cooking, care of feet, first aid, or the effect of wind, heat, and light upon shooting." This was a inauspicious start for an army intended for open warfare and built upon superior American marksmanship and maneuver.[60]

Unfortunately, as American units prepared to deploy overseas, the experiences of the 82nd and 84th Divisions were the rule rather than the exception. A War Department inspection of division cantonments in 1917 revealed serious problems with the training conducted in the new units. The inspectors noted, "Schedules of drills and instruction show an ample provision of time for this phase of instruction. Want of time, therefore, cannot be given as an excuse. *The defect lies in a want of accurate knowledge on the part of company officers and noncommissioned officers and failure on part of battalion commanders and commanders of higher units properly to supervise the drills and instruction of their commands.* . . . They fail to make satisfactory progress in drilling their commands because *they do not see* the mistakes which are constantly made, and do not, as a consequence, correct them."[61]

This unfortunate situation did not improve with time. Nearly a year after the first inspection, Col. H. O. Williams noted the same problems in training within the 84th Division. Williams reported, "The instruction of the division has not been as thorough as it should have been. Officers and noncommissioned officers are not sufficiently zealous in correcting mistakes made at drill. . . . They give command and command without any correction or any apparent effort to see that the movement is properly executed." He noted: "The greatest weakness of our system of training today is the lack of officers and noncommissioned officers who have a thorough knowledge of what they are trying to teach or who insist upon having their commands or instructions thoroughly complied with. We must have more accuracy and more attention to detail."[62] Without an adequate knowledge of the basics or a fundamental understanding of war on the Western Front, the new officers and NCOs were ill-equipped to identify and correct problems both in the training of their soldiers and in the flawed American doctrine.

The Americans did not embark into their wilderness of training alone. Both the French and the British sent a number of officers and NCOs to instruct the Americans in the mysteries of modern war. Every division had a cadre of Allied officers to teach the Americans the technical skills needed to master the new weapons of the war. For example, one officer recalled that in the 89th Division the training conducted by French and British officers was generally limited to machine guns, gas warfare, bayonet fighting, artillery firing, automatic rifles, bombs, liaison, and the Stokes mortar.[63] These technical courses tended to be the most common subjects taught by Allied officers during the stateside training of American divisions.

While the junior officers seemed to have been enthusiastic and receptive students of these technical courses, their regular army senior commanders were much less impressed. After becoming a division commander, Hugh Scott observed that the Allied officers "invariably assumed our total ignorance of everything military, and started their course with the most rudimentary subjects. I had to stop this waste of time, and told them that our regular officers needed only the newest developments as they came up, for they were otherwise as well or better trained than the officers of Europe. After this they taught us the art of throwing bombs, the use of flares, and the operation of trench mortars, *but the best thing we got from them was their new bayonet drill.*[64] The fact that he found the bayonet drill to be the most important aspect of the training that the foreign officers gave is a revealing statement on the mentality of many of the army's officers. It also shows why the training given by the Allied officers was so limited in scope and scale.

The stiff-arming of French and British tactical concepts by the Americans was not lost on the Allied instructors themselves. Their reports to the American General Staff and their own military missions bristled with anger and exasperation over their sidelining by the very people they were trying to aid. After inspecting training at Camp Shelby, Mississippi, one French officer reported, "When there is a question of drills and maneuvers in which they could readily assist, the American officers are careful not to consult them, preferring to work their own way. Our officers have no part in training except in that of specialties and do not assist at any drills." After visiting a number of American posts, he concluded that "the word 'specialist' is being deliberately exploited [by the Americans] to limit our activity."[65]

As the Americans "found their feet" in 1918, these obstructionist tendencies grew stronger. On 18 April 1918, Col. James Martin, the acting chief of the French advisory mission, reported a number of problems that his trainers were having in instructing the Americans. In one of the gravest situations, he found that his officers at Camp Custer were subject to "a certain hostility on the part of the new chief of staff." The Frenchman peevishly observed that the American "thinks that because he stayed a few days near the front that he does not need any help of foreign advisors for the instruction of the division."[66]

The Allied officers were also critical of the tactical training that the Americans were conducting. As most of these officers had personally experienced the realities of modern war, their criticism was telling. One Russian general was amazed that, given the changes in warfare that had placed increased importance on small-unit combat and had made well-trained squads "the basis of all efficiency" in higher units, he "saw practically no squad work in the entire trip." This, he believed, was "a most grievous fault" in the Americans' training. The Frenchman Major De Reviers condemned the fact that he was "unable to

FOREIGN OFFICERS WHO AID TRAINING
Left to right—Lieut. Kolb-Bernard, French; Capt. De Brissoe Owen, British; Capt. Tommy Martin, French; Major Woodcock, British;
Lieut. Michel, French; Capt. Yoxall, British; Lieut. Marquet, French.

French and British officers assigned to train the 83rd Division at Camp Sherman, Ohio, in 1917 and 1918. Allied training officers often chafed under the restrictions that the Americans placed upon what they could teach.
(From Frank H. Ward, ed., The Camp Sherman Souvenir.*)*

have our infantry combat methods accepted or to have the progressive stages of instruction directed along the lines of modern warfare" because the Americans insisted that their tactical training be based on "American Methods." He found that in the US Army "false ideas of combat prevail such as antiquated tactical theories of before the war." As the bulk of the American divisions were preparing to deploy in the spring of 1918, the head of the French military mission, General Claudon, noted that the Americans' tactical training still lacked "a programme of exercises in combat in simple but well defined and progressive steps."[67]

The Allied officers also denounced the Americans' fixation with close-order drill and waste of time on other topics of limited importance. The senior French officer assigned to Camp Sheridan was maddened by the fact that while the Americans had finally established a much-needed platoon leader school, six of its ten lessons were devoted to close-order drill. Another noted that the platoon leader school at Camp Sevier had essentially closed because senior officers had mandated that young officers attend morning close-order drill sessions rather than attend the school.[68]

Some of the Allied officers placed the blame for this squarely on the backs of the American regular officers responsible for planning the training. Of his American peers one Frenchman acidly remarked, "The field officers are very inferior as they have no special schools and do not get down to work. The efficiency of the staff is lost in innu-merable details; its idea of training is merely that of close order for-mation and outward appearance (bluff)." This officer's gloomy and prescient conclusions were, "Combat training has hardly begun and is started in the wrong direction. It is greatly to be feared that if the present mistakes continue the situation will not improve.... The di-visional staff is responsible for false orientation given to training and it is absolutely necessary, in order to climb out of this rut and avoid appalling and demoralizing losses, that measures be taken and orders given from those in command."[69] None of these comments offered any ringing endorsements of the Americans' training, doctrine, or combat leadership.

129

It could be argued that the harsh assessments of the Americans concealed a hidden agenda on the part of the Allied officers, or were merely sour grapes. There is truth in this point of view, and painting the Americans as inept amateurs was clearly a way of furthering the argument that the doughboys would be best served by being amal-gamated into existing Allied units, or at least retained under Allied command. However, one must keep in mind that the Allied officers assigned to duty in the United States understood the implications for their battered nations and armies if the US Army proved incapable of taking to the field. There is nothing to suggest that any of these offi-cers wanted to see the Americans fall flat on their faces in either train-ing or combat. The criticism that they offered was valid and reflected the veterans' hard-won battlefield wisdom.

The Americans themselves appeared at times to understand the holes in their tactical training and doctrine. Regrettably, much of this realization came only after they experienced the pain of com-bat. From division to division there were great variances in the qual-ity, quantity, and degree of realism given to the tactical training of squads, platoons, and companies. As previously noted, much of this depended upon the local realities of equipment shortages, training areas available, the weather, and troop health. When all the right variables aligned, tactical training in the United States could be quite sound and realistic. Since the time, equipment, and ammunition were available to Maj. Gen. John F. O'Ryan, he was able to have units up to the brigade level in the 27th Division practice assaults on a simu-lated enemy trench system supported by preparatory fires and a roll-ing barrage fired by the division's artillery units using live shells. The division commander noted that the "psychological effect upon units subjected to these tests was marked" and that this "rather radical training" helped to better ease the soldiers into their later shocks of combat.[70]

The available evidence indicates that the level of detail and realism in the 27th Division's tactical training was uncommon in that of most units. The field exercises in other divisions tended to range from overly scripted productions to wide-ranging free-for-alls. In most cases the training was circumscribed by the knowledge and the abilities that the unit's junior officers brought to the field. For instance, Kerr Rainsford admitted that the military knowledge of most of his fellow officers was limited to the *IDR* and *FSR* that they had learned at the OTCs.[71] As old habits in training and indoctrination are hard to break, this often meant that the officers brought the obsolete practices that they learned to the instruction of their own units.

Although the historical record suggests that officers made diligent efforts to keep abreast of rapidly changing doctrines, they seemed to be training concepts that were always two or more steps behind the realities or latest tactical developments of the Western Front. A 77th Division officer recalled that in his unit's specialist schools, the instructors "usually concluded their course with a warning that, in view of a more recent method having been ordered since the opening of the course, the methods of instruction just taught should not be practiced with the troops." Another 77th Division officer noted, "Platoons were for a time divided into grenade throwers ... rifle-grenadiers, rifle-men and liaison agents, according to the directions of the red pamphlets, recently distributed to the officers." While he was most likely referring to the red-jacketed *Supplement to Instructions for the Offensive Combat of Small Units* first published by the AEF in early 1918, by the time his unit sailed for France that edition had already been supplanted by a newer one published in April 1918. The officer also leaves the impression that even the time his unit spent on the new doctrine was rather sparse.[72]

It was not until 6 May 1918 that Payton March rescinded the hastily translated copy of *Instructions on the Offensive Combat of Small Units* that the War College Division had published in May 1917 and approved the revised edition of the work reprinted by the AEF in April 1918 (the April edition also changed the title to *Instructions for the Offensive Combat of Small Units*). Factoring in the time it took to print and distribute this manual, it is unlikely that infantry units training in the United States received it before late June or July of 1918. This meant that most of the American divisions had either already sailed for France or were in the process of deploying by the time the publication was distributed. Although the manual incorporated the AEF's *Supplement to Instructions* illustrating the combat formations and maneuver drills for infantry platoons and companies, it also contained illustrations from a British source that were not in the AEF's original publications. It is unclear why this series of British line drawings of a platoon in the attack was included in the manual. They were not referred to in the text and, more important, did not match the formations illustrated in *Supplement to Instructions*. Although the pictures are quite artistically done, they could only have complicated the efforts of junior leaders to puzzle out the already confusing American doctrine.

The lag time in adjusting emerging doctrine to training hindered the ability of stateside instruction to replicate the realities of combat in France. Most divisions spent a lot of time and effort in constructing trench systems that their units would use as part of their offensive and defensive training. Some of these systems were quite elaborate, with machine gun emplacements, troop shelters, communications and reserve trenches, and extensive barbed wire obstacles. These forms of fortification were discussed in detail in the 1917 War Department publications *Infantry in the Defense* and *Notes for Infantry Officers on Trench Warfare*. Unfortunately, the mighty constructions depicted in the publications and built at posts across the United States reflected more of the realities of 1916 than those of late 1917. By mid-1917 the German army had shifted its defensive doctrine to the elastic defense in depth. With this change the Germans moved away from massing their strength in forward trenches to a system of echelonment that employed shell holes, strongpoints, and trenches sited on reverse slopes to slowly grind down any Allied attacks as they slogged through the depth of the German defenses. The training trench system constructed by the 27th Division at Camp Wadsworth, South Carolina, had a depth of only four hundred yards—far shallower than the elastic defense systems then in use by the Germans on the Western Front.[73] Thus, the American training in no way replicated either the methods of the German defense or the Germans' penchant for well-timed counterattacks.

The training also continued to present other unrealistic visions of modern war. One officer wrote, "With the completion of the new trench system patterned after a sector of the Western Front, the regiment engaged in maneuvers which suggested something of actual war. Men leaped over or into real trenches and advanced cheering in innocent simulation of a real bayonet charge." The ease with which the Americans assaulted the trenches, unhindered by complex thoughts of artillery, machine gun fire, and supply, brings to mind the equally flawed and unrealistic training given Britain's Kitchener's Army scarcely two years before. These unrealistic visions of war even affected

131

some officers' conception of their role as combat leaders. While training at Camp Doniphan, Capt. Ernest McKeighan wrote home to his wife, "On the battlefield the captain is expected to be way back behind the lines in a 'dugout' or bombproof with a telephone at his head directing operations. They say we are not even allowed to carry a revolver but [are to be] armed with a *trench stick.*" Although McKeighan may have written this to ease his wife's worries, the passage still offers an insight into the mind-set that his training encouraged.[74]

132

It is interesting to note that the enlisted men appear often to have had a different impression of the field training than did their officers. In many ways the accounts of these soldiers offer a more honest appraisal of what they were learning. A soldier in the 77th Division observed that much of their tactical training at Camp Upton, New York, often "degenerated into nothing more than wild games of hare and hound, pursued without the slightest regard for military regulations." Similarly, another soldier maintained that during his training at Camp Funston, "Exercises in minor tactics made up in aggressiveness where they lacked in accuracy. Both sides claimed victory in many bloodless campaigns." Benson Oakley left an interesting narrative of his tactical training at Camp Hancock in April 1918. Benson wrote home,

> During the past three days we have been very busy drilling under those ignorant officers and hiking. Yesterday morning we took a ten mile hike, five miles each way to a lake and on the way we had to send out snipers, advance guards etc to watch for [the] enemy. I was one of the advance guards and it was surely a great game spying around in the woods. We were supposed to be attacked by cavalry but our companies surprised them in the woods. Our guard discovered where they were, sent one man back to the main body and then they all charged down upon them. Of course we didn't have any guns but we all went through the maneuvers just as if we were actually engaged in the present war.

These passages suggest that the soldiers involved either did not understand the purposes of the exercises and the roles they were to play in their unit's maneuvers or were blissfully ignorant of how little their carrying-on reflected the realities of combat.[75] In either case, they do draw into sharp focus the underlying problems with American stateside unit training. These naïve soldiers and officers would soon find that their German adversaries played a much rougher version of "a great game spying around the woods."

Military training, however unrealistic it may be, has the additional goal of welding all members of the unit into a cohesive whole that enables soldiers and leaders to operate effectively under the stress of combat. Although the US Army of the Great War lacked a theoretic or doctrinal basis for describing cohesion, it clearly understood that modern warfare had placed an even greater demand on small units and had a conception of unit identity. An experienced regular officer wrote, "Any group of individuals working together for a common purpose is going to establish unconsciously a group spirit of some kind. The leader knows that success largely depends on . . . this spirit. . . . By getting to know the men and 'how they feel about it,' he keeps in close touch with the spirit . . . and makes the men feel a membership in his team." In an April 1918 article in the *Infantry Journal*, a description of the psychological realities of the battlefield by William E. Hocking, a Harvard University professor of psychology, has a decidedly modern feel. Hocking described a reality of war in which "no one knows in advance how he will behave in an emergency that he has never experienced. But it may be taken for granted that everyone experiences fear." He accurately noted that the "mass attack, while expensive in terms of men, is psychologically easier, for it satisfies the protective instinct of bunching together." Hocking argued that the only way to get soldiers away from this "illusory and fatal impulse" was to train the individual soldier and officer to understand what he and his unit had to accomplish and where he was to end up at the end of the battle. As for unit leaders, "It is no longer possible for officers higher than platoon leaders to be personally in touch with their men during attack. The weight of responsibility for morale is thus thrown on lieutenants and noncommissioned officers, and under present conditions it is difficult for them to retain control. Men fight best with comrades and in units they are used to. Men are kept up to their best performance by the eyes of those who know them."[76] Thus, nearly thirty years before the publication of S. L. A. Marshall's *Men Against Fire,* the concept of the modern psychological battlefield, and the cohesion and leadership that units and individuals needed to confront it, was understood in certain corners of the American army.

Given this broad understanding of unit cohesion, one wonders why the War Department did not do more to promote it during World War I. Although the War Department was beset by a host of problems beyond its control that hobbled its wartime mobilization and training efforts, its personnel policies caused unnecessary damage to unit cohesion that seriously undermined the army's overall combat effectiveness and the leadership of its junior officers and NCOs.

Some of this was undoubtedly a reflection of the managerial approach to problem solving that accompanied the cult of professionalism.

As with other Progressive Era reforms, part of the Regular Army's move toward increased professionalism was an effort to inject efficiency into military processes and operations. Reflecting the views of the guru of Progressive Era efficiency, Frederick Winslow Taylor, the adjutant general and other members of the General Staff simply saw the manpower issue as matching human assets to the most pressing needs. Soldiers merely became interchangeable commodities that could be moved at will to enhance the overall institutional efficiency of the US Army. Throughout the conflict, the War Department displayed no reluctance in ordering massive levies and transfers of soldiers and leaders from one unit to another. These frequent and often inexplicable movements of soldiers in and out of units further crippled individual and unit training, damaged the morale of officers and enlisted men, and preempted the efforts of junior leaders to build cohesive combat-ready units.

The experience of the 82nd Division is illustrative of the continual building and tearing down of the American divisions. In August 1917 the War Department activated the division and began to man it with draftees from Alabama, Georgia, and Tennessee. Less than two months later, with the division approaching full strength and already well into the stage of collective unit training, the War Department reversed itself and ordered most of the unit's enlisted men transferred to the 30th, 31st, and 81st Divisions. The War Department's decision left the 82nd with a cadre of only 783 men to rebuild the division's organization. A regimental commander reported that these transfers left his company commanders with only five draftees in their units. Nor were the moves limited to enlisted men. The same officer noted that "each time a Signal Corps motorcycle entered camp," it caused great consternation in his regiment because it was bound to be bringing "orders for officers to proceed here, there, and everywhere for duty in the then so-called National Army." More important, these moves meant that six weeks of cohesion building and training had gone to waste.[77]

In late October and November the division was brought back up to strength with draftees from New England and the mid-Atlantic states. Col. Julian Schley remembered that the arrival of these new men was not greeted with rejoicing by his junior officers. He noted, "These men proved so poorly drilled in general that training had to commence at the beginning. Up to this time the spirit of the reserve officers had been high and the development of the men under their instruction had encouraged them. This return to the first lessons again with another set of men discouraged them and created a corresponding slump in their enthusiasm." The new levy also contained a large percentage of recent immigrants unable to speak or read English. This further hindered training and forced the division commander to organize English classes to give the soldiers the basic language knowledge necessary for combat. This problem was not limited to the 82nd Division; in 1917 one in three Americans was a first-generation immigrant, and one in five draftees was foreign-born.[78]

The division's manpower challenges did not end with the arrival of the northeastern draftees. In an effort to pool soldiers who had civilian experience in certain crafts and industrial jobs, Washington again ordered the 82nd to transfer over three thousand specialists from its ranks in November 1917. This levy fell hardest among the unit's NCOs. One bitter officer remarked, "The Division believed that the War Department had overlooked one important consideration. Although the soldier might be a very good plumber, lumberman, blacksmith or structural iron worker, a great deal of Government time and money had been expended in making him an even more valuable specialist in his present occupation: namely that of a non-commissioned officer, bayonet instructor, hand grenade expert or machine gunner." As a result of the ugly wave of xenophobia and nativism that swept over the United States in the first year of the war, the division's number of trained men was further reduced by the forced discharge of over fourteen hundred men considered to be enemy aliens by the War Department in March 1918. Although the War Department promised to refill the division with suitable specialists and replacements prior to their embarkation, the 82nd received only levies of untrained draftees. Some of these raw replacements arrived in the division only days before its departure from the United States. Pvt. Ralph Flynt was inducted into the army on April 2, 1918; eighteen days later he was on board one of the ships carrying the 82nd to France.[79]

The disruptions caused by the army's levy policy were far from isolated. The *American Expeditionary Forces Order of Battle* notes that many National Guard and most National Army divisions experienced large turnovers of personnel prior to their movement overseas.[80] Of the forty-one divisions that were sent to France, seventeen lost an aggregate of at least ten thousand men to transfers between the time they were raised and the time they sailed for France. Nearly all of the remaining divisions suffered losses that ran into the thousands, or had to cope with the constant arrival of new and largely untrained men required to bring them up to strength. Although the War Depart-

135

ment justified the constant transfers of personnel as a necessary evil that filled earlier deploying units at the expense of those sailing later in 1918, these moves appear to have been made with little or no reasoned thought or judgment.

The 86th Division was hit particularly hard by the War Department's orgy of willy-nilly transfers and illustrates the problems that the haphazard moves left in their wake. Like most of the National Army divisions, the 86th was initially filled with the first wave of draftees by October 1917. Before that month was out, however, the division was forced to send over fifty-four hundred of its newly arrived troops to the 33rd Division. Between January and April 1918, more than one hundred thousand men were sent to Camp Grant mainly to fill the division's ranks. During the same period, the division lost an aggregate of eighty thousand soldiers. On 30 April 1918 the 86th was down to only ten thousand men. These moves had a profoundly negative influence on the unit's leaders and men. In just the area of NCO leadership, the division's morale officer reported that "it is noticeable that the men are not as keen as formerly in competition for noncommissioned rank." He believed that this was due to the fear among the men that, given the recent spate of levies, they would shortly be transferred and lose any rank they had gained.[81]

The War Department's demand for men with specialized skills or those with the education and experience to attend officer training wreaked havoc on the already strained efforts of lieutenants and captains to identify and train a solid cadre of NCOs. For example, the 89th Division's intelligence officer reported, "There is some feeling of discouragement and dissatisfaction among the commissioned personnel due to the constant extraction of men from this division." He noted that the officers "feel keenly the loss of those men whom they have spent months of hard work in an effort to make them trained soldiers." On 22 February 1918 another intelligence officer complained that training in the 31st Division had been gravely disrupted by the unit's transfer of 320 "of the best N.C.O.s" to the Leon Springs OTC. Likewise, Capt. Wardlaw Miles noted that "a fierce exasperation burned in the hearts of the Company Commanders who were constantly obliged to give to other units their best non-commissioned officer material."[82]

At the beginning of the war the soldier's company commander filled out the man's qualification card listing any specific technical training, education, or experience. As these cards were the local adjutant general's only way of determining whether a soldier possessed the skills that were needed elsewhere in the army, it was easy for the officers to camouflage the man's true worth from the prying eyes of the outsiders. Going one step further, Capt. Kerr Rainsford admitted that he and his peers came to view the transfers as a "safety-valve" or quick means of ridding themselves of "the man whose face seemed irreconcilable with a steel helmet, whose name on the rollcall consisted only of consonants, or who had cast his rice pudding in the mess-sergeant's face." As it was also in their interest to have efficient small units, often the company commanders' superiors tacitly supported their obstructionism. After his unit endured a particularly large loss of current and would-be NCOs, one regimental commander later confessed that he turned a blind eye when his company commanders later listed "especially qualified men as farmers" so as to keep the soldiers off of subsequent transfer orders.[83]

137

Whatever their justification, the frequent levy of soldiers from the division caused massive and lasting disruptions to the training and cohesion of the army's small units. In the realm of training, the loss of NCOs and privates forced officers to readjust their training plans continually to account for the influx of raw recruits. With each new levy the officers and remaining NCOs also had to reassert their authority and try to rebuild the "group spirit" of their units. This constant "reinventing of the wheel" was but another obstacle to the leaders' efforts to advance their own professional development. An officer in the 305th Infantry described the process that followed after "each company had been sifted down to a mere hundred or so." As soon as new men arrived, "all over again, the company commander would have to organize his unit, re-size and re-distribute his men in order to balance the platoons; start in once more on the rudiments of drill, spending long days on the rifle range teaching the infant mind to shoot."[84]

In March 1918 the Camp Lewis assistant intelligence officer reported that in the 91st Division "the necessity of going back over close order drill for the instruction of incoming recruits caused a noticeable slump in the morale of some of the men in the infantry regiments," because the "old hands'" resented having to go again through the drudgery of close-order drill and other basic tasks.[85]

Since the War Department did not approach the issue of reassigning men to fill deploying units in any methodical manner, it exacerbated the training problems in both the gaining and the losing units. Units that had lost the soldiers to transfer were eventually refilled with partially trained men from other posts or, even worse, brought up to strength with recently drafted recruits. The levies also hindered efforts by commanders to standardize their training across all subordinate units. Col. H. O. Williams reported in July 1918 that the transfer of a vast number of relatively trained men, and their subsequent replacement with raw recruits, had greatly slowed the training of the 84th Division. Furthermore, the changes in troops had left the division with a poor balance of experienced and green soldiers within its infantry units. The failure of the division staff to reorganize these units after the War Department's levies meant that the 167th Infantry Brigade had a disproportionate level of raw recruits with "less than a week's training under arms."[86] This of course led to a situation in which training in that unit fell far behind the training of other units in the division.

The fact that large groups of raw recruits continually and unpredictably arrived in divisions throughout the winter, spring, and summer of 1918 only aggravated the challenges of the army's small-unit leaders. In the final analysis, the constant effort given to integrating the raw draftees into their units, and the time dedicated to constantly rebuilding unit NCO cadres, left even less time for the officers and NCOs to move on to more advanced levels of unit training or to devote the time they needed to improve their own competence through self-study.

The close personal bonds and social identities that link soldiers together and build ties between the soldiers and their leaders in small units are the fundamental basis for the cohesion and combat effectiveness of armies. In addition to all the wreckage of training and leadership left in the wake of the incessant transfers, the greatest damage inflicted by the moves was in these areas. Two officers from the 89th Division were keenly aware of these abiding problems. Capt. Charles Dienst of the 353rd Infantry noted that the constant turnover of the officers and soldiers of his unit "seemed at the time to be striking at the progress and efficiency of the organization. There's something in the association of men as 'bunkies' that ties them together once [and] for all." He recorded a conversation between two of these "bunkies" shortly before one of them was transferred to another post: "'I'm ready to go,' said the transferred man, 'but I should like to go with my old outfit.' And the man who was left behind answered, 'We're going to be filled up with strangers. I don't like it either.'"[87] In July 1918 the Camp Funston intelligence officer reported, "The most serious condition tending to lower the morale is the constantly [*sic*] shifting of men to other camps." He wrote, "Many friendships are no sooner made than broken. No man knows from one day to the next whether he will be in Camp Funston or sent to fill up another division."[88]

The poignancy of broken friendships masked the deeper problems of unit cohesion. With the exception of some National Guard units, upon entering military service the majority of the army's soldiers were cast into a strange world and largely cut loose from the close association of friends and family. In the early weeks of training, the soldiers established new bonds of friendship and a "surrogate family" developed within their units. As the levies caused units to be "filled up with strangers," the surrogate family was torn apart and was replaced by "one day to the next" uncertainties and worries. The importance of strong unit solidity is that it provides the essential psychological armor that the individual and the unit require to face the mental, physical, and spiritual challenges of fighting. As the transfers cut deep chinks into this psychological armor, the ability of the unit's officers and NCOs to lead in combat grew exponentially more difficult. An officer of the 307th Infantry observed that everything at Camp Upton was so hurried and confused, and his unit was so buffeted by constant personnel transfers, that it "never really found itself" until it departed for France.[89] Since having a unit "find itself" is the goal of military cohesion building, the fact that a unit was unable to accomplish this key task until it reached the war zone was an indication of the serious challenges that lay before the AEF.

139

The mobilization of the US Army was a mighty but flawed undertaking. The nation's general lack of military preparation and the press of time resulted in vast shortages of equipment and defective and incomplete plans for training the mass of new soldiers and officers. With its flawed personnel policies and other missteps, the War Department itself often tripped up unit training plans and the efforts of its junior leaders to build cohesive units. The OTC graduates were both victims and obedient minions of a training system that improperly prepared them for combat and then loosed them to spread ignorance among the draftee masses. While many motivated and well-intentioned young officers and NCOs attempted to transcend the host of training and leadership problems that confronted them, their greatest obstacles were their own limitations and inexperience. Capt. Charles Dienst recalled that as his 353rd Infantry boarded the train from Camp Funston for the embarkation ports, "its equipment was still incomplete; its training was still unfinished; and its organization was still untried. . . . Both officers and men realized the inadequacy of their preparation."[90] Regrettably, his unit was far from being alone in this sad circumstance. The inbred and ingrained flaws in leadership, training, doctrine, and organization that the army's units carried with them to France would bear bitter fruit in the campaigns of 1918.

6 "My God!
This Is Kitchener's Army All Over Again"
*Leader Training in the
American Expeditionary Forces in France*

A SOLDIER IN THE 105TH INFANTRY RECALLED that after seeing his unit arrive at a British training area in France, one of the British instructors noted with tears in his eyes, "My God! This is Kitchener's army all over again."[1] While the American chose to interpret the remark as a comment on the poor state of the British army in 1918, the statement could equally be applied to the Americans' innocence and lack of training. Like the Kitchener divisions that swelled the ranks of the British army in 1916, the Americans of 1918 were young, energetic, and woefully unprepared for the shock of battle that awaited them.

The arrival in France did not end the US Army's problems with developing competent junior leaders. Officers and NCOs arriving overseas faced new sets of challenges and obstacles to their leadership development. The AEF's own unique training policies, uncertain tactical doctrine, and mistrust of the Allies often hindered efforts to create leaders with the tactical and technical skills needed to overcome an experienced and able foe. Moreover, dramatic changes in the military situation in 1918 further sidetracked the AEF's unit and leader training. By the time the American units became involved in large-scale combat in 1918, the AEF had made few breakthroughs in improving the readiness of its junior leaders to command in war.

The training and doctrine of the AEF has long been a popular subject among historians of the era. James Rainey, Timothy Nenninger, and, more recently, Mark Grotelueschen have all examined the shortcomings of Pershing's open warfare doctrine and the problems that the AEF faced in preparing its units for combat. Rainey noted that Pershing's doctrine suffered from inherent contradictions, unrealistic visions of combat, and a rather vague looseness in its doctrinal definitions. He has also argued that this flawed doctrine, when combined with a faulty philosophy of training, shortages of equipment, and unsound personnel policies, undercut the AEF's efforts to effectively train its units for war. Similarly, Timothy Nenninger maintained that while systemic problems with fielding a large army in France, and the German offensives of 1918, dogged Pershing's efforts to weld the AEF into an efficient fighting force, ultimately, things that were under the Americans' control, such as doctrine and personnel policies, could have been better managed to increase the effectiveness of the AEF's divisions.[2]

In *The AEF Way of War*, Mark Grotelueschen expands on the arguments of Rainey and Nenninger and agrees with their negative assessment of Pershing's nebulous open warfare doctrine. However, Grotelueschen maintains that American divisions overcame the AEF GHQ's doctrinal blunders and crafted their own firepower-centered combat methods. These methods, which eschewed the GHQ's concepts of "self-reliant infantry" for ones that maximized the use of artillery, ultimately led to better combat performance in these adaptive units and fewer American casualties. He argues that despite the adherence of senior officers to the GHQ's rifle and bayonet dogma, most of the new tactics were spawned by American officers who had been heavily influenced by Allied schools and advisors.[3]

141

I have little to add to the debate over Pershing's open warfare doctrine. Rainey, Nenninger, and Grotelueschen have done a superior job of highlighting the problems with these tactics and the flawed assumptions upon which they were based. Thus, my discussion of open warfare will be limited to those points at which the doctrine intersected with the training of company-level leaders. It is interesting to note that in the War College Division's *Instructions for the Training of Platoons for Offensive Action,* the authors used the terms "trench-to-trench attack" and "attack in open warfare" to distinguish between the different methods of attack. However, they also stated, "It can not be too thoroughly recognized that although it may be necessary to slightly vary the preparations and forms of assembly for the attack in these two circumstances, the actual tactics to be employed will usually be identical."[4] There is much wisdom in this statement, and at the company level and below, much of the debate over the tactics, formations, and realities of open versus trench warfare may well have been a tempest in a teapot. For the captains, lieutenants, and NCOs, the problem remained the same: how to cross terrain covered by enemy fire and end up with enough manpower and firepower remaining to kill the defenders, seize the terrain, and hold the ground gained.

Pershing was well aware that the American units arriving in France exhibited widely varying levels of training. He was also cognizant of the fact that these soldiers would soon face a hardened and experienced enemy. Pershing believed that only a standardized training plan driven from the top down could whip the Americans into shape and overcome their glaring shortcomings. In the fall of 1917 Pershing envisioned that all of his divisions would go through a three-month program of training, with the first month taken up with individual and small-unit technical and tactical training conducted under the watchful eyes of one of the Allied armies. The divisions' second month of instruction was to be devoted to learning the intricacies of trench warfare by serving on an inactive sector of the lines, and the third month was devoted to large-unit training for open warfare. Pershing justified the plan by noting, "In order to give the troops the advantage of the latest tactical and technical developments and make up for the defects of training at home, the plan contemplated an additional period of training for divisions of about three months after reaching France. This gave us an opportunity to secure a certain uniformity in standards, and was especially valuable in affording the newly arrived troops the benefit of experience in the immediate atmosphere of war."[5] In other words, this plan would close the Americans' training gaps and also give them a degree of combat seasoning prior to any major combat operations of the AEF. Pershing envisioned that by 1919 the trained and robust AEF would be the preeminent Allied army on the Western Front.

The AEF GHQ's first effort to impose a standard training plan on its units came on 6 October 1917 when it issued the "Program of Training for the 1st Division, A.E.F." To some extent this program became the model (or at least the point of departure) for the subsequent training of all other AEF divisions. The memorandum stated that the training would be "based on developing sound leadership in succession in the squad, or group, the platoon, the company and finally the higher unit." The program called for six periods of progressive instruction totaling sixteen weeks of training. The first two periods, totaling seven weeks, focused on the training of individuals and units from squad to battalion. The third and fourth periods devoted three weeks each to the training of regiments and brigades. The fifth period, of three weeks' duration, was to train the division. The sixth period ended the divisions' training with an inspection of their readiness for combat by the corps commander or GHQ.[6] As with the stateside guide for instructing divisional units, *Infantry Training,* the "Program of Training for the 1st Division" detailed the number of hours that were to be devoted to training all of the subtasks that each echelon required to be tactically proficient. The biggest difference was that while *Infantry Training* basically stopped at the battalion level, the AEF's "Program of Instruction" extended instruction all the way to the division.

The AEF GHQ continued to refine its standardized divisional

training plan as it stood up the 2nd Division. While the "Program of Training for the 2nd Division" contained much of the same wording as the 1st Division's plan, the 2nd Division was to undergo eight periods of training lasting nineteen weeks. The 2nd Division plan devoted six weeks of training for the individual, squad, platoon, and company and only two weeks each for the training of the regiment, brigade, and division. Another difference was that during the two weeks of divisional instruction, the unit's artillery brigade would be fully integrated into the training plan. It appears that after it published the 2nd Division training plan, the AEF GHQ merely changed the name and reissued this scheme to other divisions as they arrived in France.[7]

143

All of the AEF's standardized training plans also listed Pershing's "General Principles Governing the Training of Units of the American Expeditionary Forces." These principles were intended to encapsulate Pershing's vision of the war the AEF would wage as well as his overarching expectations of his soldiers and leaders. The AEF's five training principles were,

> a. The methods to be employed must remain or become distinctly our own.

> b. All instruction must contemplate the assumption of a very vigorous offensive. This purpose will be emphasized in every phase of training until it becomes a settled habit of thought.

> c. The general principles governing combat remain unchanged in their essence. This war has developed special features which involve special phases of training, but the fundamental ideas enunciated in our Drill Regulations, Small Arms Firing Manual, Field Service Regulations, and other service manuals remain the guide for both officers and soldiers and constitutes the standard by which their efficiency is to be measured, except as is modified in detail by instruction from these headquarters.

> d. The rifle and the bayonet are the principal weapons of the infantry soldier. He will be trained in a high degree of skill as a marksman both on the target range and in field firing. An aggressive spirit must be developed until the soldier feels himself, as a bayonet fighter; invincible in battle.

> e. All officers and soldiers should realize that at no time in our history had discipline been so important; therefore discipline of the highest order must be exacted at all times. The standards of the American Army will be those of West Point. The rigid attention, upright bearing, attention to detail, uncomplaining

obedience to instructions required of the cadets will be required of every officer and soldier of our Armies in France. Failure to attain such discipline will be treated as a lack of capacity on the part of the commander to create in the subordinate that intensity of purpose and willing acceptance of hardship which are necessary to success in battle.[8]

These principles highlight a number of the tactical and leadership assumptions that the AEF's training was built upon and offer some insight into Pershing's conceptions of modern warfare.

The "principles of training" make it crystal clear that Pershing intended to train and fight using American methods, and no matter what the Americans adopted from the Allies, they would still be grounded in the concepts of the *IDR* and *FSR*. This was part of Pershing's efforts to stave off Allied attempts to amalgamate American soldiers and units into their formations. If the Americans adhered to a unique doctrine, he could argue that the mixing of American and Allied units was unfeasible.

It was Pershing's determination to make the AEF's expectations of discipline "those of West Point" that caused some of the greatest problems with junior leaders. Pershing never questioned his assumption that the near automatonlike discipline expected of the army's cadets was suited to the nation's wartime citizen soldiers and officers. Just as had been the case in the United States, leaders in the AEF found themselves in the quandary of needing to balance the call for discipline with the demand of battlefield initiative. Pershing also claimed that his concept of open warfare "demands initiative, resource, and decision upon [the] part of all commanders" and required leaders able to lead "flexible teams capable of rapid maneuvering to meet swift changes in situation."[9] But the question for the AEF's senior officers was at what point did their subordinates' exercise of initiative give way to Pershing's insistence on rigid and "uncomplaining obedience to instructions?" It was Pershing's threat that any lapses in his perfect West Point discipline would be "treated as a lack of capacity on the part of the commander" that placed a dead hand on the desire of many commanders to allow their subordinates to test the limits of their initiative and independence and make decisions on the battlefield that deviated from their last orders.

In early 1918 the AEF worked to hammer out a regular training plan for the divisions that were scheduled to arrive in France in the spring. At the Versailles Conference on 30 January 1918 Pershing agreed to an arrangement in which the British would transport to

France and train six complete American divisions in March and April. This Six Division Plan stipulated that "the training will be progressive; by platoon, company, battalion, and regiment until such times as the American division is fit to take the field as a division, when it will be handed over to the American Commanders-in-Chief." The plan gave the local British corps commander the responsibility for the Americans' training and placed American units under British command when the doughboys occupied parts of the British lines. The British also agreed to provide one officer per American infantry, machine gun, signal, and engineer battalion to assist in training. American units would be equipped with British weapons during their stint with the British army.[10]

After consultation with the AEF, on 21 March 1918 the BEF GHQ provided general guidelines to its subordinate commanders on training the Americans. The training was to occur in three periods, which roughly corresponded to the AEF's general three-month training plan for arriving divisions. The first period would concentrate on basic weapons and tactical training and provided time for American officers and NCOs to conduct visits to the front lines for twenty-four to forty-eight hours to gain some firsthand experience in the trenches. During the second period, the British would train the Americans in trench warfare by attaching US units to British formations occupying sectors of the front. As American junior leaders became more familiar with the front, complete American units would be attached to British formations in progressively larger increments. In other words, an American platoon would serve at the front as part of a British company; then the American company would do a tour as part of a British battalion, and so on up to the regimental level. The last phase of training was to last three to four weeks and was intended to provide American units, up to the regimental level, with advanced tactical training. The focus of this period was on maneuvering large units and integrating machine guns, liaison, signaling, and supply into the Americans' operations. To meet the AEF GHQ's demand for its units to be trained in "American methods," the British specified that all of the training manuals for this last period were to be supplied by the AEF.[11]

Unfortunately, the Ludendorff Offensive of March 1918 disrupted the training plan before it even got started. The subsequent German attacks in the spring and summer completely shattered Black Jack's hope for a methodical and standardized approach to American training. In the midst of the crisis the British quickly reneged on the promises they had made during the Versailles Conference. On 7 April

1918 the BEF GHQ directed its subordinate commanders to shorten their training periods for the American units under their supervision. Under the new plan, American battalions were to be sent into a quiet sector of the line after a short period of training consisting mostly of rifle firing, specialist training, and instruction in gas warfare.[12]

Unknown to Pershing, the British were also working to meet the Allies' growing manpower crisis by radically changing how American forces would be deployed to Europe. Under the terms of the London Agreement of 27 April 1918, the Americans and British settled on a plan by which the British would expedite shipping the Americans to France but would transport only American infantry, machine gun, engineer, and signal units. The intent of this plan was to get warm American bodies into the lines in France as soon as possible to relieve some of the manpower strains the Allies were feeling as a result of the German offensives.[13]

Although Pershing managed to negotiate with the British to ship divisional artillery and support units as soon as surplus transportation became available, the changes disrupted the AEF's efforts to train whole divisions as they arrived in France. This further exacerbated existing problems with American combined arms training. More important, the chaos caused by the German attacks led to a further fragmentation of the American training efforts as AEF regiments and brigades were shuttled across the front in the spring and summer of 1918 to have them close at hand in the event of any emergency. The chief of the AEF Training Branch, Brig. Gen. Harold Fiske, admitted that while "circumstances invariably prevented . . . [the three-month training program] from being carried out," the AEF at least tried to retain this basic concept for the training of newly arrived divisions. But even Fiske confessed that under the crush of events, the programs for the late-arriving divisions were severely curtailed, with the open warfare phase often reduced to only two or three weeks of training.[14]

Ultimately, the AEF's own interferences in unit training and the unsettled situation in France resulted in the situation in which no two American divisions went through the same training program in France. Although the first four AEF divisions, the 1st, 2nd, 26th, and 42nd, came the closest to achieving Pershing's three-month training plan, all bets were off when it came to the time devoted to the training of the later-arriving divisions. Due to anxious conditions in France following the Germans' March offensive, the new divisions often found their training disrupted by constant shifts in their billets based on the events at the front or the whim of far-off staff officers. The 28th Division's 109th Infantry was a case in point. The regiment arrived at Calais on 19 May 1918. When it reached its billets on 22 May, it was to begin "a long period of training" under the tutelage of the British army's Tyneside Scots. However, less than three weeks later, after receiving little more than weapons training on British small arms, the regiment was on the move to the French sector. It picked up whatever training it could along the way until finally arriving at the French lines near Montmirail on 24 June.[15]

The 33rd Division faced similar problems with seemingly arbitrary moves and likewise experienced great disruptions in its training. Upon arriving in France, the division's 131st Infantry was assigned to a training area in the British XIX Corps sector. After developing a training plan with their British cadre, coordinating for drill grounds, and establishing firing ranges, the unit "was beginning to feel that some practical work was to commence." Just as their progressive training was getting into gear, the AEF GHQ ordered the unit to move to another sector of the British front. After only eight days of training with a new cadre of British soldiers, the regiment was again ordered to change its station. The unit hurried through "classes in defense against gas, bayonet work, bombing, rifle practice, trench mortar and 37mm gun practice" and was ushered into a quiet area of trenches on the British front lines. The total time from the regiment's arrival in France to their occupation of the British trenches was thirty days, with much of that time taken up with marches from one area to another. This problem was endemic in the AEF. After inspecting units within the II Corps, Maj. Walter Short reported on 5 July 1918 that the machine gun training programs for the 27th, 30th, and 78th Divisions was dangerously incomplete and unsystematic. He blamed much of this on the fact that units were moved so often that coherent and consistent training was difficult to achieve.[16]

This lack of standardization in the training of American divisions carried with it grave challenges for the AEF's future operations. One of the main goals of unit training is to create a certain uniformity of expected performance in an army's forces. In theory, this meant that senior commanders could expect their different units to operate and respond to enemy actions in roughly similar manners. As American divisions were divided up by brigade, regiment, and battalion for training or operations with the Allies, any degree of standardization was lost. Ultimately, American units from the battalion level and above were forces of unknown quality to their higher commanders until they actually saw combat. The idiosyncrasies caused by this fragmentation of training also meant that the instruction given to infantry and machine gun companies, platoons, and squads was hit or

147

miss depending on when the unit arrived in France and the whims of the times and their trainers.

Since the training that the AEF's formations and leaders received was so distinctive from unit to unit, it is difficult to offer more than an overarching impression of the general quality and effectiveness of the Americans' instruction in France. All the American divisions that saw combat had some degree of basic instruction under the tutelage of the British or French. The first four American divisions were instructed mainly by the French and spent more time in training with the Allies than did any subsequent American divisions. Grotelueschen argues that in the case of these early divisions, this training allowed these units to shake off the obsolete practices of the *IDR* and *FSR* and, through their close association with the French, gain a deeper appreciation for the primacy of firepower in modern war.[17]

This extended training, with its focus on combining and synchronizing the new weapons of warfare now organic to infantry companies and supporting weapons, such as artillery and machine guns, also allowed junior infantry leaders to become much more technically and tactically proficient than their peers in later-arriving divisions. The officers and soldiers in the early divisions also seemed to have had a much more realistic training regimen than those units training in the States. Maj. Gen. Robert Bullard noted that the 1st Division trained with the French in a vast and detailed training trench system where "nothing was omitted or left to the imagination of the soldier." He observed, "Almost everything except the actual bursting of shells" was duplicated by the French to make the Americans' training as realistic as possible.[18]

While the early divisions may have benefited from training with the French, they also had to cope with some of the same systemic problems of equipment, junior leadership, and shortages of supplies that had dogged the training efforts of their stateside brethren. On top of this, they also suffered from having to provide the officer and soldier "overhead" for establishing the AEF's corps and higher headquarters as well as the Services of Supply (SOS).[19] The experiences of the 1st Division offer an insight into the problems faced by the first American divisions in France.

As the US Army was woefully short of machine guns, artillery, mortars, and the other weapons of modern war, the Americans relied on the overstressed Allies to provide essential war material. Over the course of the war the French alone provided 3,672 field and railroad artillery pieces and 40,884 light and heavy machine guns. Despite this largess, the AEF still faced shortages that affected soldier comfort and training in its early months in France. In the late fall of 1917, Col. Robert McCormick noted that in the 1st Division "supplies were short; clothing could not be kept up; the meat rations had to be obtained from the Canadians; payday was irregular; the mails were dilatory; forge was lacking, and the horses suffered. Artillery drivers bought oats out of their own scant funds to feed the government horses that the government did not provide for." He also noted that when the division moved to the front on 6 January 1918, "it was still short of much essential equipment, and the artillery had never been supplied with the telephone equipment needed to train its telephone details."[20]

149

Perhaps the most pressing issue that the first divisions faced was the inexperience of their junior leaders. As he prepared to depart for France in June 1917, the commander of one of the 1st Division's infantry regiments, Col. George Duncan, noted with dismay that only seventeen of the officers in his unit had at least a year in the service and that many of his best NCOs were to remain in the United States to attend OTCs. Col. George Marshall, a 1st Division staff officer, observed on 27 August 1917 that "owing to the number of officers detached from the Division . . . there exists a serious shortage of field and company officers in the command. There are companies with only two officers present for duty and the latter have rarely had more than five months' service, and frequently less." Although he supported sending officers to the front to observe the conditions of modern war, he worried that this, and the future detailing of fifteen officers per regiment to attend schools, would only exacerbate the officer shortage.[21]

The relative greenness of the American officers and the steep learning curve they faced in modern warfare frequently hindered unit training. In December 1917 the 1st Division's exasperated commander, Maj. Gen. Robert L. Bullard, wrote, "I have much difficulty in getting officers who know anything. . . . Many even of our regular officers can never be worth anything in this war [as they are] unadaptable and immovable." More cogently, he later noted, "The training also showed in American officers, except those of the very latest education, a love of tactical prescriptions, rules of thumb, a demand for orders that should fix the method of tactical procedure for all things." Even after the division had been in training for nearly six months, Bullard's comments indicate that all was not well when it came to leadership in the AEF. They also suggest that the officer corps had a long way to go to develop the initiative and flexibility demanded of combat on the Western Front.[22]

This situation was exacerbated by GHQ personnel policies that promiscuously transferred officers from one organization to another. McCormick recalled that in the spring of 1918 "there was a constant flow of junior officers through the organizations. Those who had received training at the front were ordered to other organizations, their places being taken by new arrivals from America and from the artillery schools." In an effort to overcome this problem, McCormick's commander ordered him to establish a unit school to teach the new arrivals the lessons that the Americans had learned up to that time about fighting in France. McCormick soon discovered, "Our inexperienced troops were ignorant of many military accomplishments not touched upon in this simple course. The points I sought to cover were essentials in which most of the officers who had come to the 1st division from America and from training schools in France either not been taught or had totally failed to comprehend."[23] This constant revolving door policy also meant that many junior leaders never truly mastered the tactical and technical skills of their positions, nor were they able to hone their leadership by building cohesive units. McCormick and the other old soldiers of the AEF were also learning that the units and officers arriving in France were woefully short of the basic knowledge and skills required to fight the Germans, and that time was against them.

Despite these systemic problems, the training of the early-arriving American divisions was rather long and thus allowed their junior leaders more time to learn through the school of hard knocks. However, their experience was not truly indicative of the training experiences of most of the AEF's units. An examination of the training of these more typical AEF units goes far in uncovering some of the key issues that undercut the combat effectiveness and cohesion of the US Army's small units and their difficulty in fielding combat-ready junior leaders. Most of the divisions that arrived from April 1918 to the end of the war managed to at least spend a month or more of their first phase of training with the Allies. However, the quality and quantity of this training varied greatly from division to division. If Robert McCormick is to be believed, the only thing that the arriving units had in common was their total ineptitude. He caustically observed of the new units, "They had been practiced in marching formations, and received small arms bayonet training, and had been taught infantry tactics so far as these had developed up to the Summer of 1917. They had not studied the use of the modern infantry weapons of assault, the trench mortars or infantry cannon. They have not learned the use of cover, which only comes from service at the front. . . . They were totally uninformed as to the methods of attack developed and perfected by the French at the end of 1917."[24] McCormick's trenchant remarks neatly encapsulate the challenges that the AEF faced in molding its new units, and the French and British trainers who took charge of their instruction had their hands full in trying to shape this rude clay into a fighting force.

After taking out the first four divisions and those units that the AEF skeletonized or transformed into depot divisions, roughly half of the Americans received their first phase of training with the British army. Although the British worked, in a somewhat abridged manner, within the framework of the original Six Division Plan, their approach to training contained inherent problems that tended to negate the effectiveness of the instruction that they gave the Americans. The British viewed the first phase of the Americans' training as a crash course in the new weapons of war and as a means of honing the discipline of the unpredictable doughboys.

The experiences of the 27th, 28th, and 82nd Divisions, who trained with the British in May and June of 1918, illustrate the problems with this phase of training. The first thing the British did was to break the Americans divisions down into battalion- or company-sized units and scatter them widely throughout the Flanders sector for training. An officer in the 82nd Division wrote that his brigade was so dispersed that "it takes hours to get around."[25] This scattering made training in anything larger than a battalion difficult. Furthermore, the Allies trained field artillery and machine gun battalions separately, independent of the infantry regiments that they would support in combat. This precluded combined-arms training and prevented battalion, regimental, and divisional commanders and their staffs from learning how to control and supply their units as a whole.

The actual training that the infantry units received also left a lot to be desired. In an effort to ease problems of supply and the need, after the March offensive, of having American units in training readily available to occupy a point in the line, Pershing agreed to the British desire that the Americans undergoing their instruction be armed and trained with British weapons. As such, upon arrival in the British sector, the Americans turned in their small arms and were issued British rifles and machine guns. Not only did the turnover consume valuable training time, but it also forced the Americans to learn how to operate, maintain, and employ a totally novel set of weapons. This further complicated the efforts of junior leaders to hone the technical skills of their soldiers. First Lt. Louis Brockway recalled that during the space of one year his infantrymen had been issued with four different types of weapons. He noted, "You might say our men were a little confused" by the frequent change in armament.[26]

American accounts of the first phase of training with the British frequently stressed the amount of time their trainers devoted to weapons training. Not all of this training accomplished much. As with their American cousins, the British army had a fixation with the painstaking, step-by-step introduction to weapons known as "naming the parts." This method focused more on nomenclature than on operating the weapon. The 28th Division's Chester Baker also questioned the utility of much of his weapons instruction. After he had swapped his Springfield for "the hated English Lee-Enfield," he discovered that the bore of the new rifle was "so worn it looked like a shotgun." Despite the pointlessness of trying to fire with such a dilapidated weapon, he and his comrades were duly sent to the range. After failing to do much damage to the targets, Baker gave his rifle to the British instructor who fired several shots with even less effectiveness than the American. The Britisher merely grinned and said, "Well it'll get a Jerry at 100 yards."[27]

The greatest problem with the British weapons training was that as soon as the Americans' time with the British ended, the doughboys turned in all of the British weapons they had so laboriously come to know and (except for the II Corps) never saw or used their kind again. For instance, the machine gun battalions and companies of the 28th Division were issued British Vickers machine guns and began a training program conducted by British officers and NCOs. Before the month of training was even completed, the units were ordered to turn in their Vickers machine guns in preparation for the division's move to a quiet sector of the French front. Upon arrival, the Americans were issued the French Hotchkiss machine gun and began their training all over again.[28] This sad farce happened time and time again and wasted valuable training time that could have been employed to better effect in other areas.

Despite the inherent problems associated with constantly switching weapons, it must be admitted that given the shortages of machine guns and other modern weapons in the States, some soldiers still benefited from basic training with any weapons. Charles Minder was one such soldier, but he still confessed that when he was directed to fire "twenty shots for ranging fire, and thirty for application fire," he had no clue what his instructor meant. Although Minder's firing did improve, he still faced the problem of rapidly changing out one weapon for another. Minder recalled that after training in the States on the Colt machine gun, and spending weeks with the British Vickers gun, it was not until mid-June 1918 that his unit finally received the Hotchkiss machine guns that they would use in combat. These sudden changes sowed confusion and required that he and his officers were "studying night and day, learning all we can about the new French machine gun."[29]

The remainder of the training in the first period was a rather mixed lot. A soldier of the 28th Division noted that his training "consisted of six-mile hikes each day to a hillside drill field where we practiced throwing dummy hand grenades and listened to lectures." Another of the division's doughboys later recalled, "The doughboys and engineers whiled away the long, warm days, drilling and hiking, doing much bayonet work, polishing and cleaning rifles and other equipment and putting in time as best they could" An officer in the division's 112th Infantry had a different take on the training. In a very revealing passage, he maintained, "Under British training they made surprisingly rapid progress, becoming especially skilled in bayonet work." However, the doughboys already had a bellyful of bayonet work and needed to address subjects such as tactical movement and command and control, which would serve a more useful purpose in combat. As an officer later recalled, he had "never heard an authentic and first hand account of bayonet-fighting" during or after the war.[30]

Taken as a whole, the AEF's phase one divisional training plan for infantry units was far from a resounding success. Chief of Staff Peyton March later noted that the pawning off of American instruction to the Allies had been a mistake. He stated, "The practical effect of Pershing's policy was that large bodies of American troops, divisions whose morale was at the highest point, who had four to six months of training . . . found the keen edge of their enthusiasm dulled by having to go over again drills and training which they had already undergone in America." There was much merit in March's criticism. Capt. W. A. Sirmon lamented on 23 May 1918 that "the routine is practically the same we had at Camp Gordon. No one seems to realize the training our men have had, and we are starting them over like raw recruits." Far too much of the training in the first phase was a return to the basic recruit-type training that the units had to endure in the States. More important, this training did little to address the instructional needs of the American officers and NCOs.[31] For most units there was little to no tactical unit training conducted in the first phase. However, units did send large numbers of American officers and NCOs to British schools during this period. Unfortunately, most of these schools focused more on the basic technical aspects of weapons and not enough on their tactical employment. Furthermore, the schools removed the leaders from their units at key times when the leaders should have been focusing on sorely needed unit cohesion-building.

Although the training of the AEF's infantry units was slighted, the same could not be said for its artillery regiments. Both the AEF GHQ and the Allies realized that the training of the American artillerymen had been rather sparse in the United States and took strenuous measures to ensure that the cannoneers received abundant instruction once they reached France. Pershing went so far as to demand that divisions scheduled for deployment send large numbers of artillerymen in their advance parties so they could attend French and AEF artillery schools prior to the arrival of their units. The 5th Division, for example, sent fifty officers and 350 enlisted men in their advance party so they could learn the operation and employment of the never-before-seen French 75 mm and 155 mm guns and modern indirect fire control. While infantry formations frequently had their instruction curtailed or stalled by short-notice moves, artillery units seldom faced these challenges. When the Pennsylvania doughboys of the 28th Division were sent to the trenches with less than a month of instruction, their artillery comrades spent a full three months of training before rejoining their already bloodied infantry brethren.[32]

Ironically, while American artillery units tended to receive the least amount of sound training in the United States, under the relatively short but intense training by French and British taskmasters they generally rose to become some of the most effective combat units in the AEF. The Allied instructors showed no mercy on their ignorant charges. After being sent to a French artillery school, one officer noted the trouble that he and his fellow Americans were having with trigonometry. He noted that the French cadre were amazed "that we held commissions in artillery" and ruefully confessed that "the French know this isn't an Indian war." Unfortunately, this degree of expertise in the American artillerymen came at a price. By examining the detached units listed for American divisions in the AEF Order of Battle, one quickly gets an appreciation for the amount of time that the AEF's artillery brigades and regiments spent away from their parent units for training. The 82nd Division's 157th Field Artillery Brigade, for example, was absent from the division for training from 2 June through 21 August. The divisions that arrived in the spring and summer of 1918 generally underwent their course of training (both in and behind the lines) without the presence of their organic artillery units. This led to recurring problems with infantry-artillery cooperation that plagued American operations throughout

154

the war and all too often meant that junior infantry leaders had little experience in working with artillery.[33]

Generally speaking, the divisions that trained with the French during their first phase of training had a somewhat easier time than those that trained with the British. First, there were none of the problems of training on weapons that the soldiers would never use in battle. Second, the available evidence suggests that the French-trained units underwent fewer disruptions to their training caused by frequent moves. Furthermore, the French tended to more seamlessly blend the different phases of training together, making their instruction more fluid and coherent to the Americans. The French also seem to have made more of an effort to accommodate the Americans' pride and psyche. On May Day of 1918, Marshal Pétain issued a bulletin to his subordinates that advised them how to train the Americans who were rapidly descending on them. Pétain warned his countrymen that while the Americans were only partially trained, "they have an extremely highly developed sense of amour-propre, based on their pride in belonging to one of the greatest nations of the world." As such, "an attitude of superiority over them should be assiduously avoided" and "patience and tact" should be the French trainers' watchwords.[34]

To a large extent Pétain's approach worked. American soldiers often praised the French willingness to share with the doughboys their hard-won battlefield wisdom. A soldier in the 26th Division, for example, believed that the instruction his unit's French instructor gave on trench warfare was very productive. He noted that his unit practiced both offensive and defensive tactics using a very realistic training trench system, and following each exercise, his officers would participate in detailed, but tactful, critiques of the event led by experienced French officers.[35]

Regardless of whether the French or British initially trained the Americans, the pace of training in the spring and summer of 1918 was blistering. Officers from the 4th and 91st Divisions left detailed records of their training regime during this period that highlight the bewildering rapidity of their instruction. Col. Christian Bach, the 4th Division's chief of staff, recalled that his division had to work hard in their first weeks in France to overcome the shortcomings of their initial training at Camp Greene. Although the unit initially trained with the British, it was not until it began training with the French in mid-June 1918 that its units received their Chauchat automatic rifles, had the opportunity to throw live grenades, and were able to complete the rifle training that one of the division's infantry regiments had missed in the States. Bach noted that in addition to the weapons training, the French also taught the Americans "their method of attack" and "how to pass through barrages with a minimum of loss."[36]

The record of the 4th Division's 47th Infantry illustrates the quick tempo of the unit's instruction that followed a rather pointless period with the British. The regiment arrived in France on 25 May 1918 and spent much of the next three weeks moving from one spot to another in Flanders and Picardy. It was not until 14 June that the unit began serious training with the French army. For a month the unit was put through their paces in the following events:

17–19 June. The regiment equipped with and trained on the Chauchat and Hotchkiss guns, grenades, and mortars. The unit also dug and trained in trenches.

20 June. Brigade open warfare problem

21 June. Division open warfare problem

23–24 June. Rifle and pistol practice

25 June. Open warfare maneuvers

27 June. Division open warfare problem without troops

28–30 June. "Strenuous drills"

1–3 July. Rifle practice, and on 2 July, the first group of officers and NCOs conduct a tour of the French trenches.

5 July. Regiment moves to French reserve trenches.

7–8 July. Unit training

12 July. Division open warfare maneuver

15–17 July. Regiment goes into French front lines.

The 47th Infantry's experience illustrates some of the trends and problems with the Americans' training. Given the breakneck speed of the instruction, one wonders if the regiment was truly able to overcome the unit's poor stateside training and learn the new methods being taught in France. It is clear from the schedule that the 4th Division devoted much training time to open warfare, and an officer in the 47th Infantry stated that "most drill periods were devoted to extended order drill and to the new formations which were soon to be used by the regiment in actual fighting."[37] However, given the rush to get the unit into the line, the training was far from thorough.

The infantry regiments of the 91st Division went through a similar training program. Capt. Clarence Minick, of the 361st Infantry, recounted his early training in France in his diary. Minick's unit arrived in France on 20 July 1918, and after a week of getting settled into their new surroundings, the 361st got down to training. Given the crush of events, the 91st Division, like many others arriving in the summer of 1918, had its training plans greatly abridged. Given this fact, it is no wonder that Minick's schedule for August 1918 was rather full:

2 August. Established firing range; "used cans and everything else for targets."
3 August. Battalion field problem
7 August. Company hike
9 August. Brigade maneuvers
13 August. Twelve-kilometer hike
16 August. Brigade field problem
17 August. Minick umpired a brigade field problem in the morning and a regimental field problem in the afternoon. Both were on "how to take German machine guns."

20 August. Regimental field problem
21 August. Regimental field meet competition. Events: gas mask donning, semaphore signaling, bugler match, shooting match for rifles, pistols, and automatic rifles
22 August. Brigade field problem
25 August. Regimental maneuvers
26 August. Night maneuvers
27 August. Overnight hike
31 August. Brigade field problem
3 September. Unit begins move to the front.

It is hard to gauge from Minick's terse entries the realism of the training in this vastly accelerated program. It is interesting to note that in his description of the brigade maneuvers he participated in on 9 August, he wrote, "We hiked and imagined we drove the Huns back for about 8 kilometers and returned."[38] This seems to indicate that the training had more to do with marching than with realistically fighting and maneuvering to drive back the Huns.

As the Americans entered their second phase of training, learning trench warfare by occupying sections of Allied trenches in quiet sectors, the doughboys' schooling was again rather hit or miss. Once more the American training in the line varied greatly from unit to unit depending on the area of the front the soldiers occupied, the exact time period in France in which they were training, and the predilection of their enemy to make mischief during their tour of duty. In the last, the enemy played the greatest role in the Americans' training and seasoning. When the enemy was active, the Americans' schooling could be brutally direct. An officer of the 1st Division noted of his foes, "The Germans did all they could to assist our training ... thus, American indiscretions invariably were punished. Trenches, reserve positions and batteries which were revealed by the least carelessness receive chastisement. . . . Thick heads and dull, which had failed to learn the teachings at school, had the lessons of war pounded into them by the German schoolmasters, whose model was: 'he who will not heed must feel.'" Pershing intended that the second-phase tour in the trenches "harden and accustom" his soldiers "to all sorts of fire and make veterans of the individuals." The early AEF divisions certainly accomplished this goal and made great progress in the unforgiving school of hard knocks.[39]

However, most of the divisions that arrived from the spring of

1918 onwards generally occupied less active areas of the front, usually in the slumbering French lines of the Vosges sector. That sector was often too quiet to provide much practical combat experience or give the Americans the seasoning that they so desperately needed. The lines occupied by the 325th Infantry were so quiet that its officers shot quail and gathered plums and apples in no-man's-land. For nearly two years a tacit truce between the French and Germans had kept the sector relatively calm. Neither the French nor the Germans were particularly enthusiastic at having the raw and rambunctious Americans disturbing the region's "live and let live system." This relative quietude often lacked the intensity to give the doughboys the battle hardening that Pershing sought. As Sgt. William Triplet recalled, his training in the French trenches "was a kindergarten rather than the first grade in the school of war," and he admitted that the enemy had been "too easy on us."[40]

The Americans neither understood nor respected the truce between the French and Germans. When units of the 35th Divisions were assigned to the Vosges front, the French officers in the sector told the Americans that "it is much better to lie quiet. . . . If we do not bother the Boche the Boche will not bother us." Another 35th Division doughboy contemptuously noted of this arrangement, "This ain't a war. . . . The Frogs and Krauts got it fixed up between 'em to spend their vacations where there ain't nothin' to bother 'em but scenery." The French officers assigned to the American units did everything within their power to control and divert the aggressiveness of the newcomers but were only marginally successful in this endeavor. A soldier of the 32nd Division remarked that while the French desire to keep things calm in Alsace "stifled a lot of budding initiative," the Americans were not to be denied. He boasted, "Soon shells were falling on both sides of the line, where no shells had fallen for months, and the front line trenches were no longer a place to spend a quiet evening."[41]

The Americans' opinion of their training in the trenches was mixed. A number found the experience to be a boring letdown. Sgt. Elmer Straub complained that "I am rather disappointed because we can hear only an occasional shot, and things do not seem at all lively." He later caustically wrote of his French allies, "They sure live a soft life and its no wonder they can't win the war." Although Sgt. Richard McBride and his comrades from the 82nd Division's 325th Infantry expected their time in the Allied trenches to be "our Baptism of Fire," they found it to be "a dull period as men sat for days in the trenches gazing intently at the enemy lines but seeing nothing to shoot at. Patrols were made every night without contact with the Germans." One doughboy simply noted, "The training in the Vosges did not prove of great value to the men in the Argonne battle."[42]

The commander of the 77th Division, Maj. Gen. Robert Alexander, believed that his unit's experience in the quiet sector of the Vosges had actually done more harm than good, for it failed to adequately prepare either the leaders or the men for their coming ordeals. He noted that in the quiet sector "an occasional trench raid and routine gun fire had been the limit of their warlike activity." When the division moved to the front near Vesle, "it was an entirely different war. . . . [We] were harassed day and night by shell fire and air raids, troops in [the] front line were constantly deluged with gas." It was a war in which the enemy "could, and did, make movement even of individuals most hazardous."[43]

Some Americans took a more positive view of their training and tried to milk the most from the experience that they possibly could. Their time in the trenches seemed to build self-assurance in the Americans of their capabilities and eased their soldiers into the status of veterans. An officer in the 140th Infantry stated that after his men stopped a German attack, "our success in repelling the raid helped the confidence of our men a great deal," and they also "gained increased confidence in their officers." Unfortunately, this confidence often turned to hubris. After serving in the Alsace sector, a soldier in the 32nd Division's 126th Infantry believed "the officers and men of all ranks felt certain that they could give a good account of themselves anywhere." Another soldier of the division boasted, "In short order we learned most of the tricks it had taken four years to perfect and had figured out a few on our own account." Of his time in a quiet sector of the front, a 29th Division soldier keenly observed, "While it gave the individual soldier the exalted morale so valued by military leaders, it also inclined to give him a careless disregard for the future and a contempt for enemy power that often induced him to take unnecessary chances."[44]

Other soldiers were equally incisive in their evaluation of their time in the trenches and its effectiveness in preparing them for their future combat. Shortly after the war, the 35th Division's Charles B. Hoyt recalled, "The value of the training in the Vosges . . . must be measured more by the atmosphere of war it gave rather than by the actual benefits of the training. For what the infantry learned of trench warfare was of no practical value to it in carrying on open warfare in the Argonne; for what the artillery learned in emplacing guns it had pretty nearly to unlearn in the days to come; so [too] with the field signal work and medical men."[45] Hoyt's comments neatly encapsulate the experience of many, if not most, American soldiers as they trained for trench warfare during their second phase of training.

Given the problematic nature of this training, what did it do to improve the expertise of the AEF's junior leaders? To some extent the training did provide the company-level officers and NCOs some practical experience with coping with the fear and uncertainty of combat. This often meant that the initial stint in the trenches was particularly wearing on the junior leaders. A soldier of the 305th Infantry recalled, "From the caution our platoon lieutenant took in those support trenches, and from the worried look he always wore, one would think that the fate of the army, the safety of democracy and the political freedom of the next generation depended upon our staying up all night."[46]

For the first time the officers also had to wrestle with the fact that their actions carried deadly implications and that they had to set an example for their men while controlling their own fear of the unknown. Lt. W. A. Sirmon freely admitted that during his first trench raid he was "badly frightened" and "shaking badly, but swearing to myself I would not run." Other leaders found that war was not as tidy as their training had led them to believe. After one trench raid, an officer of the 113th Infantry reported, "It was practically impossible to carry out the plan as practiced. There were no trenches. The destruction was absolute, and instead of the trench lines they had expected to encounter, the Scouts met shell hole after shell hole, heaps of earth and projecting duckboards and wire, which impeded individual progress and made extremely difficult the task of keeping groups together. Each group leader was compelled to act on his own initiative, following only as a general direction, the original plan of attack." Their time at the front and exposure to even the briefest of danger did aid officers in honing their leadership skills, developing their battlefield wisdom, and furthering the building of cohesion in their units.[47]

The Americans' time in the trenches also brought to light serious problems with the training of the AEF's junior leaders. While serving on the French front, the 82nd Division lost 44 men killed in action and another 327 wounded. Many of these casualties can be traced back to the lack of training of both officers and men. Of the 44 soldiers killed, 17 were lost in a single incident after a German shell slammed into an overcrowded trench. This occurred despite the instructions warning officers to disperse their soldiers in the trenches for "economizing personnel" and "minimizing the effects of the enemy's artillery fire."[48]

Tellingly, most of the division's wounded resulted from exposure to gas. A general lack of training and supervision by junior leaders in gas warfare plagued the AEF. These lapses also ensured that the Americans suffered inordinate casualties to poison gas when compared with those of the other combatants. Archibald Hart noted that even after his unit sent soldiers to the Gas Warfare School, his company's gas training was rather sparse. Upon his return from the gas school, the battalion's gas NCO gave a few lectures and "suggested that we don gas masks marching to and from the drill ground." After the soldiers followed this advice for a few days, Hart noted that the men grew weary of wearing the uncomfortable masks and halted the procedure. He laconically recalled, "That was the extent of our training in gas warfare." In the 89th Division an inspector discovered that one infantry battalion gas officer "had no training whatever in gas, and . . . knew nothing whatever about the subject, which may account for some lack of knowledge about gas among the Company Commanders."[49]

Far too many American officers and N C O s took a cavalier attitude toward gas training. Leslie Baker, of the 27th Division, remembered that the level of gas training within his unit was not as thorough as it needed to be. He related one incident that illustrates the lackadaisical attitude that both officers and men took towards gas warfare training. When the division gas officer queried Baker and his comrades on the extent of their gas training, "It will never be forgotten how surprised he was when we foolishly admitted we had never worn our gas marks for four hours [as required]. Consequently, the very next day we were ordered to wear our gas masks from eight o'clock in the morning until noon, which we did—most of us." However, he related that since "it was such a wonderful day . . . and we had wasted so much of it in this fashion," the company commander cut the training short and with a wink and a nudge told his soldiers before dismissing them, "I hope that everyone will remember he has had his gas mask on for four hours in case anyone should ask him." This casualness and indifference in training, sanctioned or propagated by junior leaders, later reaped its own deadly rewards. In the fighting of June 1918 one officer noted in his diary that a "combination of Yank 'take a chance' carelessness, and German gas, [was] responsible for 85% of casualties so far."[50]

Inspections of Americans undergoing their training with the Allies revealed other serious deficiencies that could often be traced back to poor leadership and instruction by company-level officers and NCOs. After inspecting machine gun units within the II Corps, Maj. Walter Short reported on 5 July 1918 that the machine gun training programs for the 27th, 30th, and 78th Divisions "had not been consistently followed" and that in those units "shooting had been largely neglected." Short noted that this fault was because commanders willfully deviated from the training program. Less than a month later, an inspection of the 27th Division discovered that the unit's officers and men lacked "sufficient knowledge of [the] properties of gas and gas defensive measures" and that its infantrymen were "generally deficient in map reading, sketching, intelligence, signaling, scouting, patrolling, grenade practice, and rifle practice." More important, junior infantry officers were "not properly instructed in [the] tactical handling of platoons and companies." Last, on 30 June 1918, Lt. Col. K. T. Riggs found that the 30th Division had serious problems with the level of gas warfare and machine gun training of the unit's soldiers. Furthermore, in the division's 119th and 120th Infantry Regiments "the methods of training used by the officers . . . [was] sketchy, with the result that men are not thoroughly trained."[51]

The third phase of training, in which large-unit open warfare maneuvers were to occur, was the most variable part of the American training plans. This phase, more than the others, was the one that was given short shrift when training time was curtailed. As Capt. Clarence Minick's diary illustrated, units made efforts to conduct this critical large-unit training, but the time and realism afforded these events were limited. Despite the importance of these large-unit exercises to the overall effectiveness of the AEF, they often lacked realism. Lt. Col. Ashby Williams sardonically recalled that just before leaving the French sector "we had our last of those famous division problems before we went into actual war. A division problem is where you imagine you have some troops and you imagine there is an enemy and you walk over a field where they are not. Very simple, isn't it? It is, however, the only opportunity the staff officers have of demonstrating to you how much they know about war. Perhaps the best thing about these problems is that they are easy to forget when you go to meet a real enemy and have to use common sense."[52] All of this sometimes appeared quite comical to the junior officers. B. A. Colonna noted that as his unit's training period came to an end, "regimental, brigade, and divisional problems began to be the rage. Since nobody below major ever got any information as to what these are all about, the troops were usually represented by flags."[53] Colona and his company-grade peers should have been more concerned over the antics of their regimental, brigade, and divisional commanders and their staffs. The failure of these exercises to address realistically the higher headquarters' fire support, supply, medical, and command and control functions would soon complicate the leadership challenges of junior officers and NCOs.

Beyond the problems of the AEF's unit training plans, overarching training issues also influenced the leadership and skills of the junior officers in Pershing's army. The first of these is the matter of doctrine. At the close of the war, Pershing boasted that the AEF had prevailed due to "its determined insistence on an offensive doctrine and upon training in warfare of movement" and its ability to overcome the soldiers' idiosyncratic stateside instruction with "a system based on correct principles."[54] Unfortunately, Pershing's doctrine was never quite as doctrinal as he believed.

163

Senior American commanders and staff officers liked to boast that the Americans had crafted the most advanced doctrine of the war. Hunter Liggett maintained, "The American High Command had made a thorough study of the experiences of the French and British during the war in the matter of organization and had, as a basis of our own organization, adopted the best of both foreign systems, with modifications to suit our psychology and problems peculiar to our own development." On 4 July 1918, Harold Fiske even conveyed to the AEF chief of staff his belief that "Berlin cannot be taken by the French or the British armies or by both of them. It can only be taken by a thoroughly trained, entirely homogeneous US Army," one, of course, trained in "American methods." This was all well and good on paper, but in execution the creation, dissemination, and training of infantry doctrine were far from clear-cut.[55]

The reliance on the Allies for conducting much of the Americans' training ensured a lack of uniformity in tactics and doctrine. American combat leaders had to reconcile the "American" tactics they had learned in the States with the new doctrines taught by the Allies. For example, Capt. Wardlaw Miles, of the 77th Division's 308th Infantry, stated that "under the tutelage of the British 39th Division, American methods were largely abandoned." Similarly, Capt. John Stringfellow noted that despite the criticism of GHQ staff officers for his unit's use of British "methods of conducting combat," the British tactics remained his primary doctrine. Capt. Paul Schmidt, a company commander in the 127th Infantry, recalled that during his month of training in France, his company underwent intensive instruction in "formations and French tactics, under French instructors." Yet despite their French training, a 5th Division officer maintained, "The Americans still clung to the idea that the rifle was the main dependence in warfare, and pushed that training with that arm to the utmost"[56]

This system of dueling doctrines certainly complicated the efforts of junior leaders to train their units for combat. In a reflection of this confusion, one 78th Division infantry officer observed, "Some men would go to a British school and qualify as instructors, only to come back and find that the American system was being used, and vice versa. Both systems might have had their good points, and did have, but the rate at which orders and instructions and ways of doing things changed from day to day was enough to bewilder old hands at the game; and we were greenhorns."[57] Even the French recognized this problem and attempted to get the Americans to take a more holistic approach to tactics. On 10 December 1917, Maj. Beaugier of the AEF's French Training Mission recommended to General McAndrews, commandant of the AEF schools, "that American officers and candidates should not imagine that there are two different ways of making war, viz: war in the American fashion, in accordance with the Field Regulations, and war 'a la Francaise,' according to the Platoon Commander's Manual and Trench Polygone."[58]

Unfortunately, the hodgepodge of doctrine was never truly resolved. The fact that American officers were making distinctions between "American methods," "British methods," and "French tactics" does question the degree of uniformity in the training and doctrine of the US Army. If the purpose of doctrine is to "get everyone on the same sheet of music," then the American practices were a discordant and inharmonious symphony.

The other challenge concerning junior leaders' absorbing a coherent doctrine was the rapid pace at which tactical methods changed in the AEF. Kenneth Hamburger maintains that the AEF GHQ went to great lengths throughout the war to adjust its doctrine based on its hard-won lessons learned.[59] Although militaries must constantly assess their tactical doctrines in wartime and make adjustments based on changes in technology or enemy tactics, these changes must be accompanied by a period of time for units to absorb and train on the new concepts. This also assumes that the units have a sound grounding in basic tactical skills and principles—a questionable assumption for many of the AEF's infantry units.

As had happened in the United States, units training in France were drowned in a flood of doctrinal manuals and other publications. In addition to the publications printed by the War Department, the AEF itself printed 154 manuals and pamphlets. It ultimately issued nearly 1.5 million copies of these publications over the course of the war.[60] This was almost enough to give every doughboy at least one doctrinal manual to carry in his pack. Most of the AEF's publications were technical manuals related to the new weapons filling the army's inventory. Many of the remaining tactical manuals suffered from some of the same problems that had plagued those of the War Department. The AEF's editions of *Instructions for the Offensive Combat of Small Units* and *Manual of the Chief of Platoon of Infantry,* for example, contained differing formations and tactics, and the GHQ made no effort to reconcile them.

Again, the officers and NCOs had to become their own interpreters of doctrine as they groped their way through their tactical training. As an officer in the 29th Division lamented, "Numerous, varied and exhaustive pamphlets were issued. . . . These were presumed to be absorbed by the officers and imparted to the men. There was much in them that was good . . . [and] much that had to be learned and forgotten, because the game of war is not constant; but they were issued in such quantity and covered so wide a scope that it was impossible for officers to master them and at the same time attend to their varied duties." Another officer noted that as his unit progressed through training, "More attention was now paid to extended formations than had been in the past." He quickly found that "no formation was standard or final." This was caused by the fact that "each new instructor and each succeeding pamphlet brought new combinations." However, he did admit that "while this instruction was indefinite and discouraging at the time," by its very fluid nature it actually "fitted well into the requirements of future campaigns."[61]

The pace of doctrinal change could be baffling to those junior officers attempting to understand and impart the new tactics and formations to their units. Robert Bullard admitted that as soon as his 1st Division came out of its time in the trenches, he instituted a hurried burst of training in which "everything new in tactics that we could hear of, whether of the Allies or the enemy, we tried."[62] One company commander found out just how quickly doctrinal change could occur in France. Just after Capt. B. A. Colonna returned from a tour in the trenches, he recalled,

> an orderly brought around late that night some red covered books and leaflets, and we were told that these would be put into effect the next day. These were the new system of combat formations, involving absolutely new extended order drill, and formation of the company. Lieut. Moore had drilled a few times in these formations; the rest of us knew no more about them than the company cooks did. So the next morning we sallied forth, books in hand, and worked the formations out step by step.

Everyone was quick to see that this was something like business, as of course our old army regulations were absurd when it came to using the new special weapons, such as automatic rifles, hand and rifle grenades, and so on.

Colonna maintained that, given this step-by-step approach, "the new formations were mastered remarkably quickly."[63] However, mastering formations and being able to match them to variations in the terrain and enemy were two separate issues. The evidence suggests that many American units became relatively competent in the former but not so skilled in the latter. This problem became evident as the doughboys of the later-arriving division entered combat.

Part of the problem of matching tactics and formations to the terrain and enemy was due to shortcomings in the AEF's "official" doctrine. The *Instructions for the Offensive Combat of Small Units* (most likely the "red covered" manuals that Colonna referred to) was the closest that the AEF ever came to a printed infantry doctrine during the war. The manual was a departure from many of the prewar precepts enshrined in the *IDR* and contained sound advice for junior leaders on how to operate in combat. It decisively departed from the army's prewar doctrine of "building up the firing line" by noting, "When a line is stopped by organized defenses, which are intact and occupied by the enemy, the reinforcement of such [a] line has little chance of producing success—it will simply increase losses." The manual also enjoined junior leaders always to seek enemy flanks rather than pushing frontal attacks, and that once platoons and companies came under fire, they should break into smaller elements and proceed forward by "employing short rushes at top speed" or "advance by filtering over ground furnishing but little cover."[64]

While Pershing and Fiske might cavil over open warfare and the superiority of the bayonet-steeled rifleman, *Offensive Combat of Small Units* advocated a close cooperation and integration of artillery, tanks, mortars, machine guns, and the other new weapons of war in assisting the infantry in killing the enemy and taking ground. Although the manual stated the obligatory mantra that "the rifle remains the first weapon of the infantryman in all the circumstances of war," it went on to describe how the infantry had to employ its other organic weapons and synchronize its operations with artillery to accomplish its missions. Furthermore, the manual directed company commanders to shatter resistance "by employing all the means at hand."[65]

What *Offensive Combat of Small Units* lacked, however, were the details of application. Although the pamphlet provided excellent general guidelines, it was much sparser in its discussion of how the junior officer was to accomplish this seamless transition from one formation to another and from the approach to the assault. While it is easy on paper to state that the officers must maneuver to attack the enemy's flank, it was quite another thing in practice. Given the uneven training of the army's junior leaders and soldiers, it was incumbent on the AEF to make the transition of doctrine on paper to doctrine in action as smooth as possible. Some of the manual's great failings were its illustrations and its texts describing how the leader combined formations with tactics. Although it adequately showed the basic formations, the way that the leader was to move his unit from one formation to the other was mostly left to the reader's interpretation of the snaking arrows illustrating the move, or his reading of the somewhat turgid accompanying texts.

The great hinge upon which the formations presented in *Offensive Combat of Small Units* swung was command and control. Although combat command and control were the perennial problem that plagued all the war's combatants, they were a particular difficulty for the Americans. In an effort to give his units staying power through the depth of the attack, in October 1917 Pershing convinced the War Department to adopt a massive structure for all AEF units. The end result of this change was the creation of ponderous 28,000-man divisions, which at the lower levels were made up of huge 250-man infantry companies and 59-man platoons. Overnight Pershing presented his half-trained junior leaders with monumental problems in maneuver, supply, combat, and command by saddling them with these cumbersome units.

Offensive Combat of Small Units tried to address these challenges by keeping the formations relatively close, thus giving the platoon and company commanders a better ability to see and direct their men, and by offering these leaders the option of organizing their platoons into "combat groups" or "half-platoons."[66] The idea was that these divisions of platoons would give the platoon leaders fewer subordinates that they would have to control directly while also creating smaller, more flexible subunits capable of semi-independent combat action as the situation demanded.

In theory, the concepts of half-platoons and combat groups were a brilliant solution to some of the problems with command and control that dogged infantry operations of the First World War. *Offensive Combat of Small Units* directly addressed the realities of confusion and friction in combat. It admitted that the use of these subunits and other "temporary groups" often "can not be avoided on account of incidents of the fight and the conditions of the terrain." As such, it was imperative that the soldiers be prepared for these unavoidable eventualities. The key to success was when these planned or accidental groups of men occurred, a leader, "whether he be non-commissioned officer or private, must rise spontaneously to direct his comrades, carry them forward when they hesitate, and prevent them from giving group."[67]

The emphasis that the manual placed on NCO leadership raised troubling issues. As previously noted, the lack of any systematic plan for identifying and training NCOs had all too often led to a culture of junior officer micromanagement of their combat units. For the vast majority of American NCOs and privates, nothing in their previous training or experience had prepared them for "spontaneously" rising to assume these combat leadership roles. While some American enlisted men later rose to this occasion in combat, they did so in spite of, rather than because of, their training in the United States and France. Thus, while the infantry doctrine promulgated in *Offensive Combat of Small Units* was often quite sound, it frequently demanded more of junior officers and NCOs than they were prepared to do.

The way the new doctrines were used during the Americans' time in the quiet sectors illustrated that while the doughboys may have understood the mechanics of the evolving doctrine, their grasp of its application remained dangerously underdeveloped. A good example of this was seen in the Americans' efforts at replicating machine gun barrages. The British had used machine guns in an indirect fire mode to create a lethal "beaten zone" of bullets falling on crossroads and other potentially crowded areas behind the German lines. The goal of this "harassment and interdiction" fire was to create casualties and undermine German morale by eliminating safe rear areas and slowing the flow of food and supplies to the forward trenches. The American machine gun officers adopted the British method with relish and at times in the war used it to great effect. However, the American officers often failed to realize the inherent problems and risks of their newfound doctrine. Sgt. Charles Minder recalled his first exposure to the concept of the machine gun barrage while in the Vosges:

> The Lieutenant came back about ten last night with a spirit level and elevated the gun a certain degree. Then he turned the gun to the left and I had to go out in front of it, about ten feet, and stick a branch in the ground about three feet high. Then he turned the gun to the right to a certain point, and I stuck another branch in the ground. . . . Then we started to shoot from right to left between the two branches, and the bullets were supposed to fall

on a cross roads behind the German lines. . . . The Lieutenant [later] told us that we did a great deal of damage. How they got this information was beyond me, unless there are spies behind the German lines.

The next night the officer again had Minder fire fifty rounds, wait a minute, and then fire fifty more for half an hour. When the officer directed Minder to fire on the same spot in the same manner that he had the night before, retribution from the Germans was not long in coming. The Germans mercilessly shelled his position and barely missed Minder.[68] Being on the receiving end of an inexpertly employed doctrine meant that Minder certainly experienced his combat "coming of wisdom" long before his officers did.

Another overarching problem with the AEF's efforts to prepare its junior leaders and units for combat was, ironically, the AEF's school system. Pershing and his staff were painfully aware of the army's inexperience in modern war and sought to mitigate this problem by creating a vast system of schools that would train the American novices in the new technical and tactical rites of combat on the Western Front. Pershing's actions in establishing the AEF's schools reflected the precepts of the cult of professionalism that had so influenced the army for the past twenty years. True to the cult's principles was the unshakable belief of Pershing, Fiske, Bullard, and other key commanders and staff officers that education would be the salvation of the American army, and they approached it with a missionary zeal. Bullard later remarked,

> Among the officers of the 1st Division there largely prevailed our old idea that experience in war was the only proper teacher of war-making, and that war having come, schools should cease: We should take to the field and learn war there. . . . The division commander and many of his officers seemed to regard the school idea as puerility, a fad of schoolmen; very troublesome and irritating at a time when everybody was getting ready to fight. These ideas remained among Americans until they had seen real war at the front. Then every commander wanted officers and men who had been through these schools. The demand for school instruction soon became so great that it could not be met.[69]

While this "road to Damascus" moment may have brought the 1st Division's officers back to the high church of education, it also blinded them to many of the problems that the schools created.

First of all, the administration and faculty of the schools required a large overhead of officers and soldiers to keep the courses running. This meant that the GHQ constantly levied AEF units for officers and men to fill these slots. Often, the schools also received permission from GHQ to retain the best students of their graduating classes to fill their instructor needs. From 5 March 1918 to 11 November 1918, the Infantry Specialist School alone retained 504 officers from its courses as instructors. This meant that approximately 9 percent of the students attending the courses did not return to their units following graduation.[70] Although these numbers seem low, their significance increased as units began to suffer losses of officers from casualties and other levies of personnel.

This means of obtaining teachers for the AEF's ever-growing schools and courses perpetuated the corrosive blind-leading-the-blind system of wartime instruction and unintentionally encouraged mediocrity in both the students and the instructors of the courses. As one infantry lieutenant commented, "Officers feared to make good grades in school because of the danger of becoming an instructor." Of his time at the Engineer School another officer recalled, "The instructors were 2nd Lts, who had finished the previous course. It was not their fault that they didn't know [the material], but it was a joke." A 126th Infantry officer noted that all company officers were required to attend I Corps schools for "a month of instruction in new formations and the use of the new weapons we received." He complained that much of the instruction "was of little account" and "the instructors generally were officers who never had active service at the front and their theories were sometimes complexing to the veterans just in from the line."[71]

To overcome the shortage of instructor, the AEF also turned to the Allies for qualified teachers. Unfortunately, the AEF's officers were just as willing as their stateside peers to limit the scope of instruction given by Allied trainers. Generally, the Americans used Allied instructors only for the technical side of weapons training and bayonet instruction. Some senior American officers feared that the Allied instructors were contaminating the Americans with their defensive mindsets and focus on "foreign methods." In July 1918 Harold Fiske wrote to the AEF chief of staff:

> Our young officers and men are prone to take the tone and tactics of those with whom they are associated, and whatever they are now learning that is false or unsuited for us will be hard to eradicate later. . . . The junior officers of both allied services, with whom our junior officers are most closely associated . . . know

little of the general characteristics of war, and their experience is almost entirely limited to the special phase of the war in the trenches. . . . The tutelage of the French and British has hindered the development of responsibility and self-reliance upon the part of our officers of all grades. All our commanders from the division down have constantly at their elbows an Englishman or Frenchman who, when any difficulty arises, immediately offers a solution. A great fraction of our officers have consequently permitted themselves to lean very largely upon their tutors with a resultant loss of initiative and sense of responsibility. The assistance of our Allies has become not an asset but a serious handicap in the training of our troops. . . . An American army can not be made by Frenchmen or Englishmen.[72]

To combat the "bad influence" of Allied soldiers and minimize the "damage" caused by Allied training, Fiske vigorously petitioned Pershing to purge the AEF's education centers of all "counterproductive" Allied influences. Heeding Fiske's advice, on August 1918 Pershing ordered GHQ slowly to shed French and British instructors from all AEF schools and units. Although these purges were never completed, they did rob the AEF of a fruitful and labor-saving source of experienced instructors.

Another problem with the AEF's schools was that they often failed to impart realistic training to their pupils or focused the instruction in the wrong areas. When Maj. Gen. Robert Alexander inspected the AEF's schools in the winter of 1918, he found numerous problems. The schools focused on the technical and specialist skills of the new weapons instead of concentrating on how best to use those weapons in combat. Alexander discovered that one six-week course was devoted to nothing but the use of the bayonet. He believed that more than two weeks of instruction in this subject was a "waste of invaluable time." The general accurately noted that "any system of infantry training must be, for the subaltern and sergeant the ability to direct platoons and sections under fire, for the corporal the same ability in the control of his squad."[73]

Alexander's observations were absolutely on the mark. The machine gun school, for example, taught "the mechanical operations of various types of machine guns; practice in known distance machine gun firing; calculations for and practice in various methods of indirect machine gun fire; a certain amount of machine gun tactics."[74] It is obvious from the machine gun school report that giving the students "a certain amount" of tactical knowledge was not the thrust of the course. The schools failed to find the balance between the "technician and the tactician" that was so desperately needed in the AEF's junior leaders.

This flaw was illustrated in the training of Cpl. Fred Takes. Takes recalled that he spent nearly a month in an AEF school learning the Chauchat automatic rifle. He wrote that the training was mostly technical in nature and that he spent only one day of the course in learning the tactical employment of the weapon. The tactical training was conducted on the drill field and consisted simply of moving "in lines of skirmish in two waves, the first wave firing as they marched. We walked about 30 yards and then lay down and fired for a while. Then we advanced about 30 yards and jumped in trenches and fired from there."[75] Again, the focus was on the mechanics of the weapons and formations rather than on a sound appreciation for the gun's tactical employment.

As Alexander also noted, in many cases the training was simply overdone. Too many of the AEF's schools demanded far too much of the student's time to train subjects that should have occupied far less space on the calendar. As one infantry officer later commented, "Three weeks courses were given in courses that any reasonable man ought to learn in three days. If he couldn't learn grenade throwing, for instance, in three days, he ought not be an officer. . . . Somebody's obsession regarding the necessity for schools kept about 50% of officers away from their units all the time, when they ought to have been giving their time to their men."[76] This officer's last observation revealed the greatest problem with the AEF's school system.

In the end, the AEF had to choose between increasing the technical knowledge of its leaders in modern warfare and allowing junior leaders to build cohesion in their units by remaining with them through the trials of unit training. Caught between this Scylla and Charybdis of training, the AEF GHQ chose to give preference to the schools. In doing so, the AEF's senior officers merely continued a precedent that had long been established in the Regular Army. In the twenty years leading up to the war, the regulars had simply gotten used to a perennial shortage of officers in their units caused by the need to staff and fill the army's various schools. But what could be safely practiced in peace could not always be so in war. This fact was not lost on the junior officers themselves. The phenomenal growth of the training courses, and their voracious demand for students, led one disgusted officer to quip, "The Germans begin a great offensive, and we retaliate by starting another school."[77]

Unit histories and personal accounts of the war are replete with tales of units losing vast numbers of officers and NCOs to the AEF's voracious schools. The AEF GHQ demanded that its units provide a quota of students for each course and brooked no rebuttal from subordinate commanders based on unit hardship or military necessity. One 7th Division officer noted that "the schools alone, prescribed by higher authority . . . threatened to absorb all the time and energy of officers and non-commissioned officers whose principal occupation should have been the care and training of their units." Frank Sibley noted that while his unit was undergoing training with the French, it had to supply a steady stream of officers and NCOs to fill the new AEF schools. He observed that given the training that needed to be conducted in the unit, the loss of these leaders was "more than could really be spared." Even George Marshall, a high practitioner of the cult of professionalism, decried the fact that the green divisions bound for the Meuse-Argonne Offensive were "absolutely scalped" of their officers "in order that the next class at Langres might start on scheduled time."[78]

Unfortunately, the AEF's draconian schools quota system yanked the leaders from these evolving platoons and companies while they were both learning the ropes of modern war and jelling as cohesive groups. Generally, a young captain or lieutenant in the AEF could expect to attend at least one school during his service in France and lose approximately one or two months of time with his soldiers. For example, after arriving in France, 1st Lt. C. E. Crane was assigned to the 55th Artillery Regiment on 18 April 1918. Crane spent all of June and half of August 1918 in various signal and artillery schools. When he went into action with his unit for the first time on 28 August, he was virtually unknown to his men and had precious little time to build a sound and symbiotic command relationship with them.[79]

As previously noted, the stateside army's personnel system had severely undermined unit cohesion through frequent and sudden transfers of solders and officers. Little in the training of junior leaders had taught them how to motivate, manage, and care for their troops. The interpersonal leadership skills that were not learned in training had to be obtained through trial and error by the leaders in their units. All too often, attendance at AEF schools prevented junior leaders from gaining the hands-on leadership experience that bonds units together. The military sociologist Darryl H. Henderson noted that to be successful, combat leaders had to establish "personal, empathic, and continuing face-to-face contact with all soldiers in the unit" to build and maintain cohesion. He noted that leadership was one of the greatest determiners of how well a unit hung together and performed under the stress of combat. Henderson argued, "Men in danger become acutely aware of the qualities of their leaders. They desire leadership so their immediate needs can be met and their anxieties controlled. In this regard, well-trained and respected company grade officers and sergeants relay a sense of competence and security to their soldiers and, if successful over a period of time, gain a degree of influence and control over members of their units."[80] Soldiers gain this appreciation of "the qualities of their leaders" through training, daily contact, and shared hardships and experiences. Unfortunately, the AEF GHQ remained oblivious to this demand, and small-unit leadership and cohesion in the US Army consequently suffered.

Time and time again, doughboys were struck by the sudden transfer or unexpected departure of their leaders at critical points in the unit's history. An infantryman in the 33rd Division recalled that his company had five different commanders during its eighteen months of service. His experience was far from rare. Connell Albertine and his comrades were shaken by the fact that just as they were departing for their stint in the French trenches, their company commander was ordered to a month-long school. As their lieutenant had already departed for another lengthy course, they were detailed an unknown lieutenant to serve as their acting platoon leader during their first time in combat. An officer in the 353rd Infantry noted that in the middle of some of the battalion's most strenuous training in France, the battalion commander, several of his officers, and "a picked sergeant from each company were called to Langres for special tactical instruction." These key leaders did not return to their unit until after it had already moved to the front. Another 89th Division officer, Lt. John Madden, recalled that just as his unit was committed to the Meuse-Argonne Campaign, he was sent to the rear to fill his unit's quota for a ten-day class in rifle and hand grenades. Because of this assignment, the young officer missed most of the fighting, and his platoon was left leaderless. The absence of the officers at these key points in their unit's history prevented them from sharing the hardships and experiences that tend to weld the leader to the led.[81]

Ultimately, the GHQ's decision to privilege schooling over troop time for its junior leaders was the wrong one. Despite the Americans' glaring lack of technical and tactical training, leadership and unit cohesion was still the more pressing issue. Some in the GHQ's hierarchy also recognized this problem. In August 1918 an AEF staff officer observed that in the 27th Division "the battalion and company commanders were frequently away on courses, thus missing the great opportunity of gaining practical experience" of serving with their commands on the front line with the Allied armies. He went on to note, "While many of the officers have attended a number of courses and have acquired a considerable amount of theoretical and tactical knowledge, their knowledge of their duties as regimental officers is not thorough. They do not often realize what their position demands of them, what their responsibilities are as regards to their men, and fail to exercise fully their powers of command."[82] Sadly, no one heeded this officer's advice, and the AEF's schools continued to demand their pound of leader flesh from combat units at the most inopportune times.

In April 1919, Pershing convened a board of officers to study the AEF's overall lessons learned from the war. This board, called the Lewis Board after its chairman, Maj. Gen. Edward M. Lewis, found much that had been wanting in the army's performance. During the proceedings, the commander of the 7th Division, Maj. Gen. Edmund Wittenmyer, commented,

> Every organization after its arrival in France was to a great extent disorganized by the system of instruction adopted by the G.H.Q., in constantly withdrawing officers and noncommissioned officers to send them to school; thus leaving the organizations entirely without their complement of instructors. While these officers and noncommissioned officers were benefited . . . the organization itself lost by their absence more than was gained by the individuals that attended the schools. . . . The action of superior authorities in taking away large numbers of officers of all grades, and enlisted men, to attend school and receive instruction absolutely destroyed all results in the way of instruction in the companies and battalions, and I consider these two organizations to be the very best schools for both soldiers and junior officers.[83]

Although Wittenmyer's penetrating observation came too late to aid the AEF, it did at least show that some of its leaders understood the price that combat units paid because of the army's school policy. The AEF's difficulty in building those vital "face-to-face" relationships between the leader and the led in small units, which Henderson claims is the key to unit cohesion and effectiveness, later bore bitter fruit in the combat it endured in the second half of 1918.

Before closing on the AEF's school system, it is important to discuss one specific school that has a bearing on this study. On 10 October 1917, AEF General Orders 46 established the AEF Army Candidates School at Langres "to provide standardized officer material to replace our losses in battle." In 1918 the AEF established additional candidate schools to train infantry, artillery, and engineers at Valbonne, Samur, and Maily. Based on Allied experiences, the AEF GHQ expected to lose 75 percent of its junior infantry officers per year. To make up for these losses, GHQ initially planned to have all divisions provide 150 enlisted men each to attend the course.[84]

In December 1917 Pershing directed that the Langres candidates school instruction in minor tactics would focus on reconnaissance, security (advance, flank, and rear guards and outposts), combat orders, marches and convoys, camping and billeting, combat operations (attack, defense, and night fighting), minor trench warfare, field sanitation, and liaison. He further stipulated that the "course will be based on Field Service Regulations and Infantry Drill Regulations, modified in detail whenever necessary to conform with the present organization of the American units in France." As a secondary text the school was to use translations of the French *Manual of the Chief of Platoon of Infantry*. As usual, the school commandant was given no guidance on how to reconcile differences between the *FSR* and *IDR* with the French manual. Ultimately, the AEF's candidate schools commissioned 10,976 officers by 14 November 1918, and were planning on having 22,000 more candidates under instruction by January 1919. Pershing later admitted, "It must not be thought that such a system is ideal, but it represents a compromise between the demand for efficiency and the imperative and immediate necessity for trained replacement officers."[85]

In theory, the curriculum of the Valbonne Infantry Army Candidates School reflected Pershing's demands for more focus on the tactics of open warfare. In late October 1918 the three months of instruction was to total 468 hours of training and study. The tactical terrain exercises were to have accounted for 190 hours of the total instruction, with 160 of those hours devoted to open warfare training, and only 12 hours on subjects related to trench warfare.[86] In other words, the tactical training was to have been much more robust and intense than that given to candidates in the United States.

Unfortunately, several factors prevented the AEF's Infantry Army Candidates School from achieving its lofty training goals. As casualties and the need to man staff positions mounted in the second half of 1918, Fiske admitted that "to meet the imperative demands for officers several courses had to be considerably shortened." For Joseph Lawrence this meant that after completing only two months of his three-month candidates school he was commissioned and immediately assigned as an infantry platoon leader in the 29th Division.[87]

The actual training conducted in the school also indicates that much of the instruction given the budding officers was problematic. The training schedules for the candidates school for October 1918 show that a great deal of the training remained at a fairly basic level, with a focus on close-order drill, bayonet and physical training, and instruction on musketry and grenade throwing. The school also lacked training aids and publications. A 20 October 1918 memorandum from Col. S. L. Pike, the school director, noted that "instruction in Infantry Drill Regulations must be given in the form of conferences as no books are available in France for issue." Pike directed his instructors to use the *Manual for Noncommissioned Officers and Privates of Infantry* as a substitute for the *IDR*. The fact that the school was using a very basic manual for instructing officers seemed to have raised few concerns among the cadre. Furthermore, Pike ordered his instructors to focus on "developing smartness and precision in close order drill . . . when adequate facilities for other instruction are lacking." Despite Pike's best intentions, it seems as though the AEF's officer training was little better than that provided by the stateside OTCs and COTSs.[88]

The AEF's candidates school suffered many of the same problems of the stateside OTCs and COTSs. Some of lapses in the AEF's officer training programs were a result of Colonel Pike's difficulties in obtaining qualified instructors. As happened with the COTSs, Pike resorted to the shortcut of retaining "a number of the smartest and most enthusiastic graduates" as instructors for new classes. In June 1918 Pike requested that thirty-one students be retained as instructors, and on 26 September he asked that eighty-two of the officers who were to graduate on 30 September be assigned to teach at the school. As was the case in the United States, there would be no end to the blind-leading-the-blind school of officer instruction in France during the war. Even Pike had to admit that his instructors, "in most cases, had never received any tactical instruction and many had received little serious instruction of any character" prior to their assignment to the school. Also, the retention of these newly commissioned officers further deprived combat units of much needed leadership.[89]

At first the AEF's candidates school employed Allied instructors to aid in the training of its future officers. On 31 March 1918 the school had twenty-three French officers assigned as instructors, with nearly all of them teaching specialist weapons such as machine guns, the Chauchat automatic rifle, the 37 mm gun, and grenades.[90] Keeping with established American practice, Pike ensured that the Allied instructors were strictly limited to technical training and generally not allowed to stray into areas of tactics and doctrine.

Just as in other AEF schools, the candidates school sought to purge Allied instructors from its cadre. As early as 26 January 1918 the commandant of the AEF's schools, Brigadier General McAndrews, instructed the director of the candidates schools to provide a list of the French instructors that they could dispense with, "without impairing the efficiency of your work."[91] However, the real push to rid American schools of Allied instructors did not become serious until the summer of 1918.

In a confidential 30 August 1918 memorandum to the directors of the AEF's School of the Line, Infantry Specialists' School, and Infantry Army Candidates School, the overall commandant of the AEF's schools, Brigadier General Smith, relayed that it was his desire to "Americanize the Army Schools in every respect." As such, he directed his subordinates to determine the absolute minimum number of British and French instructors that they needed for their courses and to provide the "names of those whose services may be dispensed with."[92] In response to Smith's directive, Colonel Pike informed his superior that the candidates school had already limited the role of the Frenchmen to merely advising the American instructors. The only subjects that the French officers actually taught the candidates were field fortifications and liaison work. Pike admitted that "while all French officers could be dispensed with," he believed that this drastic step would disrupt training. He did, however, slowly reduce his French cadre from twenty-one to ten officers.

Pike's purge of Allied instructors could not have come at a worse time. By the summer of 1918 the AEF school was having trouble obtaining qualified candidates. In a telling indicator of the overall training and experience level of the AEF's enlisted men, on 2 June 1918 one instructor reported that "no definite degree of training can be assumed in candidates entering the School." Because of this, "if any but the most elementary subjects occur early in the course, a large portion of the candidates will be unable to assimilate the work, and practically none will be qualified to act as leaders when detailed day to day." He recommended that the first month of the course be dedicated solely to very elementary subjects.[93] In the end, Pike confessed, "A large number of the candidates reporting had received very inadequate instruction [prior to attending the course], about 30% had never fired any course with the rifle or pistol, and a small percentage from the staff corps and departments had never received any infantry instruction at all. Much time, therefore, had to be spent in rudimentary work. During the short courses, the time devoted to first principles had to be reduced at the expense of smartness and precision."[94] Given that most of the candidates admitted to the course were NCOs, one would have thought that this level of rudimentary instruction would have been unnecessary. If these candidates were indicative of the overall state of NCO and soldier training in the AEF, then it was no wonder that some NCOs had trouble mastering the tactics called for in *Offensive Combat of Small Units*.

The heavy casualties of the summer and fall of 1918 only made matters worse, and the quality of candidates continued to slide. The infantry candidates class that began on 15 September 1918 contained twelve candidates who had to be removed from the course because they were illiterate and twenty-seven more who stated that they had been sent to the course against their wishes and desired to be sent back to their units. One of these men, Pvt. Howard B. Peck, maintained that he had not even been aware that he was being sent to the course until his orders arrived. When he protested the posting, his commander bluntly informed him that "he had to come as there were no other men" to send.[95] The candidates school's ravenous demand for students greatly contributed to the overall decline in the quality of its students by forcing field commanders to make a hard choice between the immediate needs of their units and the larger needs of the AEF. Although most commanders appeared to have sent their most qualified soldiers to officer training, a fair number sent their troublemakers or those men who, like Private Peck, were the most expendable.

The high demand for officers led to the expansion of the AEF's officer training system and caused the AEF GHQ to squeeze its units for more candidates in the summer and fall of 1918. The AEF expanded its original branch-specific candidates schools at Langres, Samur, and Mailly, and opened an additional infantry candidates school at Valbonne in the fall of 1918. Attendance at the Army Candidates School at Langres more than doubled from 950 in June to 2,259 in August. Unfortunately, the fighting in the Meuse-Argonne made this surge in attendance unsustainable, and by the time the 15 September course began, its enrollment was back down to 1,125. To fill all of these courses, the AEF GHQ demanded that its divisions and other subordinate units provide a monthly quota of soldiers. On 18 September 1918, for example, the GHQ informed all its divisions that they would send "twenty five suitably qualified soldiers from each infantry regiment" and six soldiers from their machine gun companies to Langres by 6 October.[96]

This constant levy to fill the divisions' quotas had immediate and negative effects on the AEF's combat units. From July to September 1918, for example, the 107th Infantry alone was tasked to provide seventy-five enlisted men to attend the candidates school. Levies such as these dealt further serious blows to the efforts of junior officers to build a competent cadre of NCOs and seriously endangered the effectiveness and cohesion of the AEF's small units. The list of the officers commissioned from the candidates school on 31 October 1918 illustrates the effect that officer training had on the AEF's pool of NCOs. Of the 847 new officers, 41 had been unit sergeants major and 195 had been first sergeants in infantry or machine gun companies. In fact, all but 26 of the graduates had been NCOs before attending the school. Capt. Wardlaw Miles noted that the commissioning of a number of 308th Infantry's NCOs "proved a great loss to the regiment" and forced him and his peers to scramble to replace these losses before entering combat. This steady drain of NCOs worsened existing problems with enlisted leadership and led to a greater entrenchment of the practice of officer micromanagement in combat units.[97]

The AEF's schools were not the only source of stress in combat units. In addition to its constant demand for students, many of the AEF's own personnel policies also obstructed training and unit cohesion. Pershing had originally intended that every fourth division that arrived in France would become a depot division to provided replacements for other AEF units. The breaking up and remaking of these units was another obstacle to effective unit and leader training. For example, the 32nd Division was informed that it was to serve as a depot division when it arrived in France in February 1918. This meant that the division's 128th Infantry had a number of its officers and nearly all of its privates scattered to the AEF's four winds. The remainder of the division was detailed to unload ships and provide other manual labor that the Services of Supply (SOS) required across France. However, after the 32nd had served in this capacity for a month, the GHQ reversed itself and returned the division to the status of a combat unit. This presented a number of problems. One division officer noted that the AEF's cavalier use of the 32nd "took the edge off the fine state of discipline" of the unit's troops. More important, the moribund 128th Infantry could only be brought up to strength by transferring men from the division's other infantry regiments. This measure further wounded the efforts of junior leaders to build unit cohesion and reduced most of the division's infantry companies to nearly a third of their authorized strength. A soldier in the 126th Infantry, one of the units "scalped" to refill the 128th, noted that when his unit was brought back up to strength by an influx of recruits, most of the new men "had received very little training before they arrived."[98]

The 32nd Division was not alone in its rather high-handed and short-sighted treatment by the GHQ. In late August 1918 the AEF GHQ ordered the 7th Division's 13th Brigade, after it had been in France for less than two weeks, to transfer an average of eighty men per company to provide replacements for the battered 4th and 26th Divisions. While in their first weeks of training in France, the 36th Division lost forty-five officers sent for reclassification and another sixty-eight officers to transfers or schools. Shortly thereafter, the division was "called upon to send a large portion of its most seasoned personnel to fill gaps in other divisions that had been fighting at the front."[99]

181

All of these changes resulted in a constant breaking and rebuilding of platoons and companies with the resultant problems in training, morale, leadership, and cohesion that came with them. In August 1918 an AEF staff officer observed that frequent changes in battalion and company commanders in the 27th Division had undermined "discipline and efficiency" within the division's units. None of this seemed to register with the AEF's senior leadership, and the GHQ's demands for levies of officers and NCOs never slowed despite what was occurring in training or battle. For example, on 26 September 1918, just as his unit was preparing to enter the Meuse-Argonne fighting, the 82nd Division's Richard McBride recorded that every company in his regiment was required to send an NCO and officer to the Untied States to serve as instructors. These types of transfers bit hard into the AEF's small units. On 5 August 1918, Sgt. William R. Phillips wrote in his diary, "My Lieut. Niel of the third platoon was sent back to the U.S. to train a new bunch of men. I sure hated to see him leave." Although Niel was replaced with a new platoon leader, Phillips noted that he too was transferred within a week.[100]

Taken as a whole, the AEF's attempts to prepare its junior leaders for combat floundered in the wake of doctrinal uncertainty; rushed, incomplete, or unrealistic training; and ill-conceived personnel and training policies. Although the German offensives in 1918 caused or exacerbated some of these problems, many of them were the result of the Americans' own miscalculations, hubris, and blunders. Pershing's efforts to correct the training deficiencies that he believed his divisions brought with them from the United States unintentionally damaged unit cohesion in the AEF. The AEF's elaborate schools system offered scant improvements in the tactical competency of the army's leadership at the cost of great disruptions to the "team building" of its small units due to its insatiable demand for manpower. The fragmented state of the Americans' infantry doctrine added levels of confusion and uncertainly in their training and operations that would only be resolved in the crucible of combat. Ultimately, the failure of the AEF's training plan left its ill-prepared junior leaders with no other option than to muddle through their combat preparations as best as they could with units that often lacked the corporate spirit that was so essential to success and survival in battle. As the AEF prepared to enter major combat in the summer of 1918, the unheeded ghosts of Kitchener's army warned the Americans of what lay ahead.

7 "Gone Blooey"
The AEF's Systems for Addressing Officer Incompetence and Inefficiency

JOHN J. PERSHING WAS A HARD MAN. He was exacting in his expectations of efficiency and discipline and strictly weighed the ability of his subordinates to achieve results on and off the battlefield. He had an unbending concept of duty and was seldom swayed by friendship or long-standing personal relationships when it came to accomplishing missions. Maj. Gen. Robert Bullard, one of the AEF's senior officers who consistently lived up to the Iron Commander's rigorous standards, noted that when Pershing arrived at the front he was often "good-humored" and "agreeable." However, Bullard knew "that underneath his easy manner was inexorable ruin to the commander who did not have things right. He shows the least personal feeling of all the commanders that I have ever known, and never spares the incompetent." Maj. Gen. Robert Alexander echoed these remarks and observed, an "individual, whatever his grade, had only one chance to demonstrate his capacity or incapacity. In the latter event there was no alternative but to relieve the individual at once."[1]

Pershing, a man already predisposed not to suffer fools lightly, fully realized that his personal reputation as well as that of the army and the nation were inextricably linked to the results attained by the AEF. Given these high stakes, Pershing saw to it that no underperforming Regular Army general, over-the-hill National Guard major, or wet-nosed OTC lieutenant was going to hinder the performance or efficiency of his army. The system that the AEF established to deal with unfit or incompetent leaders tells us much about the wartime officer corps and the command climate in Pershing's army. Although the GHQ designed its reclassification system to weed out incompetent officers, it also created a climate of fear that further undercut the initiative of junior leaders and the overall effectiveness of the AEF.

On 16 November 1917 Pershing issued General Order 62 directing commanders at the division level or higher to establish local boards for the "examination of officers who have demonstrated their unfitness." When the divisional boards proved cumbersome and ineffective, Pershing established standing AEF-wide reclassification and efficiency boards as part of the Casual Officers' Depot at Blois.

At first, the boards could only recommend that the officer in question could be returned to duty or be cashiered and returned to the United States. However, due to the continuing shortage of officers in the AEF, on 11 April 1918 GHQ authorized the reclassification boards to reassign inefficient but otherwise capable combat arms officers to positions within the SOS. If those officers redeemed themselves by the satisfactory performance of duties in the SOS, their discharges would be voided. The Blois depot also held boards for 1,078 officers rendered unfit for combat duty for reason of poor health or wounds to determine if they could be used in some other capacity within the AEF.[2]

Over the course of the war the AEF sent more than 1,081 officers before reclassification boards at Blois after they had failed in their original units. An additional 270 officers appeared before the Blois efficiency board after their conduct or performance had raised questions about their fitness to remain in the service.[3] However, these 1,361 officers listed in the final report of the Blois Casual Officers' Depot do not tell the whole story of the AEF's reclassification system, and the actual number of AEF officers cashiered or reassigned was much higher. For example, while the records of the board proceedings for 50 African American infantry officers in the grades of major through second lieutenant are in the files of the Blois Reclassification Depot, only 31 of these officers are listed (and thus counted) in the Casual Officers' Depot final report. Furthermore, a file of "proposed eliminations for inefficiency" in the National Archives contains the record briefs for an additional 349 officers sent for reclassification. Of the 349 officers who appeared before these boards, only 46 had any case files in the Blois records or were listed in the Casual Officers' Depot final report.[4] Although the circumstances of these boards are unclear, the addition of their records boosts the number of officers sent for reclassification to 1,682. In addition, on 18 December 1918 the AEF GHQ established another officers' reclassification depot at Gondrecourt. By the time the Combat Officers' Depot, at Gondrecourt, closed on 30 April 1919, it had reclassified another 161 officers.[5] This brings the minimum number of officers sent for reclassification after being relieved from their positions to at least 1,843 men.

In May 1919 the SOS deputy chief of staff estimated that approximately 82,000 officers served in the AEF from 1917 through 1919. Given this figure, he also estimated that "approximately, 1 regular officer in 40, and one temporary officer in 80 were found unsuited for the duties they were performing and had to be reclassified."[6] However, even using 1,843 as the number of officers sent for reclassification, that still meant that only 2 percent of the AEF's commissioned ranks were boarded during the war. Still, these minuscule numbers do not accurately represent the powerful influence that the reclassification system held over the AEF's officer corps.

Maj. Gen. James Harbord claimed that Blois was a "Human Salvage Plant" that reclaimed "human beings to an untold and incalculable value." Those sent for reclassification did not see it in this same positive light. Being sent for reclassification was a humiliation for the officers involved. Brig. Gen. L. M. Nuttman, commander of the Blois depot, recalled that the officers awaiting judgment "arrived in various states of mind which ranged from extreme anger, through a feeling of injury and a passive acceptance of fate, to an entire loss of self respect." To the Regular Army officers, being reclassified was the shipwreck of their careers and left the taint that they had failed the highest trial of their profession. For National Guard officers, being removed from their units meant the ultimate embarrassment of having to return to their home communities with their reputations sullied by the stench of failure. To the many young National Army officers who had so eagerly filled the wartime OTCs, reporting to Blois indicated that they had failed as men by being tested and found wanting in the Rooseveltian world of the "strenuous life." Lt. Harvey Harris had a chance encounter with a group of captains who were in route to their new assignments after their reclassification boards. They told Harris that they had been treated as "privates in every sense" at the depot, and one stated that "he would have [to work] for 5 years to get his self respect back." As these officers could attest, Blois was the bogeyman that haunted the psyche of the American officer.[7]

During the war, the term "gone blooey" or "blooeyed" entered the American lexicon as slang for a failure or a colossal malfunction. In the AEF it carried the same meaning as the British army's "Stellenbosched" or the French army's "degommes": an officer cashiered in disgrace. Even though 882 of the 891 officers assigned to the SOS after their Blois boards later "made good" in their new positions, they never truly shook off the impression that somehow they were "damaged goods." Soon after becoming the commander of the SOS, Harbord noted, "The spirit of the S.O.S. has been rather low. All officers who fail at the front are sent back to be utilized in the myriad activities of the Service of Supplies where something can be found for one of almost any profession or trade. This record of failure has had a depressing effect on the spirit of the important work of the S.O.S."[8] Thus, while he could gush of Blois being a "Human Salvage Plant," in his more honest moments Harbord admitted that reclassifying an officer at Blois was the AEF's version of making him wear the scarlet letter.

Given the perception that American officers had of being bloo-eyed, the threat of being sent to Blois was a cudgel commanders could use to keep their subordinate officers in line. Capt. John Castles re-called that the commander of his infantry regiment launched a tirade against his officers and "ended by saying that twenty-seven new offi-cers from the Reserve were coming to join the Regiment and that any one of the old ones who didn't attend to business would soon go back to the United States 'with his tail between his legs.'" Another officer remembered that a division commander threatened to send one of his brigade commanders "to Blois in disgrace" after a failed attack during the Soissons Offensive of July 1918. Capt. Coby McIntyre stated that one of the few times that he witnessed any nervousness on the part of Col. Frank Hume, the commander of the 103rd Infantry, was during field problems in France. McIntyre noted, "In the problem the cards were stacked against Colonel Hume, and he felt, rightly or wrongly, the high command, never considered too friendly to National Guard officers, might look upon any failure on his part as ground for re-moval from his command." Thus, while the actual number of officers sent to Blois for reclassification was rather small, the fear that the re-movals inspired rippled through the AEF and influenced the behav-ior of American officers throughout the war.[9]

Although Blois dealt with the cases of officers from throughout the AEF, officers from combat units made up the vast bulk of those sent for reclassification. Only eighty-three officers were sent to Blois from the vast expanse of the SOS. Infantrymen, artillerymen, and aviators alone accounted for three-quarters of all reclassifications. Nearly 44 percent of all the Blois boards involved infantry officers, and this branch made up the largest number of men sent by any single arm of service. Artillerymen were a distant second, comprising over 23 percent of all reclassifications. Table 7–1 illustrates the statistics from the boards based on service branch and rank.

A number of misdeeds or missteps could lead an officer to Blois for reclassification. Given the rising importance of technology in warfare, and the trend towards professionalization, there was often a direct correlation between the specific expectations and demands of the officer's arm of service and his removal from his position and unit. In April 1918 GHQ directed that officers who failed out of AEF or Allied schools would be considered unfit to serve in their respective branches and were to be sent to Blois for reclassification. This school provision fell hardest on artillerymen and aviators. For example, of the 318 records for National Guard and National Army artillerymen in the ranks of major through second lieutenant in the Blois case files,

Table 7-1. Reclassifications at Blois by Branch of Service and Rank

Branch	Brig. General	Colonel	Lt. Colonel	Major	Captain	First Lt.	Second Lt.	Total and %
Infantry	8	39	29	93	188	143	98	598 (43.8%)
Artillery	2	7	11	21	58	80	141	320 (23.4%)
Air Service	0	0	0	6	7	62	26	101 (7.4%)
Engineer	0	2	3	13	22	18	12	70 (5.1%)
Medical Corps	0	1	3	17	16	28	4	69 (5%)
Quartermaster	0	0	2	10	14	10	14	50 (3.6%)
Coast Artillery	2	2	1	3	9	16	11	44 (3.2%)
Signal Corps	0	0	0	0	15	18	8	41 (3%)
All Others	0	3	2	10	25	19	11	70 (5%)

Note: This table is based on information found in Tables 1 and 2 in *The Reclassification System of the A.E.F. (Blois)*. It does not include the seven Marine Corps officers and one Navy officer sent to Blois, nor are these men counted in the table's percentages.

160 (50 percent) were sent for reclassification as a result of their failure at the DeSourge, Samur, or Coetquidun field artillery schools or after demonstrating that they could not master the technical and mathematical skills required of modern artillery officers.[10] The great number of artillerymen reclassified due to course failures reflects the massive technical and tactical changes that accompanied the rapid evolution of artillery during the war. The enforcement of rigid standards in the artillery schools also helps to explain why the AEF's artillery units became so lethal and effective over the course of the war.

The largest group of officers sent to Blois was made up of infantry majors, captains, and lieutenants. The 515 case files for these National Guard and National Army officers give us insight into the tactical and leadership skills and abilities that Pershing and his commanders expected their junior combat leaders to wield. By examining which infantry officers were ordered to Blois, and the reasons given for their reclassification, one also gains a greater appreciation of the general shortcomings of the US Army in the Great War.

The Blois boards shoehorned the officers appearing before them into four broad categories: "misfit," "temperamental," "inefficient," or "physical." These categories were subjective and rather ill-defined. In general, a "misfit" was an officer whose skills and abilities were poorly suited for the position, unit, or branch in which he served. McAdams added that these officers also included those who were in staff departments or branches "whose civilian vocational training did not fit them for such assignment."[11] The boards classified over 68 percent of all officers into this category.

The officers in the next largest group were those that the boards classified as "temperamental." These officers supposedly possessed personalities, quirks, or dispositions that made it difficult for them to work with others or otherwise hindered their effectiveness as leaders or soldiers. This group accounted for 16 percent of all reclassifications. The "inefficient" category was the murkiest of an already nebulous system of classification. It appears to have been a catchall category for officers whose performance or general incompetence made them unfit to serve in their grade or position. They made up 10.5 percent of the Blois reclassifications.

Last were those officers classified in the "physical" category. The boards considered these men as physically unfit for combat positions due to age, physical limitations, or their inability to take the mental strains of combat. This included a number of officers sent to Blois with shell shock. This reclassification category should not be confused with the physical disability boards also held at Blois. The officers who received a "physical" designation from the reclassification board were those who field commanders deemed unable to accomplish their duties because of their psychological or physical infirmities. Only 5.5 percent of all reclassifications resulted from physical inability.[12]

As Blois's four categories failed to describe adequately the detailed reasons for the officers' being reclassified, I have further sorted the infantry officers into fourteen more descriptive groupings. I freely admit that this system suffers from some of the same subjectivity that marred the original. In many cases the officers were sent to Blois for a multitude of reasons that often cut across the categories listed in Table 7–2. In these cases I attempted to place the officers in the category that seemed most to indicate their alleged failing, or that which most directly led them to Blois. Despite the inherent shortcomings of this system, it does provide a more nuanced view of the reasons that the officers were relieved of their positions than do those given by the original boards.

Some of the categories in Table 7–2 require some explanation. Officers who lacked force, energy, initiative, or aggressiveness, or were too slow or hesitant for combat, reflect characterizations used by the senior officers in their original reports. These characterizations also denote those leaders who lacked the personal presence to inspire soldiers as well as those unable to act independently of direct orders. This category is closely linked to "lacks leadership." "Lacks leadership" signifies those officers who demonstrated an inability to train, control, or direct their soldiers to accomplish their tasks or missions. The category "temperamental" retains the original meaning as used at Blois.

Table 7-2. Reasons for Reclassification of Infantry National Guard
and National Army Officers by Rank

Reason for Reclassification	Major	Captain	First Lieutenant	Second Lieutenant
Lacks, force, energy, initiative, or aggressiveness	17 (22%)	34 (20.4%)	39 (25.8%)	33 (27.2%)
Lacks leadership (ability to control, train, discipline troops)	12 (15.5%)	28 (16.8%)	27 (17.8%)	20 (16.5%)
Lacks tactical skills or professional knowledge	10 (13%)	11 (6.6%)	15 (10%)	13 (10.7%)
Too old or unfit for combat	14 (18%)	10 (6%)	16 (10.5%)	5 (4.1%)
Temperamental, argumentative, or insubordinate	5 (6.4%)	9 (5.4%)	7 (4.6%)	3 (2.4%)
Lacks basic education or mentality	3 (4%)	8 (4.8%)	8 (5.2%)	4 (3.3%)
Too nervous or lost nerve in combat	1 (1.2%)	14 (8.4%)	7 (4.6%)	3 (2.4%)
Personal moral failing	5 (6.4%)	8 (4.8%)	4 (2.6%)	10 (8.2%)
Poor unit administrator	1 (1.2%)	23 (13.8%)	15 (9.9%)	12 (9.9%)
Too slow or hesitant for combat	2 (2.5%)	4 (2.4%)	4 (2.6%)	1 (0.8%)
Poor combat performance	7 (9%)	14 (8.4%)	7 (4.6%)	10 (8.2%)
Suspected disloyalty to USA	0	2 (1.2%)	1 (0.6%)	0
Too immature for combat duty	0	1 (0.6%)	0	3 (2.4%)
Unknown	0	0	1 (0.6%)	4 (3.3%)

"Personal moral failing" describes those officers whose conduct was at odds with the army's expectations of gentility, morality, or standards of behavior. Reflecting the mores of the time, this category also includes those cases in which intemperance led to the officer's boarding. It should be pointed out that the reclassification system was not linked directly to the system of military justice. If, during the course of their investigation, the board found that the officer in question violated any of the Articles of War, it could merely recommend he be sent before a court-martial. While the AEF tried 1,093 officers from June 1917 to June 1919, most of those cases had nothing to do with the reclassification system.[13] Most of the officers in this category at Blois had a string of minor moral infractions that never quite added up to a level that required court-martial. In some cases, once an officer was cleared by court-martial his commander still sent him to Blois just to be rid of him.

The officers listed as "poor unit administrator" failed in the areas of supply, personnel, and messing or in the additional noncombat duties that were a part of regimental life. Officers classified under "poor combat performance" were those whose relief from command was directly tied to their, or their unit's, failure in action. Finally, under "suspected disloyalty to the U.S.A." were those officers sent to Blois based on the accusation that by their deeds or words they had exhibited pro-German or anti-American sympathies.

The reasons that the commanders gave for cashiering their Blois-bound officers offer a unique perspective into the major pitfalls that the AEF faced in combat leadership. It is significant that the largest percentages of officers in all four ranks were reclassified because they lacked force, energy, initiative, or aggressiveness. These figures support the argument that the army as a whole had placed too much emphasis on strict obedience of orders and too little on training its junior officers on how to act when they were not directly under the supervision of their superior. Too often the officers sent to Blois fell victim to senior officers who criticized them for failing to use initiative while simultaneously giving them little space or latitude to develop a knack for independent action. Hunter Liggett later remarked that "the failure of more than one unit" in the AEF "could be traced directly to the inability of the officer in command to delegate authority." He also noted the sad reality that senior commanders tried to "do a sergeant major's, lieutenant's, or an adjutant's" job rather than focusing on their own.[14] If Liggett's observations are accurate, the AEF's junior combat leaders frequently found themselves between the rock of the demands for initiative and the hard place of their superiors' micromanagement.

One of the greatest ironies of the reclassification system was that while officers were being sent to Blois because of their lack of aggressiveness and initiative, the fear that the depot inspired actually worked against encouraging these attributes within the AEF's junior officers. The fear of being blooeyed motivated many commanders to keep their subordinates firmly in line by limiting any of their actions or activities that might reflect badly on their commands.

This fear of removal, and their lack of confidence in the training and abilities of their junior leaders, encouraged the micromanagement by senior officers that Liggett described. As an officer in the inspector general's office later noted, "Officers commenced to exhibit a degree of fear and apprehension lest some unavoidable event, something which they could not control, might operate to ruin their careers." Some officers went as far as to prevent their more talented junior leaders from attending needed technical schools because of "the danger to themselves of being relieved of command for some error made by the less efficient officers." With the specter of Blois never far from their minds, regimental and higher commanders seldom allowed their subordinates the latitude to make or to learn from their mistakes prior to going into combat.[15]

The fact that failure in leadership was the second largest reason given by commanders for the relief of their subordinates points to another flaw in the AEF's combat leadership. As noted in previous chapters, the army itself realized that it had no method for training and developing leadership in its new officers. The Old Army system of having officers learn the leadership side of their trade during a long-term apprenticeship was unworkable in World War I. Maj. Gen. David Shanks warned in 1918 that "the weakest point in the training of our young officers is their lack of knowledge and experience in the handling and management of their men." He sadly noted that the army still expected an officer to learn leadership "by intuition and observation . . . feeling his way along, profiting only by the mistakes he may make."[16]

191

Despite the fact that the army had no coherent plan for teaching young officers how to be leaders, the records of Blois show that it could be unforgiving to those who failed to acquire these skills on their own. Some failed because the job was simply too big for their limited training and abilities. When 250 men were placed under his command, Capt. James Archer was found to be unable to "handle a large company under the existing conditions."[17] In other instances the officers seemed to be at a loss to know what to do as leaders. For example, there is no question that Capt. Wilbur N. Farson, a company commander in the 135th Machine Gun Battalion, was a poor leader. He was relieved of command on 6 September 1918 because he failed to exercise even basic leadership while his unit was occupying trenches in a quiet sector. He never visited his platoon gun positions and "did not make platoon commander[s] properly instruct [their] men in their duties," and as a result, "an inspection disclosed men in both platoons almost wholly ignorant" of what they were expected to do in combat.[18]

Those officers reclassified because they lacked the requisite tactical skills or professional knowledge had often been ill-served by the army's chaotic training system. Lt. Albert C. Pate was packed off to Blois after having been in his battalion for less than a month and in command of his platoon for less than a week. He admitted in June 1918 that "he knew nothing about drill." The board's investigation showed that after arriving in France as a casual officer, his assignment as a town mayor and regimental billeting officer had hindered his tactical training. Pate noted that most of his training in France had consisted of "bayonet drill, close order drill and this new formation."[19]

Closely tied to both leadership and initiative was the need for the officer to be a sound administrator. Part of the social contract that exists in all military organizations is the soldier's expectations that his leaders will look out for his general welfare and provide for his basics needs in return for his military service. Officers who failed to live up to their side of the bargain by failing to provide the basic food, clothing, shelter, medical, and personnel support to their men were fatal to their unit's cohesion and combat effectiveness. Officers who failed in these tasks comprised approximately 10 percent of those sent to Blois. Good examples of the officers reclassified for their failures as administrators were Captains Mirandon and Copeland. The 113th Infantry's Joseph E. Mirandon took so little interest in the "care and feeding of his men" that his regimental commander was forced to "devote his personal attention to the matter several times."[20]

The commander of the 813th Pioneer Infantry noted that Capt. Raymond E. Copeland had "tried to do everything himself" and, as a consequence, "succeeded in doing almost nothing." The colonel found that Copeland was hopelessly inept at accomplishing the routine administrative tasks of running a company and had reported two men as being AWOL for three days when in fact they had been on a kitchen detail.[21] Sadly, while Copeland and Mirandon represent the worst cases, their problems were tied to a larger issue within the AEF of junior leaders having only a sketchy grasp of logistics. Without a strong corps of NCOs to help in these matters, the AEF often found its tactical operations dogged by missteps in feeding, supplying, and providing medical care for its soldiers. Failings in these areas also undercut the paternalistic underpinnings of small-unit leadership and subsequently reduced unit cohesion and effectiveness.

It is interesting that relatively few officers were sent to Blois for reasons that were directly related to their failures in combat. Perhaps the Great War battlefield was so unforgiving that those who made the greatest blunders never survived to be boarded. The fact that few of the officers sent to Blois had been wounded in action may also point to a reticence on the part of commanders to sully the reputations of men who had nobly sacrificed for the cause. Another possibility was that the units had managed to weed out their least competent officers prior to combat. A case in point was the experience of the 37th Division. During its time in France the division commander, Maj. Gen. Charles Farnsworth, ordered nine majors, fourteen captains, and thirty-one lieutenants to Blois. Farnsworth sent the bulk of those officers (four majors, eight captains, and twenty lieutenants) for reclassification while the 37th Division was training behind the lines or occupying trenches in quiet sectors of the front. This suggests that while much of the training with the Allies and in the quiet sectors may have been flawed, it at least allowed commanders to identify and remove some of their commissioned deadwood.

An analysis of the number of officers sent to Blois by division illustrates that the removal of an officer was guided more by the individual commander's idiosyncratic vision of "good and bad" leadership rather than any objective standard. Although the 37th Division had the dubious distinction of sending the largest number of infantry officers for reclassification, its mobilization and training differed little from that of units that sent far fewer officers to Blois. As a point of comparison, the 82nd Division, which both arrived earlier and saw more combat than the 37th Division, sent only three officers to Blois during its time in France. The sad fact was that some divisions and regiments simply "ate their young" at a greater rate than others.

The long shadow of Blois and Pershing's intolerance of failure created an atmosphere of fear within the AEF. Some commanders certainly sacrificed officers to reclassification to place the blame for their unit's failures on others, or at least show that they were being proactive in correcting any of their unit's shortcomings. It is not surprising that the number of officers relieved from the 37th, 79th, and 92nd Divisions spiked after their lackluster performance in the Meuse-Argonne.

Certain division or regimental commanders sought to show their toughness to their superiors while also dealing with "problem" officers in their ranks. The 36th Division appears to have had an overabundance of these cases. One of the division's officers recalled that soon after arriving in France, the division's staff was reorganized and a general resifting of the unit's senior officers occurred. He noted that the first act of the new staff was to go after with a vengeance those officers within the division that it deemed unfit for service. It made a great impression on him that in a matter of days the commander of the 71st Infantry Brigade, Brig. Gen. Henry Hutchings, two colonels, two lieutenant colonels, five majors, and a number of captains and lieutenants were relieved of duty and sent to Blois. As the division got settled in, more removals followed. One of the purged officers, Lt. Mancel Coghlan, claimed that he was sent to Blois merely to fill his battalion's quota for reclassifications. He maintained that his division commander, Maj. Gen. William R. Smith, "made a statement he was going to have a board, and if it were necessary to have a man before the board, the officer would go back to the States whether or not he was inefficient."[22]

Officers from other divisions also claimed to have been scapegoats for the failings of superior officers who were later relieved themselves. The 140th Infantry's Capt. Henry E. Lewis noted that the colonel who sent him to Blois had since been relieved in the short time that it took him to report to the depot. While serving as a battalion commander in the 308th Infantry, Capt. Charles H. Harrington refused to attack the Château du Diable on 6 September 1918 because he believed the assault "could not be carried out because of the demoralized condition of the men" in his unit. Harrington argued that he stopped his scheduled attack when a promised artillery barrage failed to materialize. The board believed that Harrington had been wrongly removed for the incident and subsequently returned him to a combat assignment with the 77th Division.[23]

Many, if not most, of the officers sent to Blois probably deserved their fate, but they were merely the worst, or the most unlucky, of a corps of junior leaders whose training and experience had not adequately prepared them for the tactical, technical, and leadership challenges they faced. However, in seeking perfection in the leadership of the AEF, Pershing had also encouraged the creation of a command climate based on the fear of being relieved of command. This climate permeated units from their very arrival under his command. One artillery officer reported that a steady winnowing of new officers began soon after his unit landed in France in June 1918. In a very short amount of time he witnessed "two Lieutenants now paymasters. One Captain [now a] mess officer for life. One Major simply relieved from duty. Lots of others will see Front only via movies." He noted that officers were being subjected to "efficiency exams daily" that were resulting in "heads falling."[24] While the climate that this officer observed pushed commanders to accomplish their missions and demand results from their subordinates, it also encouraged them to micromanage their units, reduce the initiative of their subordinates, stifle the development of their junior leaders, and heedlessly push attacks after it was clear that such efforts were not worth the cost of the gain.

194

8 Noncoms, Doughboys, and the Sam Brownes
The Relations between the Leader and the Led in the US Army

On 20 August 1917, General Pershing issued General Order 23 establishing the standards for the wearing of uniforms in the AEF. The order stipulated that "when in uniform outside of their own quarters all officers will wear the Sam Browne belt except when actually serving in the trenches."[1] In one fell swoop Pershing established a physical manifestation of the differences between officers and enlisted men in France. The belt was merely another brick in the regulatory and customary wall that created and maintained the gulf between the leader and the led in the US Army. The fact that the term "Sam Brownes" became an enlisted man's slang for the officers in the AEF shows that the common doughboys and their NCOs were well aware of the importance of symbolism and the realities that governed their interaction with the commissioned leaders appointed over them.

When examining the relationship between the leader and the led in the AEF, it is important to note that much of the evidence from memoirs and other sources written after the war tends to suggest that officers, NCOs, and soldiers generally got along well. Between 1975 and 1983 the US Army Military History Institute received approximately fifty-three hundred responses to a survey it sent to veterans of World War I. Two of the questions on the survey related directly to the subject of leadership and asked the veteran to comment on the quality of leadership in the military as well as any examples they had of particularly good or bad leadership they had encountered. A review of the surveys indicates that a majority of the veterans believed that their leadership was good. For example, a 28th Division infantryman, Cpl. Alonzo LaVenture, stated that on the whole "our officers and noncoms were capable of doing [their] duties," but he also noted, "Once in a great time we would get a dud." Another veteran, Sgt.Maj. Mervyn F. Burke, maintained that "leadership was uniformly good, particularly in the 1st Div, as the majority (at least in 1917) had had good training."[2]

The doughboys' generally positive view of their leaders was also shown in the reports of base censors soon after the war. In December

1918 and January 1919 the AEF censor's office opened and read the homeward-bound mail of troops in the 78th, 79th, 80th, and 81st Divisions to gauge the soldiers' morale and concerns. In all of the units except the 81st Division, the soldiers were usually satisfied with their leaders. The examination of the 81st Division's mail, however, revealed that there was "a decided dissatisfaction over the officers at present assigned to them." The revelation about the morale in the division led the AEF GHQ to launch a more thorough investigation of the doughboys' complaints. A further examination of 8,485 letters, and interviews of 150 of the 81st Division's soldiers and officers, discovered that "the feeling of the enlisted men toward their officers is very satisfactory and all are working in harmony."[3]

Other veterans were less complimentary of their leaders and made it clear that the space of several years after the war had not diminished their resentment over the treatment they had received in the service. Howard Supple, an infantryman in the 35th Division, praised senior Regular Army officers but noted that "in the lower grades, with a few outstanding exceptions, [there was a] rather mediocre level of leadership." In a similar vein, the 1st Division's Pvt. Elmer Stovall believed that the "ninety day wonders" that came to his unit "lacked the leadership ability needed to inspire the respect of the men, especially the older career men." Pvt. Harry King of the 2nd Division's 23rd Infantry acidly observed that some of his officers were so bad that he wondered why they were leaders as opposed to members of a garbage detail.[4]

Even if the veterans who believed that their wartime leadership was good represented the majority view, the ones who presented a more jaundiced view of their leaders point to tensions in the relationship between the leaders and the led in the AEF. Many of the criticisms given by the latter group were also reflected in wartime records and the memoirs and diaries published by soldiers after the war. These sources show that tensions often sprang up in AEF units due to the failure of leaders to live up to their soldiers' expectations and the belief that senior officers did not understand the physical realities and hardships endured by their soldiers. Part of these problems could also be traced to the fact that the hierarchical nature of military service and the reality that "rank hath its privileges" was at odds with the larger society's political and social conceptions of egalitarianism. Even though the nation's traditional concept of egalitarianism was under siege by the ongoing consolidation of wealth and power into the hands of fewer and fewer people, the notion of social equality remained strong within the ranks of the white native doughboys.

Upon enlistment or conscription, the soldier became a ward of the

"I know we're fighting for democracy, but next time the Colonel comes round salute, you ——— son of a ———!"

This 1918 cartoon from noted British artist Bruce Bairnsfather highlights some of the cultural challenges American officers and NCOs faced in instilling military discipline in American doughboys.
(From Bruce Bairnsfather, Fragment from France, Part Six.)

state. In return for the man's military service, the army, as the state's agent, became responsible for his health and welfare. This meant that the society and the soldier expected the army to provide for the individual's food, clothing, shelter, and medical care. At the company level and below was the crucible where this exchange transpired. A unit's NCOs and junior officers were the leaders that the army held responsible for providing the vital goods and services their soldiers needed.

198 From the company commander to the squad leaders there existed a social contract between the leaders and the led. The soldiers expected their leaders to provide for their comfort, welfare, and heath, and in return the men more or less willingly agreed to follow the orders of their leaders and place their lives and labors at the disposal of their commanders. As an institution, the army understood this social contract and had long made paternalism, rather than iron-fisted coercion, its preferred method for regulating the relationships among officers, NCOs, and soldiers.

As has been shown in previous chapters, as well as official publications, regulations, and the semiofficial writings of Regular Army officers, the concept of paternalism was stressed to the officers commissioned during the war. It is clear that many officers understood and internalized the paternalistic social contract that governed their relations with their soldiers. Capt. Carroll Swan recalled that in his unit he treated his men like "a son or brother" and viewed his company as a "family." He maintained, "The Captain's responsibility is a great one. Every one of those two hundred and fifty boys look to him for everything. Their morals, their discipline, their training, their joys and sorrows, their health, their very life and death are in his hands." Less prosaically, Hervey Allen, an infantry officer in the 28th Division, simply declared, "The men expected to be fed, and they looked to the officers to feed them. To feed, clothe, equip, and pay the men—that is about all a line officer can do anyway. . . . Excuses make cold fare."[5]

One of the basic requirements of combat leadership is the ability of junior officers and NCOs to "do routine things, routinely" in a competent manner. To the soldiers of the AEF, the routine things they expected of their leaders were the officers' and NCOs' abilities to live up to their end of the social contract by consistently providing the basic goods and services the doughboys required. As George Marshall accurately noted, "War and training here is mud and rain and cold. The officer, platoon chief, who can keep his men's socks and shoes greased and dry and his horses groomed and the picket line above the flood of water and mud—he is the greatest contributor to our success in this war." The leaders who could deliver the goods not only tended to build cohesive units but also were able to draw upon a reservoir of good will with their soldiers when combat situations demanded hardships and privations. For example, the 165th Infantry's Sgt. Tom FitzSimmons was seen by Pvt. Albert Ettinger as "one of those natural leaders who men would follow anywhere." Ettinger claimed that the men respected and followed the sergeant because "they knew that he would never let them down" and because "he always made certain that his men had dry quarters, plenty to eat, and that their boots and uniforms were in good condition." When the 165th Infantry became mired in a brutal two-day attack to take Landres-et-Saint-Georges on 14 and 15 October 1918, FitzSimmons's Stokes mortar crews stuck by their guns and their sergeant despite frightful casualties.[6]

199

Unfortunately, not all of the AEF's leaders were up the challenge of "doing routine things, routinely." On 30 June 1918 the II Corps inspector, Lt. Col. K. T. Riggs, found that in the 30th Division the officers were not adequately caring for the welfare of their subordinates. An inspection of the 27th Division in July 1918 revealed that junior infantry officers were not "sufficiently instructed and zealous in providing for the health and comfort of [their] men, especially in matters of rations, bathing, and clothing."[7]

The inspections illustrated one of the great cracks in the army's system of paternalistic leadership. There was a great difference between leaders' understanding the importance of caring for their soldiers and actually knowing how to accomplish the routine things that ensured the delivery of the required goods and services. Very little of the training that officers received in the OTCs and COTSs was related to the arcane arts of feeding, supplying, and administering small units. The army's attempts to correct these shortfalls by expanding instruction in these areas in the winter of 1918 were overtaken by the dire shortage of officers in the following spring and summer. This problem was slightly less acute for National Guard officers. Whatever the flaws of the tactical training they received on the Mexican border in 1916, the deployment at least provided guard officers with much-needed hands-on experience in feeding, sheltering, and supplying soldiers in both garrison and field environments.[8] On the whole, however, most junior officers still had to rely on hard-earned experience to learn how to live up to their end of the social contract.

In *Cohesion: The Human Element in Combat,* military sociologist Darryl Henderson maintains that the most effective leaders were those officers and NCOs who were competent in the military skills required of their positions and who also had the ability to create "personal, empathetic, and continuing face-to-face contact with all soldiers in the unit."[9] The junior officers and NCOs played the vital role of establishing the norms of behavior that governed the unit's day-to-day operations.

Henderson maintains that the leader derives the authority and influence to establish the unit's norms and achieve the goals of the higher military organization in his unit by drawing upon and wielding the four sources of power available to him. These sources are reward and coercive power, legitimate power, referent power, and expert power. Reward and coercive power gives the leader the ability to build and direct group norms by giving the individual positive and negative incentives to conform to the unit's expected behavior. Rewards and punishment target the individual's self-esteem, sense of security, and acceptance within the framework of the unit as a whole, thus giving the leader a great source of authority while also reinforcing the unity and loyalty of the group. Legitimate power is derived from the culture, laws, and values of the larger society. It gives leaders the official and legal right to exercise the authority of their position. Referent power is the leader's ability to control others based on the respect and affection that the leader receives from the led. The referent leader has built within his unit an "intense identification" between himself and his soldiers based on his intimate knowledge of his subordinates, his proven ability to deal with difficult situations, and his willingness to share the hardships of his men. Expert power is given to the leader when he is "perceived as having superior knowledge and ability important to the soldier and his unit" that improves the group's effectiveness or survival.[10] Although Henderson argues that units led by officers using referent power tended to be the most cohesive, all of the sources of influence were critical to linking the goals of the unit to those of the larger organization. Henderson's thesis and leadership model were borne out in the experiences of the AEF.

Time and time again American soldiers praised those officers and NCOs who truly lived the paternalistic ethos and demonstrated by their actions a concern for their men that transcended that required by regulation and custom. Those officers able to draw upon this referent power through their personality and actions established deep bonds of mutual respect and affection that provided a tough psychological armor for their units and its individual members that aided them in coping with the strains of campaigning. This referent power required leaders to sacrifice some of the distance and privileges of rank to build face-to-face relationships. Charles Minder noted that during a particularly arduous march, "The Captain did other wonderful things for the fellows. I saw him give a couple of fellows some water from his own canteen. He walked his horse instead of riding him, and that is something that few officers would do." He also noted that the captain even carried a soldier's pack for him when the man seemed to be overtaken by heat and fatigue. In another example, Pvt. Ray Johnson remembered that after the 145th Infantry's machine gun company was marched to a desolate village behind the lines after being relieved from combat, the company commander discovered that his higher headquarters had failed to coordinate for billeting and other accommodations. The captain and his remaining lieutenant knew that they had "a bunch of men on their hands who were foot-sore, underfed, in low spirits, and on the verge of serious illness." After finding shelter for their soldiers, the two officers "took turn standing guard . . . so that every man could snatch some sleep. Their self-imposed duty lasted until the next morning."[11]

Regrettably, there was also a downside to leadership based upon referent power. The death of a respected and beloved leader, an event that happened quite frequently during the war, could cause a great slump in unit morale and effectiveness and create a great obstacle for the officer or NCO who replaced the lost man. For example, Pvt. L. V. Jacks recalled that a "deep gloom settled down" upon his battalion during the Meuse-Argonne fighting after the death of their battalion commander, Major Thompson. The private later wrote that the major "was the only officer who was universally liked and admired by the enlisted men, and his demise produced an instant and serious depression." Another soldier recalled that when a popular officer in the unit was killed, his "battalion lost heart."[12]

Soldiers also expected their leaders to share the privations, hardships, and dangers that they lived with on a daily basis. Leaders who endured these challenges built morale and cohesion by physically proving to the soldiers that their leaders had a concrete understanding of the conditions under which they labored and had a solid conception of the abilities and limitations of their men. The presence of leaders during times of stress and danger also linked them to their soldiers through common experiences and demonstrated to the men that their leaders could completely perform their duties under the worst of conditions. Those officers and NCOs who did so strengthened their referent power within their units. Cpl. Chester Baker was pleased to note that during his unit's first stint in the trenches "I noticed that the dugout assigned to Lieutenant Thompson remained empty. During the entire engagement that was to come, I never saw

201

him take advantage of its greater safety; he stayed in the trenches with his men." The 26th Division's Lt. Walter O'Donohue also won the affections of his men by not standing on rank and always sharing the discomforts and hazards of the long marches and battles with his men. This affection and loyalty later paid the officer a great dividend when the dugout he was in collapsed during a heavy enemy bombardment. One of his soldiers saved O'Donohue's life by braving the German shelling to dig the officer out of the collapsed shelter.[13]

202 Conversely, those leaders who failed to share the privations of their soldiers or to build the close face-to-face relationships touted by Henderson often faced a rocky road when they tried to command their units. One of Thomas Barber's fellow pioneer infantry officers undermined the discipline of his company by never allowing his men to stop their road repair work when they came under enemy shellfire, while the officers always took shelter themselves. The 42nd Division's Charles MacArthur was incensed when his officers ordered him and his comrades to work through the night while they went off to sleep in a comfortable dugout.[14]

When faced with their leaders' failure to abide by the unspoken code of shared hardship, the soldiers often refused to be bound by the leaders' orders, worked to undermine their authority, or rejoiced in (and sometimes abetted) the leaders' downfall. In the case of the pioneer infantry company mentioned above, the soldiers in the unit simply waited for their officers to disappear into their bomb shelters before they stopped working and sought cover themselves.[15]

In another instance a company commander in the 26th Division became angry that some of his men had fallen out of a cold and dreary march and were riding in the unit's wagons. The captain, who was riding a horse, ordered his first sergeant to use a pistol if necessary to keep the men out of the wagons and moving forward. When the top sergeant tried to tell the officer that the men in the wagons were too sick and weary to march, the officer abruptly cut him off and informed the NCO that he would hold him personally responsible for seeing that his orders were carried out. When the company reached its destination, the captain found that the first sergeant and most of the company's NCOs had left the formation and headed for the town of Toul in protest of what they saw as the officer's high-handed actions. Similarly, Sgt. Elmer Straub noted that some men grew so tired of the attitudes and actions of their officers that they staged mini-mutinies against their leaders. After repeatedly being berated by his horse-mounted lieutenant for moving too slowly on a march, one soldier in Straub's unit threw his equipment into the mud, refused to move forward, and informed the imperious officer that "he wasn't a pack horse."[16]

The AEF's enlisted men enjoyed seeing a hated officer or NCO brought low for their transgressions against the social contract or the code of shared hardships. William Clarke recalled that his company commander "looked and acted the aristocrat," and "was aloof and not easily approachable." He was taken aback by the officer's unwillingness either to mix with his soldiers or to get his hands dirty while training in the United States. The captain's actions led Clarke to conclude, "How or why he was ever cast in the role of a commanding officer of a company of combat troops, I don't know." The officer got his comeuppance during his unit's passage to France. When word reached the captain of a possible submarine attack on the troopship, the officer rushed on deck in a panic wearing an inflatable lifesaving suit that was not issued to his soldiers. There he was met with "a great woop and holler of scorn and derision" by his unforgiving troops for his ridiculous and frightened appearance. Clarke noted that the officer's "usefulness to his men had ended on the deck" of the ship, and from then on he remained a "Captain in name only."[17] All of these incidents illustrate that despite Pershing's efforts to instill the unquestioning discipline of West Point into the AEF, at the lowest levels the concepts of obedience and discipline were governed by ongoing negotiations between the leader and the led based on concepts of reciprocity and expectations of shared hardships.

Although army regulations and military and civil law gave all officers and NCOs the ability to use what Darryl Henderson termed "legitimate" and "reward and coercive" powers, in the AEF these were both rather brittle reeds on which its leaders could base their authority. On paper, military law, especially for offenses conducted in wartime, was draconian and allowed for the use of imprisonment with hard labor or the death penalty for a vast array of offenses. Yet while the raw number of general and summary courts-martial increased between 1915 and 1918, this increase was not proportional to the massive expansion of the army. The actual percentage of soldiers being brought before the courts actually plummeted during the war. In 1916, for example, over 4.7 percent of the army's enlisted men were tried by general court-martial. In 1918 the percentage of soldiers tried by general military courts dropped to 0.87 percent. During the war the AEF executed only ten soldiers. All of the men executed were charged with the rape or murder of civilians rather than offenses directly related to military service, such as desertion or misconduct in the face of the enemy.[18]

203

A number of factors tended to limit the coercive power that leaders could wield over their soldiers. Under the press of events in 1918, the AEF GHQ tried to limit the numbers of military trials and avoid removing soldiers from the ranks for lengthy trials or discharges. On 13 April 1918, GHQ strictly limited sentences that removed soldiers from France by dishonorable discharge or by imprisonment at the disciplinary barracks at Fort Leavenworth. Its order also urged commanders to use a court-martial only as a last resort. The following month, the AEF GHQ issued General Order 78, allowing division commanders to establish "provisional disciplinary detachments" within their battalions as a means for maintaining order without losing time and combat power to trials or other punishments.[19]

The guidance from higher headquarters was clear: Pershing and his senior staff did not want time and effort wasted on a lot of military trials. However, by so ordering, the GHQ also severely restricted the punishments its subordinate commanders could use to discipline reprobates or set examples that might deter misconduct. Although the sentences of summary courts and nonjudicial commanders' boards could be onerous to the soldiers, few seemed to be worried by the threat of these punishments. The easy come–easy go nature of promoting NCOs, and the lack of privileges associated with those ranks, meant that the threat of demotion was hollow. Even worse, some officers found that they had to reappoint busted NCOs quickly because no other men in the unit were qualified to hold the positions.[20]

Another factor limiting the effectiveness of punitive measures as a tool to reinforce authority was the general lack of knowledge of military law by many of the new officers. When the army moved to the COTS system for commissioning officers, the War Department specified that the candidates would receive thirteen hours of instruction on military law and the *Manual for Courts-Martial*. Although this move doubled the amount of instruction given on these subjects in the previous OTCs, when the COTS classes were condensed to make up for officer shortages, military law was one of the classes that was curtailed. Thus, new platoon, company, and battalion commanders often had only a most basic and sketchy concept of the coercive powers at their disposal.

The new officers' lack of knowledge of military law also led to several occasions in which the leaders exceeded the punishments allowed by army regulations and the *Manual for Courts-Martial*. Capt. Edward Tanner's commander sent him before a reclassification board in August 1918 for ordering his NCOs to beat or whip five men he believed were drunk, ill-disciplined, and insubordinate. At his hearing the captain maintained that he was not "going to hand these cases over to a summary court" and had "wanted [the NCOs] to tend to these things themselves" to build a spirit of responsibility in his subordinate leaders.[21] It was clear that Tanner did not believe that his actions were either wrong or out of his purview as a company commander.

In another case in which leaders exceeded their authority, the ending was much more tragic. After Pvt. Julius VanCamp was found guilty of insubordination and sentenced to extra duty by his company commander, he refused to work or obey the orders of the sentry placed over him. When informed of VanCamp's refusal, the company commander and first sergeant assigned a corporal to watch over him and "make the man work." The corporal interpreted his orders literally, and when VanCamp again refused to obey, he began prodding him with a bayonet. When VanCamp turned on the corporal in anger, the startled NCO mortally wounded him with a stab to the groin. The regimental commander ultimately preferred charges against the corporal and ordered an investigation of the company commander.[22] In both the Tanner and the VanCamp cases, the junior leaders believed that they were well within their rights to impose physical punishments on soldiers for the good order and discipline of their units. In an army in which senior leaders were fearful of relief and subordinate leaders were poorly instructed on the powers and limitations of military law, it was natural for senior officers to restrain their subordinates' punishments and use of military courts.

The soldiers were quick to pick up on the limits of their superiors' coercive power and sometimes were more than willing to call the leader's bluff. Cpl. Paul Maxwell was disgusted when his commander was replaced by "a martinet with a sadistic complex." He recalled that from the new officer's first appearance, the officer undermined the morale of the unit. Maxwell later wrote,

> Introducing himself to the Battery he denounced us as a bunch of spoiled sissys and said he hated our guts but he would convert us into a tough fighting outfit or kill us in the attempt. . . . His favorite maneuver was Summary Court Martial. . . . To some unfortunate individual who made a trivial mistake that meant three days in the Guard House with suspension of pay. My mistake was, at inspection of quarters he spied a small piece of lint between my bunk and Corporal Jones.' He held us both responsible, and gave us the choice of Company Punishment or Courts Martial. Jones went down first and chose Company Punishment.

When I got the same proposition I chose Courts Martial. The Captain jumped to his feet, shook his fist in my face and called me a damned lousy soldier, who would voluntarily besmirch his record with a Courts Martial.[23]

As the general court-martial that Maxwell demanded would have brought the commander's actions under the scrutiny of his superiors, the flustered captain angrily dismissed the charges against Maxwell.

The longer that soldiers served, the less willing they became to accept leadership based on coercion. This was more so in the ranks of combat veterans. Pvt. John Barkley noted that it was tough to be a replacement officer assigned to a veteran unit. The officers themselves sometime exacerbated this problem by their words and actions. He recalled that during the Argonne drive, "The new officers began making themselves unpopular as soon as they arrived. They were replacements. Not a front-line officer in the lot. The men were all too desperate to be bothered with forms and they weren't very respectful." After one of the new arrivals yelled, "What's a matter with this goddam hard-boiled outfit?" and accused the men of acting like babies, the veterans informed him in no uncertain terms that they weren't "in the mood for parade ground stuff." Although the officer threatened to have the men arrested, the soldiers' hoots of derision drove the man to realize the weakness of his hand and beat a hasty retreat. The threat, "I'll put you under arrest," became a running joke among the veterans.[24]

Charles MacArthur recounted a similar event in his unit. Soon after they were pulled from the Meuse-Argonne attack for rest and refitting, a new and unpopular officer demanded that the men make an extra effort to clean up their surroundings in a badly damaged French village. MacArthur jokingly recalled, "All this was woman's work, and we told the lieutenant so. We told him to go away and not do that any more. At first we were very polite, and it wasn't until Lieutenant Wegner got cross and bad-tempered that we were the least bit severe with him. Lieutenants are like children that way. They begin by talking back. The next thing you know they are out of hand and have to be sent to military school." Likewise, during the Saint-Mihiel Offensive, Leslie Langille and his comrades grew so tired that they preferred to be left alone to sleep through German barrages rather than being rousted out of their slumber by their officers and forced to take shelter. When the men refused to move, Langille recalled, "the officers threatened to court-martial the whole outfit, but a court-martial would be a welcome relief to the mud and rain and shells, so nobody pays any attention to their threats and goes on sleeping." The men's insubordinate attitudes were not helped by the fact that their company first sergeant had taken up residence in a large dugout and even insisted that his meals be brought to him there. The experiences of Maxwell, Barkley, MacArthur, and Langille all point to the fact that the relationship between the leader and the led in the US Army was far from being a one-way street. Leadership was based on a system of ongoing negotiations among officers, NCOs, and their soldiers in which the character, ability, and willingness of the leader to share his men's hardship counted much more than army regulations or the president's signature on a commission.[25]

207

The ability of AEF junior leaders to base their authority on expert power was also limited. As has been discussed in previous chapters, the degree of training and experience between junior leaders and their soldiers was rather slim. Shortly after the war, Raymond B. Fosdick, the director of the Commission on Training Camp Activities and the War Department's special consultant on troop morale, conducted an inspection tour of the AEF to gauge the overall morale of the army in France. Fosdick reported to the secretary of war:

> In our army both officers and men are drawn from a common economic and social reservoir. There are plenty of men of superior education and high mental and moral qualities in the ranks of the A.E.F. Conversely, there are plenty of commissioned and non-commissioned officers who have none of these attributes. I do not believe that an army was ever recruited in which the common soldier possessed such a high average and social experience as in the American Army of 1917 and 1919. By the same reasoning it must be admitted that in no army have the officers been superior to their men by so small a margin.[26]

Fosdick's rather stinging indictment of leadership was also shared by some of the officers themselves. Maj. Robert G. Calder wrote, "In this war our men in the ranks have been superior to our officers, that is as soldiers they were better than the officers were as leaders." The AEF's soldiers were clearly "thinking bayonets" that quickly came to realize the personal cost of their officers' lack of tactical knowledge or skill. As one captain confessed, "It is useless to try to fool the American enlisted man: he soon loses respect for his officers when he observes their lack of experience, gained through the school of hard knocks."[27]

The enlisted men also expected that their leaders display coolness and courage in combat. This expectation was grounded in long-standing American concepts of manliness as well the more recent ideas that war and the strenuous life were the ultimate test of character and moral fiber. If an officer or NCO proved himself brave and unflappable under fire, the American soldiers seemed willing to overlook many of the man's minor shortcomings. One 27th Division machine gunner praised his company officers because "they were with the men at times, and their quick decisions, involving as they did all our lives, were such as to steady the men and give them confidence in the success of the operations." Albert Ettinger's faith in Sgt. Tom Fitz-Simmons was based on the fact that the NCO was also cool under fire, led men in combat from the front, and "was courageous, intelligent, considerate of his men, and lucky—an unbeatable combination."[28]

Leaders who failed the test of courage quickly lost the respect and obedience of their soldiers and became the objects of scorn and ridicule. After all his officers had been wounded and evacuated in the Argonne fight, Horace Baker's battalion commander sent a new lieutenant to take command of the company. Unfortunately, Baker recalled, "this worthy stayed in the dugout the two days he was with us . . . and I never saw him." By hiding out in the bunker, the hapless officer lost all moral suasion with the soldiers. The men simply acted as if he did not exist and went to their remaining NCOs for orders and guidance. A soldier in the 42nd Division recalled that two of the soldiers in his company got back at an unpopular and somewhat nervous lieutenant during the Meuse-Argonne Campaign by firing their .45 pistols over the dugout where the officer had hidden himself and yelling loudly in "bum German" to increase the man's distress.[29]

Of course there was a downside to the soldiers' expectations of steadfast bravery in their leaders. Keeping up the outward appearance of daring and courageousness was a great psychological burden to many officers. As earlier noted, during his first patrol into no-man's-land, Lt. W. A. Sirmon admitted that he was "badly frightened." He later wrote, "I was shaking badly, but swearing to myself I would not run." Others were pushed to recklessly expose themselves to danger in order to "prove" their merit to their soldiers and peers. As his company was entering its first stint in the trenches, a fellow officer noted that Lt. Wendell Westover did not look well. Although the regiment's doctor had diagnosed the officer as unfit for duty due to a high fever and possible tonsillitis, the officer refused to be hospitalized. Westover replied that his refusal was because "if I fail to go in the first night the platoon will think I'm yellow." The desire for officers to prove themselves in combat led Capt. John Stringfellow's regimental commander to issue an exasperated directive to the unit's officers stating, "I have had a number of requests from you for permission to lead squads in raids upon the enemy's trench. It is properly the duty of a corporal. If you still desire to do this, I will remove your shoulder bars and place corporal's chevrons upon your sleeves and put your bars on the corporal." These expectations of bravery and the need to prove oneself later contributed to the AEF's high officer losses in combat.[30]

One of the other expectations that AEF soldiers had of their officers was their ability to balance the needs of the mission with the needs of their men. The army and its senior commanders desired that their subordinates achieve their missions and contribute to the success of the overall operation. The individual soldier's desire, however, is to survive the mission while achieving the greatest possible physical comfort in terms of safety, food, clothing, and shelter. While the former's view is shaped by calculations of strategy and tactics, the latter's view is governed by Abraham Maslow's hierarchy of needs. The junior leader was the one who is caught between these two often diametrically opposed realities and forced to reconcile the needs of the individual with the demands of the larger organization. This balancing act required the leader to understand the ever-changing limitations on military actions imposed by the physical and mental needs of the soldiers and also his unit's role within the mission of the higher organization.

A number of the AEF's junior leaders seemed to have understood this need to balance mission and men. Some also recognized the fact that this effort to establish a workable equilibrium was often a thankless job that pleased neither party. Lt. Wendell Westover, of the 2nd Division, recalled that lieutenants held the key position in the army between the individual soldier and the impersonal calculations of the rest of the chain of command. He neatly summarized the psychological strain that this balancing act placed on these key leaders by noting that to an officer fell "the direct and all important task of leadership, of understanding, of living the human and inhuman days with the Men who occupy the ground and gain the decision in battle. . . . To him first comes responsibility—responsibility for other men's lives. His the closest contact; his the greatest grief. The Lieutenant is taught that it is necessary to sacrifice men in the attainment of a battlefield objective—then asked to lead those men into battle." These strains were also felt at the lowest levels of leadership. Cpl. Charles Minder of the 306th Machine Gun Battalion mused in a letter home that "ever since I was drafted into the army, I never had any use whatsoever for Sergeants," but as soon as he became an NCO he had come to understand the pressure they were under. He also lamented that "the disadvantage of being a Corporal" was that "you've always got to act as if you side with the officers, otherwise you would have no control over the men at all."[31]

The pressures and stresses felt by leaders such as Minder and Westover were increased when orders by superior officers or staff members seemed detached from the realities of life in small units. Lt. Hervey Allen perceptively noted "one of the hardest things for an officer to do is to enforce a stupid order when the men are intelligent enough to know better. This is where 'discipline' generally and fortunately breaks down in the American army." In his farewell address to his company, Capt. B. A. Colonna recalled, "The Co. Cmdr. is the one man who can't pass the buck on responsibility. We had to take the bushels of orders we received, eliminate those utterly impossible, select from those remaining what seemed essential . . . and then get those things done by the company. . . . And then one usually amasses a balling out for something or other that he has left out." These officers all realized that when units went from being colored flags on a map to thinking, flesh-and-blood soldiers, executing orders and missions became a much more complicated undertaking than senior officers could calculate.[32]

Those officers who were able to find the proper equilibrium between mission and men were able to reap the rewards of referent power among their soldiers while avoiding the ire of their superiors. Chester Baker recalled that his company commander, Lieutenant Thompson, had gained the respect of his men by sharing their privations and issuing his orders more as "polite requests" than barking directives. However, it was the officer's behavior in a single incident that won Baker's undying affection and loyalty. During a stint in the trenches, Thompson ordered Baker to deliver a message to the battalion command post through heavy shellfire. Upon arriving at the battalion headquarters, the major ordered Baker to deliver a reply to Thompson and then to return back through heavy shellfire to the battalion command post. After returning to Thompson and informing the lieutenant of the major's directive to return to headquarters, Thompson told Baker, "Forget it, Corporal. . . . If he wants another report on conditions here, he can come for it himself." Baker later recalled, "I thanked God that Lieutenant Thompson had his men's welfare more at heart than the Major did."[33] In this case Thompson believed that the needs of Baker simply outweighed the somewhat lesser needs of his battalion commander.

By establishing a close personal relationship with their soldiers, some officers also gained the insight to know when they should crack down on discipline and when they were best served by allowing certain transgressions by their men to go unpunished. Lt. Jeremiah Evarts discovered that sometimes good leadership entailed serving as a heat shield to protect subordinates from the wrath of their higher commanders. While his company was occupying trenches near Cantigny in May 1918, one of Evarts's soldiers had a dud artillery shell land between his legs, and the man "only missed being castrated by about ten inches." After this close call the soldier went AWOL and was found the next evening holed up in cave, drunk and threatening to shoot anyone who tried to roust him. Evarts managed to convince the man to drop his weapon and talked him out of his hole. When Evarts's company commander wanted to have the man brought up on serious charges, the lieutenant convinced the captain that the man "was too good a soldier to lose or to break his spirit" and gave the wayward soldier only company punishment for his transgressions. In the end, Evarts's ability to shield the soldier paid off, and the man later did well in combat.[34]

211

A number of sources point to the fact that despite the efforts of many officers to live the paternalistic ethos of leadership and establish strong bonds with their soldiers through face-to-face relationships and shared danger and hardships, junior leaders often fell short of these lofty goals. In his report to the secretary of war on the morale of soldiers in the AEF, Raymond Fosdick noted that far too many officers tended to act in a manner considered "galling to the democratic spirit of the troops." Although the paternalistic ethos demanded that officers place the welfare and comfort of their soldiers above their own, he discovered that "the possession of a Sam Browne belt in the A.E.F. has carried with it advantages out of all proportion to disciplinary requirements or the needs of the occasion, and officers have been allowed and encouraged to claim and even monopolize such advantages in ways that have shown a total lack of the spirit of fair play." Fosdick noted that while it was easy for officers to obtain passes to local towns or to see the sights, it was nearly impossible for enlisted men to gain the same privilege. He even observed incidents in which officers had ordered enlisted men to give up their seats at the Y.M.C.A. or other morale-building performances. He argued that to the average American soldier, "these privileges suggest a caste system which has no sanction in America and against which they instinctively rebel."[35]

Fosdick's accusations were echoed by a number of doughboys. The 1st Division's Sgt. Charles Strikell pointed out, "One thing not understood by the enlisted man was the great gulf that existed between a commissioned officer and an enlisted man. . . . The enlisted man could never understand why an officer should have better food, more leave, better quarters than he did. He could not understand why the officer was always the boss when often he did not know what he was talking about." Even some officers shared Strikell's critique. In a survey of officers awaiting demobilization, a number of them expressed regret at their own, or their peers,' failure to better safeguard the welfare and just treatment of their soldiers. An infantry officer shamefully admitted the "tendency of officers to always consider their own comforts and pleasure rather than that of their men." Another denounced officers who were "able to pull stuff . . . that the same officers would court martial an enlisted man for."[36]

From their first step onto a troopship going to France to their last step off of the ship returning them home, the actions of the AEF's officers drove home to the enlisted men the great gulf that existed between the ranks. Onboard ship, the officers enjoyed sleeping in cabins and eating at their own well-supplied mess. On 10 April 1918, Lieutenant Harris wrote to his parents, "The trip over was wonderful. . . . No trip could be more enjoyable. The scotch was plentiful and the food excellent. . . . the damnedest pleasure trip I ever took!" Things weren't quite so nice in the soldiers' berths below deck. One 42nd Division doughboy recalled that soldiers were "crowded like horses into narrow bunks, with the plainest of food, in total darkness at night, denied even the solace of a cigarette except by daylight." Food lines were long, washing was impossible, and restrictions often prevented men from getting any fresh air topside.[37]

Although there was little that most officers on board could do about conditions on the troopships, that was scant consolation to the soldiers who saw how the other half lived. One officer sadly recorded that "the officers were well fed and in a civilized fashion in the cabin, which didn't help matters much for the men."[38]

The inequalities of enlisted life were far from over when the soldiers reached France. To maintain the social distance between officers and their men, the AEF followed the European practice of messing and billeting officers separately from the troops while in training camps or garrisons. When occupying a new town, an officer from the regiment or battalion preceded his unit to arrange with the local French authorities for quartering the unit's officers and men. This officer also ensured that the lodging for his peers and superiors was the best that the town could offer. While officers were quartered in inns or local residences, the soldiers were billeted in schools, churches, or, more often, the haylofts of local barns. Hervey Allen rather shamefacedly admitted that while his soldiers made the best of sharing shelter with the locals' farm animals, he and his fellow officers slept in a "real bed, a deep feather bed, in an old peasant's house." The officers' quartering arrangements frequently included meals with the host family, while the men subsisted on the pedestrian bounty of the company kitchen or whatever delicacies they could buy from the townspeople. However, one second lieutenant acknowledged that given these enforced distinctions, "officers do not live close enough to their men and therefore do not learn the personality of each."[39]

As the American enlisted men came into contact with soldiers in the other Allied armies, some became painfully aware of the AEF's class system. This especially applied to NCOs. While observing a British unit in the trenches, Cpl. Joseph Lawrence noted the higher degree of respect and authority that British NCOs were given in comparison to American sergeants. He saw that a British company sergeant major was not only granted more authority and privileges than an American first sergeant, but also "as a rule he has more influence with the men than do their immediate officers." Likewise, Sgt. Richard McBride observed, "The British sergeant is a highly respected individual and of course is accorded privileges beyond those of the soldiers in the ranks." He noted that unlike the situation in the American army, British NCOs were assigned quarters in billets rather than barns when in the rear area.[40]

Given the strength of the separate and unequal status between the ranks of the US Army, it is not surprising that some officers continued to assert their privileges even while at the front. Charles MacArthur recalled that one new officer "suffered from acute sensibilities and a perpetual desire to assert his rank" and that it "was rumored that he had caused himself to be tattooed with gold bars" in case he was ever found by his soldiers bare-chested. Pvt. Jonas E. Warrell noted that soon after he found a relatively comfortable place to spend the night after a hard day of fighting in the Vesle sector in August 1918, an officer commandeered his cozy quarters. During the Meuse-Argonne battle, Capt. Thomas Barber saw nothing wrong with making a "snug little home" in a safe and dry German dugout while his troops slept out in the open in pup tents. He justified his actions by noting that while the men all had shelter tents, he only had a blanket and poncho.[41]

Despite all these transgressions against the paternalistic code, an officer really had to step out of line before he was brought to heel by his peers or superiors. Capt. John Stringfellow wrote that after one of his fellow officers received a slight wound he appeared at the forward dressing station and demanded that the crowd of wounded enlisted men awaiting treatment make way for him because of his rank. The medical officer in charge of the station would have none of the man's assertion of privilege and promptly ordered the captain to wait his turn.[42]

It is hard to say how great the gulf was in the relationship between the AEF's officers and men, but these incidents all point to the existence of a vast gap between the rhetoric of paternalism and the realities lived by the army's enlisted men.

There were other indications that the American officer corps, especially among the ranks of the wartime OTC and COTS graduates, had failed to internalize the Regular Army's prewar precepts of paternalism. Wartime reports and postwar memoirs and diaries relate incidents in which junior leaders had abused their power or had used physical force when dealing with their enlisted men. In January 1919 the Morale Branch of the General Staff directed the morale officers at stateside posts to conduct surveys of enlisted men awaiting demobilization. The surveys were to gauge the soldiers' attitudes about their military experiences and lasting impressions of the service.

The soldiers interviewed at Camp Grant, Illinois stated that the things that most galled them in their service were the attitudes and unfair actions of their officers and NCOs. They stated, "The man who feels himself humiliated by swearing, punished for an unknown offense, or under the control of 'hard-boiled' non-commissioned officers is so engrossed with these ideas and resentful of them that it is almost impossible for him to go into training with enough spirit and attention for him to learn either well or rapidly."[43] As in Fosdick's report to the secretary of war, the soldiers in the stateside survey indicated that they chafed under their leaders' demands for absolute authority without any apparent military necessity and the tendency of officers and NCOs to use harsh and abusive language toward them.

The soldiers surveyed at Camp Devens, Massachusetts, held similar attitudes to those at Camp Grant. When asked of their opinions of the relationship between officers and men, most stated that the dictates of military discipline required that there be a strict separation between the ranks. They also acknowledged that discipline that might have been viewed by civilians as harsh and undemocratic was still necessary to keep the army from turning into a mob. However, a large number of respondents qualified their statements by noting that OTC officers tended to abuse their authority and take an unnecessarily strict and unbending approach to discipline. One soldier noted, "We must have discipline in the Army, but not like some of these 90-day lieutenants think." Another stated, "There is such a thing as carrying [discipline] to extremes, which I have noticed most of the National Army officers, who never had a man under them before in their lives, practiced."[44]

Some soldiers maintained that they had not given up their rights as citizens when they entered the service and were quick to denounce leaders who they believed had contravened army regulations. The records of the Morale Branch of the General Staff contain a number of letters written by doughboys to political, civic, and military leaders complaining of their treatment at the hands of officers and NCOs. For example, on 27 August 1918 a soldier assigned to the 153rd Depot Brigade at Camp Dix, New Jersey, sent an anonymous letter to the national headquarters of the American Red Cross complaining of the treatment that he and his comrades were receiving in the camp. The man claimed that the unit's leaders "talk to us like dogs," and when a group of the recruits spoke to the major about their concerns, they found that "he is just as bad as any one else here he curse us out about coming to him complaining about food."[45]

The groups who tended to be on the receiving end of much of the abuse by officers and NCOs were non–English-speaking immigrants and African Americans. Although historian Nancy Gentile Ford has noted that the army, as an institution, made great leaps during the war to accommodate the needs of the immigrant soldiers, and that the soldiers themselves were active agents in carving out their own space in the military, these efforts had done nothing to eliminate the prejudice and xenophobia of the average Anglo-Saxon native-born officer or NCO. The Morale Branch and its agents received a number of complaints against native-born junior leaders from immigrant soldiers. A group of immigrant recruits informed the Camp Devens morale officer that they were glad that "We have escaped the clutches of some non-commissioned officers who continually humiliated us, put us to work at hard labor and often assaulted and kicked us. The hardest and dirtiest work was performed by details of us non–English-speaking soldiers. We are a laboring party, instead of soldiers."[46]

On 20 July 1918, the Camp Devens intelligence and morale officer, Capt. Ernest Wood, reported to his superior on the General Staff on the problems between the camp's officers and NCOs and its large number of foreign soldiers. He believed that NCOs were not "treating all those of foreign birth with consideration" and that constantly calling them "foreigners" and "guineas" was turning a "patriotic and loyal soldier" into a disciplinary problem. His pleas had little effect, for nearly two months later he was still reporting that the camp's officers held negative attitudes toward "non–English-speaking selectives"

215

and often referred to these soldiers as "'Guineas,' 'wops,' 'squareheads,' etc." He also noticed a tendency for officers to use personal violence to get the immigrant soldiers "to perform promptly commands which they do not understand."[47]

The verbal and physical abuse of foreign soldiers was not limited to Camp Devens. The morale officer at Camp Gordon, Georgia, noted that several of his sources on the post reported that "trouble and ill feelings" were being created by the propensity of the post's native-born leaders calling foreign-born soldiers "various epithets such as 'wops,' 'dagoes,' etc." He recommended that the use of such terms be prohibited.[48]

The relationship between white officers and NCOs and African American soldiers was a bit more complex. On one hand, there were white officers such as Col. William Hayward, the commander of the 369th Infantry, and Col. James Moss, commander of the 367th Infantry, who consistently displayed a concern for the welfare and success of their black soldiers. In a May 1918 article in *Outlook* magazine, Moss wrote that when dealing with black soldiers, "Make the colored man feel that you have faith in him, and then, by sympathetic and conscientious training and instruction, help him fit himself in a military way to vindicate that faith, to 'make good.' Be strict with him, but treat him fairly and justly, making him realize that in your dealings with him he will always be given a square deal."[49] Although their paternalism had a twinge of condescension, both Moss and Hayward seemed to believe that they were contributing to the "uplift of the race." On the other hand were the officers who came to their assignments fully expecting their black soldiers to fail or believing that, given their soldiers' innate racial flaws, they had to be driven instead of led and controlled by draconian discipline. Maj. Gen. Robert Bullard stated that much of the problem of the 92nd Division stemmed from the prevalence of the latter type among its white senior staff officers and commanders. Bullard noted that, with exception of its commander, Charles Ballou, "not one of them believed that the 92nd Division would ever be worth anything as soldiers" and that the white regulars "would have given anything to be transferred to any other duty."[50]

The AEF's African American soldiers had the same expectations of their leaders as did white doughboys. Black soldiers expected the junior leaders, of whatever race, to demonstrate courage in battle and to care for their health, comfort, and welfare. However, the black doughboys also seem to have expected their leaders to work on their behalf to eliminate or minimize the inequalities that they faced due to their race in the use of their units, their access to recreation facilities, and their access to justice. When one of their soldiers was kicked by a white NCO for "talking back" while the 371st Infantry was assigned to stevedore duty at Saint-Nazaire, the unit's black officers confronted the post commander and demanded that their men be treated with the respect due all American soldiers. With this accomplished, "the men went back to work overjoyed to know at last that officers of their own color had intervened in their behalf." On a similar occasion, Maj. Arthur Little, a white battalion commander in the 369th Infantry, insisted that an MP officer discipline one of his white policemen for beating one of Little's black soldiers. The MP officer explained that the "niggers were feeling their oats a bit" and that he had been given instructions to "take it out of them as quickly, just as soon as they arrived, so as not to have any trouble later on." Little's refusal to accept this explanation, and his efforts to have the policy overturned, helped him to gain the respect and support of his soldiers.[51]

217

Like other American soldiers, the black doughboys were quick to criticize their leaders when they felt that the officers and NCOs had acted unfairly, violated army regulations, or failed to live up to their end of the social contract. In October 1918 a group of African American soldiers training at Camp Gordon wrote letters directly to Secretary of War Baker complaining of the treatment they received at the Georgia post. One soldier wrote that not only did they not get enough food, but also, "we are in tents [with] no flooring in winter clothing just thin blankets no wood to burn." He also noted that black soldiers were worked seven days a week and were never given passes to Atlanta. Another complained of inadequate food and medical care and the fact that the men had to perform hard labor while sick. On 3 November 1918 a black soldier training at Camp Grant, Illinois, reported similar abuse in a letter to his hometown pastor. The pastor in turn forwarded it to Emmett J. Scott, the War Department's special assistant to the secretary of war. The soldier echoed many of the same complaints as had the men at Camp Gordon and noted the propensity of white southern "cracker officers" to punctuate their commands with curses at the troops. The man noted, "They treat us as though we are dogs. We are cursed and dogged around just as though we are not human."[52]

The Camp Gordon soldier's complaint against white southern "cracker officers" highlights one of the greatest problems in the relationship between black soldiers and white officers and NCOs. Many senior War Department officials and officers generally accepted the assumption that white southerners made the best possible leaders for African American units because they had more experience in "bossing Negros" and supposedly had a better understanding of the psychology of black people.[53] Concern over the treatment of African American soldiers in a number of stateside camps led the Morale Branch to have its post representatives survey and question white leaders of "colored troops" about the state of race relations in their units and their attitudes toward, and condition of, their soldiers. The survey also addressed the question of the effectiveness of white southerners as the leaders of African American units.

The morale officer at Camp Pike, Arkansas, reported on 6 November 1918 that, contrary to some beliefs, white officers and NCOs did not need to have prior experience in "handling negroes" to be successful leaders. However, he warned that white leaders who felt themselves "especially capable of handling negroes because of previous experience among them in labor work, and who feel they can only be led by driving," should not be placed with black troops. Unfortunately, the morale officer's warning seldom seems to have been heeded. The African American journalist Monroe Mason described the white commander of the 371st Infantry, Col. Perry Miles, as "a southern gentleman with the well-known proclivities of the 'South-in-the-Saddle' in his veins," and the colonel of the 372nd Infantry as an officer with "a leaning toward southern prejudices." Mason argued that the leadership of white southern officers was irreparably tainted with prejudice and racial assumption of the unsuitability of black men to serve as officers or soldiers.[54]

Statements made by white southern officers tend to back Mason's assertion that they had little faith in their soldiers and consciously or unconsciously undermined the morale and effectiveness of their units. One white battalion commander noted that "the general attitude of white officers over negro troops is one of desire to educate and help the negro—an attitude almost of pity for his ignorance and hopelessness." One wonders why any officer would want to lead hopeless soldiers into combat? These attitudes even reached the highest levels of command in the AEF. Maj. Gen. Robert Bullard, an Alabaman with experience leading a black regiment, wrote, "Having passed a pleasant boyhood with the Negroes and had this satisfactory experience with them in my early military life, I found myself with most kindly feelings towards them." Yet for all of his "kindly feelings," Bullard admitted in his diary of the 92nd Division, "They are really inferior soldiers," and "Poor Negroes! They are hopelessly inferior."[55]

It seems that the actions and attitudes of white officers had to be particularly harsh against their black soldiers before the army took any action against them. In one instance, Capt. Timothy Mahoney was investigated by the AEF's Intelligence Section in May 1918 for comments he had made about his black soldiers. Mahoney was alleged to have stated before leaving the United States, "All I hope is that when they send me across I can line up this bunch of niggers of mine into some formation and that a German shell will get them and me." At least two white officers were sent to Blois for reclassification because of their "unsuitability to command colored troops." Capt. William Caldwell was relieved of command in the 92nd Division because he lacked the patience and "sympathy for colored troops" required of his assignment. The other officer, 1st Lt. Julius Rogovin, was sent to Blois because of his "natural dislike for colored troops." Rogovin believed that his relief stemmed from the fact that he was a southerner whose northern superiors "didn't agree with his mode of handling them." He stated, "My personal conviction after years of intermittent observations in various parts of [the] south" was that one had to "rule a negro with a firm hand." Between the overtly crude racism of Mahoney and Rogovin and the paternalistically clad and more subtle racism of Bullard, it was no wonder that African American officers and soldiers often had little faith in the social contract that supposedly linked the white leaders to the black led.[56]

Although immigrants and African American soldiers had special cause to doubt the commitment of their white native-born officers and NCOs to the army's high-minded concepts of leadership, why was this also a problem across the army? Fosdick blamed the failure of the AEF's leaders to live up to its paternalistic ideas on the inability of Regular Army officers to understand that citizen soldiers required different handling from that needed for the hardened men who filled the army's prewar ranks. He also maintained that the regulars had done too well at teaching the army's new officers that rank hath its privileges and the need to guard against overfamiliarity between the ranks and had done too poorly at instructing them on their duties and obligations to their soldiers. Of the OTCs, Fosdick complained, "These schools with their hasty training too often turned out officers with no well-developed sense of responsibility, officers to whom the Sam Browne belt and the epaulets were merely the badge of a superior social class, the symbols of rights and privileges jealously to be guarded even at the expense of the welfare and morale of the men of their commands." Given the actions of some of the AEF's officers, Fosdick's assessment seems to have had merit. The regulars' disdain for the National Guard, and their efforts to imbue the graduates of the OTCs and COTSs with their own Uptonian visions of "a proper military policy" for the nation, gave the new officers a skewed vision of the proper relationship between officers and their men. A number of OTC graduates recalled that they left officer training believing that the National Guard's concepts of leadership were fundamentally flawed. After graduating from OTC, Lt. Maury Maverick was assigned to the 157th Infantry, formerly the 1st Infantry Regiment of the Colorado National Guard. He recalled, "Many of us arrived with big ideas—that we would reform this 'militia regiment,' with our superior training." Col. Robert Morehead of the 139th Field Artillery noted that soon after his unit began training, he received sixteen new officers fresh from the first OTCs. He was also taken aback by the fact that "in the training camp they had been led to have an unfavorable impression of the National Guard and for a long time many of them were dissatisfied with their assignment to a National Guard regiment. This feeling on the part of Reserve Officers, I subsequently learned was very general throughout the army at the time." An infantry captain reported that "Regular Army officers as a general thing, openly belittle the National Guard" and worked to hold back guard units and officers. The success of regular officers in molding the graduates of the OTCs in their own image exacerbated or created tensions between the leaders and the led in AEF units.[57]

The history of conflict between regulars and guardsmen, and the various routes to a commission, led to a factional strife within many units that undermined morale and cohesion. In October 1917, Will Judy recorded that "I had not been in Camp Logan longer than forty-eight hours when I was knocked down by army caste. The regulars speak slightingly of the national guard and the guard calls the national army conscripts. Even within the regular army is a caste of castes, the West Point group; these conspire against their fellows who have come up from the ranks." At times, National Guard officers returned the alleged contempt of the OTC graduates with disdain of their own. When Lt. Hugh Thompson reported to the Rainbow Division's 168th Infantry, the unit's National Guard officers gave him and his fellow reservists a very chilly reception. His new commander told in no uncertain terms that his unit was formerly of the Iowa National Guard, and it had traditions that the new officers were expected to uphold. One of the guard officers loudly proclaimed that the new officers were so ill-trained that they "couldn't turn a platoon around in a ten-acre field," and Thompson's first impression of the battalion commander was that he seemed to like only two things: "cognac and cursing the helpless reservists."[58]

One of the greatest sources of factionalism in the AEF's officer corps was the differing views of discipline and the proper relationship between officers and their men held by guardsmen and regulars. Regulars often criticized National Guard officers for failing to understand that discipline required the maintenance of a sharp line of separation between officers and their soldiers. A Regular Army corps inspector informed officers of the 35th Division that they were "too familiar with their men" and sniffed, "This divisions bears all the ear marks of a National Guard Division." In a 6 November 1918 report on the conditions within the 26th Division, Colonel J. A. Baer stated, "This division has babied its men heretofore and as a result there has been continued complaining that the soldiers are tired and overworked. This does not come from the soldiers but from a few officers who have brought with them from the National Guard a paternal attitude toward the men, and the feeling that their men should be rested and spared further hard work." The regular officers often believed that what they saw as lax discipline in the National Guard units stemmed from too much fraternization between the ranks. This was a concept that they also drilled into the heads of their students at OTCs and COTSs.[59]

221

To drive home their version of proper officer–enlisted man relations, regular officers were quick to discipline officers who appeared to cross the boundary between paternalism and shared privations and fraternization. Second Lt. Homer Davis's commander sent him to Blois for reclassification because he was "lax in discipline" and "mixed" with his soldiers. Davis also drew his commander's ire for wishing to be billeted with his men rather than occupy separate quarters. Another officer, Arthur Campbell, was boarded because he "had a distinctive and bad effect on the discipline of enlisted men, not only for failing to maintain discipline but [also] for permitting and encouraging them to familiarity." The message of these actions was clear: "proper" officers built and maintained a strong wall of separation between themselves and their men.[60]

The regular officers' efforts to shape the opinions and leadership of OTC graduates seem to have borne fruit. Some of the new officers often commented on the negative effect of hometown ties and politics on the efficient operation of National Guard units. One newly commissioned officer noted that when he was assigned to his unit, there was "subtle antagonism between N[ational] G[uard] and us of the Reserve." After his National Guard superior recommended that he "go easy with the boys," he wrote, "He's a druggist back home [and was always] thinking of post-war business. That's the weakness of the National Guard. Too many personal relations." He later be-wailed "rotten Nat'l Guard discipline," which he believed was inevitable, "with men and officers so chummy."[61]

On 22 February 1918, the 31st Division's intelligence officer reported that the unit's officers were angered at the replacement of senior National Guard officers with regulars. The staff officer downplayed the problem and stated that the guardsmen would be more efficient if not assigned to units with men under their command "with whom they had been 'buddies' in civil life."[62]

Unfortunately, the regular officers failed to see that the guardsmen's approach to leadership and discipline was often much more effective in building cohesive units than those they advocated to their OTC protégés. Much of the available evidence points to the fact that the willingness of guard officers to mix with their soldiers, share in their hardships, and overlook minor lapses in discipline led to a degree of harmony and understanding between officers and men that was sometimes muted in Regular Army and National Army divisions. Capt. Colby McIntyre argued that Col. Frank Hume, the commander of the 103rd Infantry, possessed an understanding of the citizen soldier that was often lacking in Regular Army and OTC officers. He recalled that when Hume discovered a sergeant of his unit sneaking liquor into camp on Christmas Eve 1917, with a wink and a nudge the colonel told the NCO, "Well go ahead and enjoy yourselves. And tell the boys I wish th' hell I could be with 'em, but I can't."[63] Hume tried to drum into his officers that "a soldier was better all around" when he was given a bit of freedom and understanding. The National Guard colonel's personal ties to his men created a tight-knit unit able to withstand the battering the unit received at Soissons and the Meuse-Argonne.

Other National Guard officers also saw the benefit of building close personal ties with their soldiers. Capt. Ben Chastaine noted that there was much anger in the 36th Division when regulars or wartime reserve officers supplanted guard officers. He believed that the National Guard system helped to reinforce unit cohesion and combat effectiveness because the men and officers respected each other. Despite the regulars' fears, he argued that "discipline was not marred" by the fact that the guard's officers and men had known each other socially at home before the war. Another doughboy wrote that in his National Guard unit, "Instead of the recruit being frightened or timid in the presence of the officer and proverbial 'hard-boiled sergeants,' they were made to know that their troubles, questions, or what-not would be considered thoroughly by both noncoms and officers." The unit's company commander saw it as his mission to "make the Company one big family."[64]

Some National Guard officers maintained that what the regulars saw as the guard's greatest weaknesses, the unit's attachments to their home communities and prewar social ties between officers and men, were actually the guardsmen's greatest strengths. Maj. Emerson Taylor noted of National Guard units,

> In a peculiar sense the regimental commanders were looked upon by the thousands of good men and women whose boys were with the troops as the guardians and friends of those lads as well as their leaders in battle. In every case they were daily subjected to a very heavy and continual pressure, in the form of direct personal appeals, from their own intimate friends, from men of high position and influence, as well as from pathetic hundreds of anxious, proud fathers and mothers, "to look out for my boy," "to bring Joe home safe," "to see that he behaves himself," "to give Bill a chance," and so on.[65]

223

National Guard enlisted men understood that when the war was over, they would have to return to small towns or neighborhoods where their civilian standing would be influenced by their reputations as soldiers.

However, Emerson Taylor also noted that the benefit of close unit cohesion based on a regiment's special bond with the local community also placed a great strain on the guard commanders. He recalled that his regimental commander had to "assume a position of responsibility to the community which was the home of their respective regiments," and "he was expected to bring glory and renown to his home town." That these community leader-officers were expected by the folks at home to "bring Joe home safe" added another source of anxiety for guard commanders. Several men in Col. Frank Hume's 103rd Infantry stated that they were moved by the fact that when their commander "read each day's casualties report he would stand there and cry" because he realized that he now had to write letters home to his friends and neighbors informing them of the loss of their sons, fathers and husbands.[66]

While many American soldiers appreciated the technical and tactical expertise of Regular Army officers, a number of National Guard and National Army soldiers criticized their leadership and standoffish behavior. The junior leaders were also critical of their Regular Army superiors' abilities to motivate and lead citizen soldiers. Cpl. Edmund Grossman, for instance, believed that his National Guard commanders were much better than the West Pointers who replaced them. Unlike the guardsmen, he found regulars officers "snobbish and distant" and unable to inspire soldiers by their personality. One officer stated, "Regular officers [were] too slow to realize that they were dealing with a citizen army." Company officers accused their regular superiors of "Prussianism": a haughty and arrogant disregard for American soldiers and democratic ideas caused by the assumed superiority of the senior officer's worth, prestige, and position. One officer complained, "Regular officers failed in many cases to get the best work out of the new men, because they treated them like niggers. No man keeps his self-respect when bullied, ragged and brow-beaten."[67]

Another fallout from the regulars' approach to discipline and officer-enlisted relations was the unintentional attitudes and skewed perceptions of leadership that it often inculcated within its corps of junior OTC and COTS officers. There was no monopoly on poor leadership in the AEF. The ranks of National Guard, Regular Army, and National Army officers all had their share of martinets and incompetents. However, the OTCs appear to have produced more than their share of officers with a strict conception of discipline and obedience and inflated sense of their positions and prerogatives. As the curriculum, tone, and leadership of the OTCs were provided by the Regular Army, then regular officers must accept much of the blame for this problem. After the candidates were pumped up with Uptonian denouncements of lax National Guard discipline and released to lead units with only the sketchiest concepts of leadership, it is no wonder that the new officers sometimes failed to grasp the responsibilities of their positions or the customary and regulatory limits on their powers. Despite the regulars' precepts of paternalism, one OTC graduate went so far as to claim, "The officer in charge of the training company I was in at Plattsburg, said that soldiers should be treated more as dogs than men."[68] Although the candidate may have misunderstood what his Regular Army instructor was teaching, the fact that the statement stuck with the new officer is an indication that the camps' training of leadership was problematic and contradictory.

A number of training camp officers believed that the army's system of training and acculturating junior commissioned leaders was flawed and failed to prepare new officers for their role as combat leaders. These critics were often enlisted men who had been commissioned during the war. Their ability to view the army's leadership from both sides of the fence gave them unique insights into the problems of officer-enlisted relations. An infantry lieutenant observed, "[There is] too large a gulf between officers and enlisted men. I've been both

and know it is not necessary. It is feudal in tendency and undemocratic. It does not make for better discipline in most cases. Familiarity breeds contempt, but the chasm is too great at present. It should
not be possible for an officer to deal in personalities of a belittling and
inhuman kind." In the Morale Branch Officers' Survey, several officers commented that the army could overcome its leadership problems if it required officers to spend time in the ranks before being
commissioned. One officer noted that "this experience is necessary
to teach a young officer to understand his men." Another stated that
this practice would be beneficial because "the great source of friction
was ignorance of the men's conditions and attitudes" by their junior
officers.[69]

    The infantry lieutenant's condemnation of an officer with "a be-
littling and inhuman kind" of personality points to one of the major
shortcomings of the new officers. The historian and World War II
veteran Paul Fussell complained that one of the most irritating, soul-
crushing, and morale-sapping elements in the relationship between
the leader and the led during the Second World War was the over-
weening presence of what he termed "chickenshit." In defining and
describing chickenshit, Fussell notes,

> It does not imply complaint about the inevitable inconveniences
> of military life: overcrowding and lack of privacy, tedious insti-
> tutional cookery, depravation of personality, general boredom.
> Nothing much can be done about these things. Chickenshit
> refers rather to behavior that makes military life worse than it
> needs to be: petty harassment of the weak by the strong; open
> scrimmage for power and authority and prestige; sadism thinly
> disguised as necessary discipline; a constant "paying off of old
> scores"; insistence on the letter rather than the spirit of ordi-
> nances. Chickenshit is so called—instead of horse- or bull- or
> elephant shit—because it is small-minded and ignoble and takes
> the trivial seriously. Chickenshit can be recognized instantly be-
> cause it never has anything to do with winning the war.[70]

Chickenshit was a concept that would have been instantly recogniz-
able to the American veterans of the Great War. For the fathers and
older brothers of the "G.I. Generation," the source of chickenshit was
those young officers of "a belittling and inhuman kind" produced by
the officer training camps and schools.

    The waging of chickenshit was the source of frequent comment
by the war's doughboys and revealed the deep current of resentment
that some of the new officers intentionally or inadvertently built up

in their units. The following examples provide an indication of the scope and impact of chickenshit in the AEF. Pvt. Albert Ettinger recalled that his company of the 165th Infantry was unlucky enough to have assigned to it an OTC graduate who was "meaner than hell and disliked by most of the men." This second lieutenant "thought himself a combination of Caesar, Napoleon, and Hannibal" and refused to listen to the experienced NCOs of the unit. Soon after arriving in the unit, the officer kicked Ettinger when he found him sleeping after drill and humiliated him in front of the company. The unit's enlisted men grew so tired of the lieutenant's behavior that they constantly sought ways to bring him down and enjoyed playing him off against the company's other officers, who equally loathed the martinet.[71]

While billeted in a shell-torn French town close to the front, Cpl. Frank Faulkner of the 23rd Infantry used a lull in the fighting to wash and dry his mud-covered leggings. Leaving the leggings to dry, he went to the nearby company kitchen to get grease to oil his boots. The officer of the day spotted Faulkner near the kitchen and placed him under arrest for appearing in the cratered streets out of proper uniform. In Archibald Hart's company, one lieutenant developed a reputation for both meanness and pettiness. In front of one of soldier's visiting parents and girlfriend, the lieutenant ordered the man to give him his rifle for inspection. Since the man was on guard duty, he was not permitted to relinquish the arm to anyone. After the officer repeated the order and the soldier gave him the rifle, the lieutenant placed the man under arrest and marched him to the brig—thus eliminating the possibility of the soldier's receiving a pass to spend time with his family.[72]

The attitudes and posturing of young officers quickly alienated Pvt. Paul Maxwell from his first moments in the army. As his train arrived at Petersburg, Virginia, "Before anyone could move, a Second Lieutenant, fresh out of training school, entered each end of the coach barking orders a la Black Jack Pershing." Upon reaching Camp Lee, the group of recruits were met by a long-service regular NCO who, Maxwell noted, was much more courteous and professional and "not at all like the strutting, newly commissioned 2nd Lieutenants who were literally bursting at the seams with self importance."[73]

It did not take Maxwell long to become familiar with the inanity of army life and the advantages enjoyed by his officers. Along with twenty other recruits who arrived with him in camp, his first duty was to move a pile of lumber from one location to another location a few hundred yards away. It was pouring rain at the time, and they had yet to be issued any raincoats or ponchos and were thus "soon soaked to the skin." Maxwell could not fail to notice that the lieutenant supervising his rather wet detachment was "wearing a rubberized hat cover, long rain coat and rubber boots" and gave no thought to the men's discomfort. The young soldier generally praised the long-service Regular Army senior officers and NCOs he encountered. He stated that those notables "far outweighed the disdain I feel for the small minority of junior officers whose conceit and lack of ability should disqualify them for positions of Leadership."[74]

Soldiers resented the petty indignities that poor or thoughtless junior leaders could heap upon them. In one unit the company commander would curtly remind a soldier, "You're at attention," if the man eased his body after saluting and reporting to him. Furthermore,

> When the captain entered the lower hall and "*Attention!*" was not called, or was called in a weak voice, the man who failed in his duty was made to repeat "*Attention!*" until he could call it out and call it out loudly. For him who failed to leap to his feet at the same warning, there were a few jumping lessons until he acquired the agility of a jack-in-the-box. If a new recruit, after a week or two in the battery, was asked by the captain for his name, and answered, "Brown," all the officials from the lowest ranking corporal to the ranking lieutenant were brought to account for that man not knowing enough to affix his title "Private."

The other extreme of such pettifoggery was the tendency of officers to see their soldiers not as individuals but rather as an anonymous mass. Charles MacArthur chafed at the practice in which "Hey you" became the universal nickname given to soldiers by officers. Capt. Will Judy, of the 33rd Division, shrewdly observed that a soldier preferred "to be called by his name rather than private," because his name was "his last hold on individuality."[75]

Officers who stood on their dignity, demanded undue privileges, or enforced petty and trifling regulations left a lasting and negative impression on the soldiers they encountered. The actions and attitudes that Lucian Truscott witnessed in junior officers in World War I influenced his leadership even after he became a general in World War II. Shortly after being commissioned from the ranks in 1917, Truscott noted,

> Young officers were impressed with the concept that it was the responsibility of every officer to enforce all orders and to maintain the customs and traditions of the service. Some young officers seemed to regard this as almost a recreational activity. . . .

Woe betide the hapless recruit who passed one of these "ninety day wonders" without rendering the appropriate honors or with blouse unbuttoned or uniform otherwise awry. "Well soldier, where do you think you are going?" "Down town." "Don't you know you're supposed to say 'Sir' to an officer?" "Yessir." "Well, let's see you salute properly." Then repeated salutes and corrections until the officer was satisfied . . . and then the soldier would go on his way, his evening ruined. Then the young officer would seek another diversion. No doubt such incidents were repeated in every town adjacent to military camps. It is not surprising that so many men ended the war with a hatred of things military, for which they blamed the Regular Army.[76]

Luckily, a bit more experience with troops seems to have weaned most of the new officers from their attachment to the chickenshit that Truscott and the other soldiers named above had witnessed or endured.

It is impossible to know how prevalent the abuse of power and status was in the AEF's officer corps. The number of references made to it in the Morale Branch Officers' Survey and other accounts by participants indicate that it certainly happened more frequently than it needed to. The fact that a number of junior officers recognized and condemned the practice points to a probability that it was common enough to merit their comments. For example, one officer remorsefully noted that "there is considerable 'bullying' and 'lording it over'" enlisted men by officers "that accomplishes no useful purpose."[77] It should be hoped that Corporal Maxwell was right in his estimation that it was only a "small minority of junior officers" who caused problems for their soldiers. However, that minority of leaders who continued to revel in the assumed power and glory of their august positions remained a drain on small-unit leadership throughout the war and contributed to the litany of other problems faced by the AEF.

Sadly, the Regular Army's efforts to instill a great social distance between the ranks, and insistence upon absolute discipline and obedience by soldiers to officers, was so pervasive that even those officers who rejected the notions often found it difficult to challenge those institutional norms and expectations. The shared danger and conditions of combat tended to bring officers and men together. Unfortunately, the institutional expectations sometimes meant that this meeting of the minds was short-lived. One soldier recalled with sorrow that soon after his unit was pulled out of the Argonne fighting, his commanders returned to "tin-soldiering." He noted that "officers

This comic take on the officer–enlisted man relationship also reflects the doughboys' general distaste for officers who stood too much on their dignity and rank.
(*From* Stars and Stripes, *22 February 1918.*)

so lately snuggling by the most humble privates in shell holes, were once again bedecked in their Camp Gordon dignity."[78]

Aside from the issues of officer–enlisted man relations, the AEF also faced friction among its company, field, and flag-grade officers. One of the major criticisms of the senior officers by their captains and lieutenants was that the majors, colonels, and generals were out of touch with the new realities of war. Capt. Mark Clark, who later commanded the Fifth Army in World War II, noted that a number of old Regular Army officers were out of their depth when confronted with the realities of modern warfare. He recalled that one old major's professional horizons seemed to be limited solely to inspecting his soldiers' feet during marches and training to make sure that they were following proper procedures of podiatric care. Clark ruefully noted, "So many of these old-timers were just out of it when it came to the tactics of the type of war we had in France."[79]

Likewise, 2nd Lt. Herman Dacus recalled that shortly before the Saint-Mihiel Offensive, his regiment received a new colonel who had spent most of the war in the Philippines. The colonel tried to convince his veteran junior officers that "the best way to wipe out a machine gun was to crawl up on it a half mile across an open field." At the end of the first day's fighting at Saint-Mihiel, the same officer ordered Dacus's company commander to have his company dig complete trench lines instead of foxholes for its overnight stay, even though the unit was in reserve and not near the front lines. After the colonel left, the captain and the lieutenant "forgot" to issue the order.[80]

Junior officers were even more critical of staff officers. One soldier denounced the fact that "the staffs in the rear had no experience at the front" and "were bothersome in their ideas." An infantry officer in the 82nd Division was equally irritated at the interference of the staff officers of his higher headquarters. On 6 September 1918, he noted, "Worse than a German offensive is the drive of staff officers launched against us by someone higher up. I spent several hours listening to staff officers who have never had even an ambush patrol beyond our own wire, tell how an enemy machine gun nest should be cleared out." In a similar vein, Lt. Hugh Thompson wrote that while his unit was serving its tour in the frontline trenches of a French sector, a senior inspector from GHQ ordered another officer to fire the flare signaling the artillery to fire a protective barrage. The colonel wanted to test the artillery to see if they were on their toes and ready to fire. Although the young officer tried to explain to his superior that such tests were ordered via telephone, and that the flares were only for real emergencies, the colonel gave the officer a pointed lecture on insubordination and ordered the man to fire the flare. Thompson recalled that "the colonel had then demanded, during the resulting volcano" of fire, that the junior officer "have the barrage stopped and had been dumbfounded to learn that no one could halt the whiz-bangs once the signal for the barrage had been given." The colonel's rather silly actions unmasked the supporting artillery's gun positions and fire plans and forced much reorganization and movement the next day.[81]

Captains and lieutenants were also angered by discourteous or high-handed treatment at the hands of their superiors. The junior officers, who were in many cases well-educated business or professional men, were not used to being treated in a patronizing or curt manner. Dissatisfied with his brigade commander's lack of tact and leadership, one captain indignantly noted, "I appreciate the difference between disciplinary reprimand and a cursing out. The General isn't careful which he uses these days." As his company passed his car-borne regimental commander while marching up to the lines, Capt. B. A. Colonna was greeted by the colonel's impatient exhortations for his unit to "step out." Since the soldiers had been marching for hours on roads that were little more than patches of deep and glutinous mud while weighed down by full packs and ammunition loads, Colonna recalled that his men "showed our military discipline and Christian forbearance by not saying what we thought of this request." One officer complained that "the majority of work in the army is obtained by nagging at subordinate officers and by threats of various kinds," while another noted that "general officers [were] usually childish and autocratic in minor details."[82]

S. L. A. Marshall was correct when he noted, "The battlefield is cold. It is the lonesomest place which men may share together."[83] The ability of soldiers to overcome the isolation and inertia that this unique human environment creates is unit cohesion that is built, nurtured, and maintained by the organization's junior leaders. The leaders that are most effective in building this small-unit solidarity are those who are consistently competent in their duties and are able to a build relationship with their soldiers based on the close bonds of mutual understanding and respect. When examining combat leadership in the AEF, we find many obstacles that hindered the creation of this type of leadership. Many, if not most, of the army's officers, especially from the National Guard, were able to overcome these pitfalls to lead their units with grace, competence, and skill. Unfortunately, a number of other officers fell victim to the egotism of power and privilege and the Regular Army's strict vision of discipline, obedience, and the separation of the ranks. These officers proved themselves to be burdens to their soldiers and detriments to the effectiveness of their units. Furthermore, backbiting and mistrust among Regular Army, National Guard, and National Army officers, and also between the leaders and the led of all ranks, undermined cohesion and further damaged the fighting power and morale of the AEF by spreading discontentment and uncertainty within the army's units. When these leadership issues were added to the AEF's other problems in training, doctrine, and personnel policy, the army faced a rough and bloody road as it entered major combat operations in 1918.

9 Combat Physics and the Ugly Realities of Attritional Warfare

BEFORE EXAMINING THE COMBAT PERFORMANCE of the
AEF's junior leaders and the challenges they faced in battle, it is important to discuss the nature of combat that they faced on the Western Front. In describing the conditions under which the AEF's junior leaders operated, it is possible to gain an understanding of the battle-field realities that circumscribed the range of their tactical options and decision making.

During the Great War the confluence of a number of factors created a "perfect storm" of attritional realities that tended to favor the conflict's defender over its attackers. Attacking infantry faced grave challenges in bringing forward enough firepower to suppress the enemy defenders long enough for the assault to cross "the fire swept ground." Tactical communications, especially those used for controlling artillery fires, also favored the defenders. The attacker's cumbersome means for communicating with his superiors or supporting artillery had to be carried forward with him during the assault. Any failure in communications might delay needed reserve forces and precious supporting fires or prevent senior officers from exploiting success or avoiding disaster.

The war was above all an artillery war. Few attacks on the Western Front gained any degree of headway without a thorough and concentrated use of artillery fires. Unfortunately, artillery presented the attacker with a great dilemma. The attacker had to use large amounts of artillery to enable the infantry to "break in" to the defender's lines. However, the use of that artillery so cratered and scarred the landscape that it became exceptionally difficult for the attacker to move forward the reinforcements, supplies, and cannon that enabled him to turn the break-in into a rapid and decisive break-out.

With the attacker lacking the mobile communications, logistics, and fire support to turn tactical success into an operational victory, combat became a grinding attritional contest. This contest pitted the staying power of the defender against the will and ability of the attacker to accept the cost of coming to grips with the enemy. The 2nd Division's Lt. Wendell Westover offered a vivid example of these sad attritional realities in his description of an infantry platoon's efforts to reduce a German machine gun position during the attack on Blanc Mont Ridge in early October 1918. He observed,

> The section attacking a Boche machine gun, deployed in a long thin line; first a few men on one flank would rush forward a short distance, then, as the fire was directed at their attack, those on the other end would make a quick advance. The flanks were creeping outward and the line developing into an arc which would eventually envelop the gun position—that is if a sufficient number of survivors was left to reach it. This was no parade ground demonstration. . . . Already five men lay still on the ground over which they had come. The right squad sprang up and forward. One, two, four men dropped; staggering, slumping forward to the ground. The rest threw themselves into shell holes and paused. They were close now, but where there had been thirty men a few moments before, only nine were still able to move. Of these, two more fell in the final assault.[1]

Despite the efforts of all the major combatants to change the tactical equation illustrated in Westover's passage by fielding new technologies, developing new tactics, or adjusting the weight, mix, or method of artillery fires, none were able to break the attritional realities that bound them during the war. The odds remained stacked against the attacker, and his successes, like that of the 2nd Division's platoon, were generally pyrrhic in nature.

None of the war's leaders were able to change the Great War's ugly and fixed attritional core. Tactical, operational, and strategic victory came only when the enemy reached the point of exhaustion. As no one could change this core reality, the attacker could only adjust the variables of combat on the margins of this type of war in an attempt to tip the attritional balance ever so slightly in his favor. It was on these margins of attritional warfare where combat leadership came into play. On these margins a small-unit leader could try to wring every possible advantage out of a given combat situation by adapting his tactics, employing his firepower, and using his manpower to inflict relatively more damage on the enemy than was absorbed by his unit. In other words, the battalion, company, and platoons were still going to bleed men, but the question was, in that bleeding, could the leader compel the enemy to hemorrhage his human and material resources at a unsustainable rate?

The margins of combat in attritional warfare were related to both the material and the moral. The material aspects dealt with numbers:

numbers of men, numbers and quality of weapons, amount and quality of food, and amount of ammunition. The moral aspects were tied to the training and confidence of the soldiers and leaders, the willingness of soldiers to sacrifice for their comrades and cause, the ability of the leader to match tactics, formations, and weapons to variations in the terrain and enemy, and the ability of the leader to inspire, over-awe, or cajole his troops into follow him.

In examining small-unit combat leadership in the AEF, we must start with an understanding of the tactical world in which the junior leaders lived and the "physics" that governed their combat. Combat physics is about matter and energy: those physical realities that defined the junior leader's range of options in combat. An example of this is the weight and ammunition consumption of a machine gun. The combat physics of an M1914 Hotchkiss machine gun informed the leader that regardless of any desire he might have to deploy this weapon with the forward assault elements, its 109-pound mass generally meant that its overburdened crew would fall behind as the less encumbered riflemen pushed forward, or that its struggling and easily identifiable crew would draw the attention of the enemy long before the gun could be brought into action. Unlike those of scientific physics, the laws of combat physics are not fixed. Thus, while the weight of the machine gun did not change, the ability of the crew to bring the gun into action could be governed by intangibles such as the crew's physical and mental state, their morale, the inspiration (or lack thereof) of leadership, or the skill or luck of the crew in finding a safe route to the objective.

The first thing to note of the junior leader's tactical world was its chaotic nature. This was not just the typical and all-important fog and friction of combat, but also the overarching chaotic atmosphere that characterized life in the US Army during the war. As has been noted in previous chapters, junior leaders were constantly buffeted by change—change in weapons, change in doctrine, and change in personnel. Each of these changes brought with it technical, tactical, and leadership realities that increased the complexity of leading small units.

Shortly after the war, Col. John Parker noted that with the war-time explosion of support weapons, "the infantry organization has now reached such a complexity that the infantry regiment in combat is difficult to manage[,] from the unit command of a platoon leader up to include the regiment as a whole." He pointed out that "the platoon leader has at least seven different weapons, four of which are organized in teams that lose their efficiency with the loss of a single expert. The company commander has the same problem, and the battalion commander has it in a still greater degree."[2] When the Americans entered major combat in the fourth year of the war, it was an army of 1914 thrust with bewildering rapidity into 1918. As Parker suggested, in very short order the Americans had to adopt new organizational schemes, weapons, tactics, and concepts of waging war that were alien to their previous ways of thinking and acting. Leaders at every level were confronted with new realities for which nothing in their previous education, training, or experience had prepared them.

When examining the tactical world of the AEF's infantry captains, lieutenants, and NCOs, it is wise to start with the role that artillery played in the lives of their units. Maj. R. C. Birmingham, of the 11th Infantry, flatly declared that "it is always necessary to have artillery preparation in order to make a successful advance without ruinous losses."[3] Unquestionably, artillery was the king of the junior leader's tactical world. It was the war's greatest casualty producer. Artillery voraciously gobbled men during large-scale battles and constantly nibbled away at them during periods of relative quiet in the lines. Brig. Gen. Beaumont Buck recorded that in the weeks leading up to the Cantigny battle, his brigade lost an average of fifty men per day, mostly to artillery.[4] For the attacker, artillery offered four great hopes: first, that its preparatory fires would destroy enemy troop concentrations, strong points, and barbed wire; second, that its counterbattery fires would stave off the fury of the enemy's artillery; third, that its rolling barrages would suppress the enemy's defense fires and allow the attacking infantry a relatively easy crossing of the deadly ground; and last, that the shelling of the enemy's rear areas would impede the flow of reinforcements and supplies to the front line long enough for the attackers to consolidate their gains. Without the powerful support of artillery, the attacking infantry seldom had any chance to achieve their objectives unless they were willing to undergo a dreadful bloodletting.

Historian Mark Grotelueschen notes that some of the AEF's senior leaders, such as Maj. Gen. Charles Summerall, recognized the attritional nature of the Great War and embraced a doctrine based on the seizing of limited objectives by attacks using a liberal amount of artillery. Furthermore, as American divisions came to appreciate the firepower realities of modern war and woke to the fact that the GHQ's "open warfare" was not a viable solution to the tactical challenges confronting them, they devised their own firepower and maneuver solutions to minimize casualties while still destroying the enemy.[5] Allowing for detailed planning, logistical preparation, and good liaison between the infantry and the artillery, the American cannoneers were a lethal bunch.

There can be no question that the American artillery grew to be a fearsome weapon in the last six months of the war. Time and time again in the fall of 1918 German officers commented on the effectiveness of the doughboy artillery. The Germans facing an attack by the American 2nd Division at Blanc Mont in early October 1918 noted that the American preparatory fires knocked out 50 percent of the machine guns of one of the machine gun companies of their 74th Reserve Infantry Regiment and caused 25 percent casualties in another regiment. An officer of the German 102nd Infantry regiment reported, "Only when there is an incomparably strong artillery preparation, as on 23 October 1918, when our forward garrisons lacked sufficient shelter, can an [American] attack gain ground. The artillery preparation was effective because a monstrous amount of artillery was expended." Although this German officer was dismissive of American infantry and disdainful of the AEF's prodigious use of shells, his comments also betrayed an element of fear and foreboding.[6]

237

Although the artillery was inextricably linked to the American infantry's chance of success, it was far from being a tactical panacea. The effectiveness of artillery in World War I was tied to a number of factors governed by combat physics and the Clausewitzian fog and friction of war. The greatest problem of the artillery was responsiveness: the ability for the guns to provide immediate fires for the attacking infantry to destroy or suppress the enemy wherever they were encountered. American artillery was most effective when it had the time to identify enemy defenses, had amassed enough shells to destroy those targets, and had seamlessly coordinated its fire plan with the infantry's scheme of maneuver. It was generally in the initial bombardments of an offensive, or the preplanned preliminary bombardments supporting planned attacks, that the gunners, commanders, and staff planners had the time needed to bring these requirements into alignment.

It was after the initial planned bombardments that the infantry and artillery ran into problems. The major obstacle to responsive fires and good liaison between the infantry and artillery was communications. A board of senior officers convened by the AEF GHQ shortly after the Armistice to assess the future of American organizations, tactics, and doctrine based on the AEF's combat experience reported that the artillery's greatest shortcoming was the lack of "a fully organized system of observation and quick communication adapted to the new conditions" of the attack. The board's officers wrote that "the position of the enemy guns and of his infantry were not known as had been the case in trench warfare, and the machinery for getting this information, for quickly transmitting it, and for promptly securing adjusted fire were not adequate."[7]

The need for "securing adjusted fire" was predicated on the changing nature of the enemy's defense. On 5 September 1918 the AEF GHQ published *Combat Instructions* in the hope of preventing American commanders from repeating some of the costly mistakes that AEF units had made during the summer battles of 1918. *Combat Instructions* bluntly informed the American commanders that "the German machine guns constitute the chief weapon to be combatted by our infantry."[8] The heavy losses of their infantry increasingly led the Germans to substitute machine gun firepower for diminishing manpower. From the Aisne-Marne to the Meuse-Argonne, the American infantry faced enemy defenses that consisted largely of well-sited, protected, and concealed machine gun positions. These positions were generally tied in with flanking units with mutually supporting, interlocking fields of fire. As the Germans went to great lengths to conceal these positions, the attacking American infantry often did not find them until the Germans opened fire. The most effective means for countering these positions was to destroy them with artillery. However, this required that someone at the point of the attack had to be able to contact the artillery and then adjust its fires onto the enemy locations. This is where the system broke down.

At the infantry battalion, company, and platoon level, tactical communications were limited to field phones, runners, visual signals, or carrier pigeons. Each of these means of communications had built-in flaws that limited the responsiveness of supporting artillery fires. Wire communications, the linking of telephone cables between field phones and switchboards, were the quickest and most responsive means that forward commanders had at their disposal for staying in contact with their higher headquarters and artillery units. Unfortunately, this wire had to be spooled out by signal parties who moved with the infantry as they pushed forward. This was a laborious and dangerous task. One signalman recalled, "The St. Mihiel affair was a veritable nightmare for the telephone men in the artillery. After the first jump-off we were advancing so rapidly that it was nearly impossible to maintain communications." He also lamented, "We would lose men and equipment so fast that I wonder we did anything at all."[9] Although the pace of the advance slowed during the Meuse-Argonne Campaign, the problems described in this passage continued to plague wire communications to the end of the war.

Wire communications were also vulnerable to shell fire and accidental disruption. In the offense the line was laid on top of the ground or hastily strung from trees or overhangs. This meant that any random shell, passing vehicle, or clumsy doughboy could sever the connection with ease. A lineman later wrote that the "universal lament" of signalers was that "artillery, with their heavy gun carriages and cumbersome equipment, were everlastingly ripping up their field wire and imposing upon the already overburdened signal troops a never-ending task of maintenance."[10] If the line was broken between the infantry and the guns, someone had the time-consuming and dangerous task of following the wire back and repairing the break.

239

The infantry could also use visual signals or carrier pigeons to contact the artillery. Visual signals could be sent by means of rockets, flares, flags, or flashing electric lights. The use of visual signals required that both parties see and understand the meaning of the signals. While this sounds quite simple, it was not always so in practice. A battalion commander in the 353rd Infantry noted that during the Saint-Mihiel operation the division and regiment's system of communications was so convoluted that just minutes before the unit was to go over the top, he and four of his company commanders were "earnestly engaged in trying to get an understanding of the signals to be used and the meaning of them." Hervey Allen recalled that just as his unit was going into its attack on Fismes in August 1918, each platoon was issued rockets to use to signal for an artillery barrage. Unfortunately, the newfangled rockets came with no instructions, and none of his men "had the slightest idea how to use" them.[11]

Another trouble with visual signals was that it was difficult to convey anything more than a simple message. Although a signal flare easily indicated that the artillery should lift its preplanned fires, it was quite another thing to use flares for the involved process of calling for and adjusting fire on unplanned targets.

The use of carrier pigeons was also problematic. The Signal Corps trained over fifteen thousand birds during the war and claimed that "they delivered more than 95 per cent of the messages intrusted to them." However, the birds were susceptible to gas and could be difficult to manage. As his battalion was about to launch an assault in the Argonne, one infantry officer recalled that the regimental signal officer gave him a number of pigeon cages to carry forward as well as "a lengthy discussion as to the care and keeping of that particular species of bird and the method of sending messages and the like." The battalion officer was dismissive of the bird's capabilities and pushed them off on a signal sergeant rather than taking them to the front. The officer also noted that when supplies ran low, the pigeons made a convenient and tasty meal. Limits in the number of available birds, and the number of cages that the infantry could carry in the front line, restricted the number of messages that could be sent by carrier pigeon by any combat unit. The units of the Lost Battalion, for example, carried forward only seven birds during its memorable attack into the Argonne Forest.[12]

Last, the leaders at the battalion level and below could use runners to keep in touch with their higher headquarters and artillery. Runners were the slowest and most vulnerable means of communications. As a result of the 1st Division's experiences during the Aisne-Marne Campaign, Maj. Gen. Charles Summerall attempted to prohibit the use of runners for tactical communications because they were "slow and uncertain and the casualties among them are out of proportion to the service that they render." Runners were also susceptible to a wide range of human frailties. In the midst of the Argonne fighting, Lt. Joseph Lawrence sent his two runners back to warn his superiors of a possible enemy counterattack, only to have the pair disappear. A search for the missing men later found them sleeping in an abandoned German dugout. When questioned, they confessed to having failed to deliver the message and "protested that they were not yellow—they were worn out, dead tired." Most important, sending runners back and forth to report required adjustments to artillery fire was ineffective.[13]

All of these issues exacerbated problems of fire support and command and control. Clare Kenamore argued that at least some of the confusion that hobbled the 35th Division's operations in the opening days of the Argonne Offensive was a result of its chaotic signals and communications system. Wire was broken by fire, runners were killed and wounded, troops were uncertain of the meaning of signal flares and rockets, and some of the flare cartridges issued did not fit the flare guns of the division's officers. When asked the location of one of the infantry regiments, one senior division staff officer could only say, "I wish I knew . . . and I wish I knew even where the brigade headquarters are." Kenamore was not alone in pointing to the inadequate means for combat communications. An officer in the 140th Infantry recalled of his unit's experiences in the Argonne and its baleful influence on fire support, "The artillery had been furnished with excellent wireless equipment—but none was given to the infantry, therefore it was useless. The telephone equipment was lacking, and the rocket-signals were of such a character as to be useless. The artillery . . . was ready to do its duty. . . . But it was hampered by lack of information, as it was compelled to depend entirely on liaison through runners, and the information provided was not always correct . . . and many of them gave their lives on the field." Although the Americans made heroic efforts to correct the immediacy and responsiveness of artillery, the AEF was never truly able to overcome the technical barriers that prevented rapid communications from the front line to the guns.[14]

To offset the artillery's lack of responsiveness, the fire plans of the American divisions focused on providing heavy preplanned preparatory fires and rolling barrages. The preparatory fires were intended to destroy or suppress enemy positions. The rolling barrage was based on the assumption that the preparatory fires were never going to kill all of the defenders, but by having a curtain of fire land just to the front of the advancing infantry, the enemy would be forced to seek shelter and thus be unable to fire on the attacking Americans. A rolling barrage required that the infantry and artillery adhere to a centralized and strictly regulated plan that was difficult to change after the fires began. As the rolling barrage proceeded at a fixed rate, the onus was on the infantry to stay as close to the falling shells as possible. When all went well, the effect of the rolling barrage was decisive to the success of the infantry. An infantry officer in the 36th Division remembered that during a 27 October 1918 attack on the Forest Ferme, his soldiers "followed the barrage so closely that they were almost 'leaning against it.'" The tactic worked so well that they caught the Germans while they were still in their shelters and took their objective with light casualties.[15]

Regrettably, the rigidity and complexity of the fire plan frequently left the infantry without the vital support of suppressive fires at the very moment they needed them most. The divisions that arrived in the summer of 1918 faced a very steep learning curve when it came to planning and coordinating artillery fires. In his official report of his unit's action in the Saint-Mihiel Offensive, the commander of the 353rd Infantry, Col. James Reeves, was critical of the artillery fire plan. Not only was it late in arriving, but it was also so complex that it was impossible to be understood by "even the company commanders, much less the platoon commanders." More important, unless there were very close liaison between the infantry and the artillery, anything that delayed the infantry advance, even momentarily, soon left the doughboys without protection as the barrage moved forward on its set timelines. Delays in orders to attack, unexpected enemy contact, uncut barbed wire, or merely the difficulty that combat-worn and heavily laden soldiers had in crossing muddy and shell-torn ground could all result in the dreaded loss of the barrage. This sad fate even struck seasoned units with much experience in infantry-artillery cooperation. On 15 October 1918, officers of the veteran 38th Infantry complained "that there was no way of holding the barrage on an objective when the infantry front line has been held up."[16]

Other factors also limited the effectiveness of the artillery support given the American infantry. The IX Corps commander, Maj. Gen. Henry Allen, argued that when American units took the time to plan their operations with a view of synchronizing the firepower of all the arms with the maneuver of the infantry, they did usually accomplished their missions "with relatively small losses." However, "when orders were given to exploit the ground to the front . . . the losses were often greater." He maintained, "The cause of this was the continuation of the advance without proper preparation due to want of time to study the new front and to bring up the artillery to where it could be used."[17]

Artillery was not an asset that a division or brigade commander could turn on like a tap. It took time to move the guns into position, to plan targets, and to build stockpiles of shells for prolonged bombardments. As early as August 1918 the AEF GHQ reported in its study of the Aisne-Marne battle that "in several instances barrage orders were not received by the units charged with their execution in time to permit the numerous calculations necessary." Two months later GHQ was still lamenting the difficulty in using divisional artillery after the initial advance due to the need to find targets and move forward. However, recognizing the problem and being able to fix it were not the same. Some things, such as moving the guns up over shell-torn mud and overcrowded roads, were simply beyond the ability of the artillerymen to remedy.[18]

There were limitations to what even the heaviest American artillery bombardment could do. If the timing of the fires was off and the infantry "lost the barrage," if the fires did not neutralize the defenders, or if the infantry was simply unlucky, the effectiveness of the artillery was muted. If, for whatever reason, the coordination between the infantry and the guns broke down, infantry commanders were generally left with no other option than to push forward without support. Following the attacks in the Meuse-Argonne from 26 to 30 September 1918, the AEF general staff admonished field commander that "advances were generally too slow and too cautious. The fruits of victory were, therefore, not what they might have been." GHQ criticized the tendency of infantry commanders to "remain inactive in the presence of relatively small hostile forces while waiting for orders, or for artillery support, or for machine guns, or missing grenades, etc."

The staff's solution was simple: "It is seldom wrong to attack. The best way to clear up a doubtful situation is to advance. In the attack it is better to lose many men than to fail to take ground. Inaction is the worst military crime."[19]

The veiled threat in this message to infantry commanders was clear. Despite the AEF's acknowledgement that artillery played a key role in the infantry's success, the failure of the artillery was no excuse for the infantry to delay an attack. This point was made abundantly clear to Capt. Charles Harrington, an acting battalion commander in the 308th Infantry. He was relieved from command and sent to Blois on 5 September 1918 for refusing to attack during the Oise-Aisne drive after a promised artillery barrage failed to materialize. This ugly reality also confronted the commander of the 126th Infantry on 28 August 1918 when the unit's planned rolling barrage never occurred. After waiting fifteen minutes, the commander made the agonizing decision to move forward only supported by the fire of his own machine guns. After two days of battling without much artillery support, the regiment was reduced to an effective strength of less than a thousand men.[20]

For all the power of the American artillery, it could be a very fickle and temperamental tool. Infantry leaders at the company level and below had little to no say in the planning of artillery fires and very limited means to request artillery support after the battle began. Thus, for infantry captains, lieutenants, and NCOs, either the artillery fires came or they didn't. In either case, H-hour was H-hour, and the infantry was going over the top with the fires or without them. Although it is easy to deride Pershing for his desire to create "self-reliant infantry," the sad truth was that at some point in every one of the AEF's engagements the terrain, the proximity to the enemy, or the breakdown of the coordination between the infantry and its supporting arms threw the American infantrymen upon their own resources. At these points the doughboys could rely only upon the weapons that were organic to their companies, platoons, and squads.

The world of the infantry small-unit leader was generally bounded by the confines of his regiment. The war brought massive change to the size and composition of the regiment, as it did to all organizations in the US Army. Before the war, a full-strength infantry regiment consisted of 51 officers and 1,500 enlisted men. With the exception of a provisional machine gun company containing six guns, the unit's remaining soldiers were armed with rifles or pistols. By June 1918 the infantry regiment had grown to 112 officers and 3,720 enlisted men. The regimental commander also controlled an organic machine gun company with sixteen guns, a Stokes mortar section of six tubes, and three 37 mm guns.

The regiment was divided into three battalions each under the command of a major. Each battalion contained four rifle companies. The battalion contained no organic support weapons except automatic rifles and rifle grenade launchers within their companies. However, it was standard practice in the AEF to attach the regiment's machine guns, Stokes mortars, and 37 mm guns to the lead assault battalion in the attack. Given the relative scarcity of these heavy weapons, there were seldom enough to go around.

When the battalion did receive these support weapons, the challenge for the commander was how to integrate them into his plan of maneuver to best aid his infantrymen in accomplishing their missions. To achieve this, the battalion commander needed a firm grasp of the inherent technical capabilities and limitations of these weapons as well as an idea of how they needed to be employed and supplied to be effective in combat. If the commander understood these realities, the employment of support weapons was one of those variables on the margins of attritional warfare that could materially aid him in reducing friendly casualties while still accomplishing his unit's mission.

Let us examine these weapons and the challenges and benefits they brought. Machine guns were the most common support weapons used in the AEF. The AEF generally used four types of machine guns: the French M1914 Hotchkiss, the British Vickers gun, the M1915 (a .30–06–chambered copy of the Vickers gun), and the M1917 Browning gun. Each of these guns had inherent advantages and disadvantages that affected their tactical use.

The M1914 Hotchkiss gun had a maximum rate of fire of four hundred rounds per minute but was fed by inserting twenty-five-round metal strips into the side of the weapon. This meant that the gun's actual rate of fire was determined by the assistant gunner's dexterity and quickness in feeding the fragile strips into the weapon. The gun was air-cooled and thus was liable to overheat during prolonged firing, and its fixed-height tripod gave the weapon a high silhouette, making its crew more vulnerable to enemy fire. The greatest problem with it was that gun and tripod together weighed a whopping 109 pounds. This made the guns' crews slow-moving targets to the enemy as they lugged that ponderous weight forward.[21]

The British and American Vickers gun fired a maximum of 500 rounds per minute using 250-round cloth belts. Although lighter than the Hotchkiss, its 98-pound weight was still quite a load. The

M1917 Browning proved to be one of the best machine gun designs of the war. John Browning's weapon matched the rate of fire of the Vickers but weighed only 84.5 pounds. Although the Ordnance Department intended to replace all of the AEF's machine guns with the M1917, the pace of operations gave them no time to accomplish the changeout, and only 1,168 guns actually made it to the front before the Armistice.[22]

On paper, both the Vickers and the M1917 Browning had very high rates of fire, but to keep from overheating even these water-cooled guns, the actual rate of fire in combat was generally 250 rounds per minute. Both the Vickers and the Browning suffered from the inherent limitations of water-cooled machine guns. The water jacket that surrounded the gun's barrel had to be kept full to sustain prolonged firing. The weapons were equipped with a condensation can to catch the steam created by the hot barrels during firing. Unfortunately, water in the jackets was still lost in the process, thus requiring crews to have ready access to relatively clean water sources during combat, and if the crews brought the water with them in the condensation cans, it added an additional burden on the already encumbered men.

All of the era's machine guns were prodigious consumers of ammunition. The iron laws of physics dictated that there was only so much ammunition that the crew could bring with them into action. This generally forced machine gun officers to rely on ammunition carriers detailed from the infantry to move the weighty burdens forward into battle.

When properly employed, machine guns offered the infantry immediate firepower to deal with a host of combat challenges. Much like artillery, the machine guns could fire over the heads of the attacking infantry to suppress enemy defenders. They could also fire to the flanks or rear of the infantry attacks to prevent the enemy from bringing forward reinforcements or supplies. If rapidly brought forward to objectives recently seized by the infantry, the machine guns could break up the German counterattacks that invariably followed an American tactical gain.

To get these effects, however, the battalion commanders had to overcome a number of obstacles. First and foremost was their own ignorance of how to use these guns properly in combat. The summer battles of 1918 highlighted this major problem. At Chateau-Thierry in July 1918, Maj. C. A. Dravo, a machine gun officer in the 42nd Division, observed that "the battalion commanders had their hands full with a thousand infantrymen and had neither the time nor the opportunity for any study of the situation as might be applied to the machine guns attached to his command." Likewise, Capt. A. M. Patch recalled that "there was no machine gun plan" for the machine gunners supporting the 1st Division's infantry attacks during the Soissons drive. Patch noted that few infantry battalion commanders gave the machine guns any missions, and thus, "the assistance [the guns] really rendered the infantry, constituted a deplorable and disastrous spectacle." The outcome of these missteps was that after four days of fighting, 80 percent of the machine gunners were casualties and approximately 85 percent of their equipment was destroyed or abandoned." Another participant in the Aisne-Marne battles argued that the machine gunners suffered heavy casualties because they had no orders and "simply followed the attacking waves."[23]

The AEF GHQ and some senior commanders tried to address these problems. In the *Notes on Recent Operations, No. 1,* issued after the July battles, the GHQ castigated the infantry battalion commanders for not adequately directing their attached machine gun companies and berated the machine gun company commanders for their lack of initiative in employing their weapons and advising their superiors of the guns' best use. Maj. Gen. Charles Summerall, the commander of the 1st Division, demanded that his subordinates improve the responsiveness of their heavy machine guns to the needs of the infantry by establishing a stronger system of liaison between the two units. He also required that machine guns and other support weapons "must at all times be under the eye of a known leader, who is able to preserve cohesion and compel energetic action."[24]

Both the GHQ and Summerall identified faulty training and leadership as the key problems with efficiently using these weapons. It was clear that neither the infantry nor the machine gun commanders had the requisite training to employ the guns effectively in combat. Part of this was the fault of the AEF's own schools. Maj. Robert Calder, commander of the 1st Division's 3rd Machine Gun Battalion, argued that the army's support weapons "were seldom used to best advantage" because "the personnel was trained in the technical side of the weapons, they were not sufficiently familiar with their tactical employment." Col. Joseph Sanborn noted that the machine gunners of his 131st Infantry were given very good instruction on how to operate the Vickers guns at an AEF school, but the school "had no competent instructors in tactical employment of guns." Looking back to the scant amount of time spent on training candidates on machine guns and the other support weapons in the OTCs and COTSs, it is little wonder that their graduates often lacked a basic understanding of these tools.[25]

After enduring the school of hard knocks during the Aisne-Marne and Oise-Aisne Offensives, some of the AEF's units puzzled out ways to better coordinate supporting machine gun fires with the infantry's maneuver. After a week of fighting in the Meuse-Argonne, the AEF General Staff reported that "the use of machine guns generally showed a decided improvement over previous operations" but went on to warn that "much more remains to be done before they exert the continuous influence throughout the fight." As GHQ pushed green units into their first major actions, the novices too were schooled by the taskmaster of experience. The 29th Division's Joseph Lawrence noted that after the attack of the 113th Infantry stalled on 10 October 1918,

> An attempt was made to break the German lines with machine guns, and as I lay in my hole I saw the machine gunners rush forward through the infantry and mount their guns. I do not believe that they fired a shot; the gunners were mowed down before they could pull a trigger. . . . Our colonel was later criticized by General Morton for the loss of life . . . for he contended that if the infantry could not break the enemy lines, machine gunners who were handicapped with heavy equipment and guns that could only be mounted in exposed positions should not be expected to do so.

Even the seasoned A E F divisions had difficulties in this area. A General Staff officer inspecting the 3rd Division on 15 October 1918 reported that officers in the 38th Infantry "found it impossible in their regiments to have close liaison with the machine gun commander," and it was better to leave the machine gun units to "function independent of instructions of infantry battalion commanders." Despite the improvements that some units made in infantry–machine gun liaison, overall the AEF inspector general found that throughout the Meuse-Argonne Campaign the AEF's infantry leaders still lacked the know-how to employ their support weapons and were thus "unable to derive much benefit from these arms."[26]

In addition to machine guns, each infantry regiment also had a Stokes mortar platoon equipped with six mortars and a 37 mm gun section of three cannons. Both of these weapons were intended to give the infantry quick and responsive tools to destroy or suppress German machine gun nests or strong points. When properly handled, these weapons were very effective in accomplishing their intended missions. Unfortunately, they also suffered from inherent limitations.

The 3-inch Stokes mortar was the "trench artillery" most used in American infantry regiments. It had a maximum range of around eight hundred meters and fired an 11-pound high-explosive shell. The barrel, bipod, and baseplate weighed 108 pounds. To make matters worse, its ammunition was even more cumbersome and difficult to bring forward into battle than that of the machine guns. The 37 mm gun, or "one-pounder," fired a flat-trajectory 1.2-pound high explosive shell with a maximum effective range of one thousand to fifteen hundred meters. Its accuracy made it ideal for eliminating enemy machine guns and pillboxes. Unfortunately, the gun and tripod for the weapon weighed a hefty 170 pounds. It goes without saying that both the regimental Stokes mortars and 37 mm guns were difficult to manhandle over anything but the flattest and smoothest terrain, features sorely lacking on most of the AEF's battlefields. Furthermore, the Stokes mortars and 37 mm guns suffered the same problems of employment as did the machine guns: if the infantry officers did not know how to use them, they were of little value in combat. Col. Robert McCormick noted that in the divisions that arrived in the spring and summer of 1918, "knowledge of modern battle conditions was wanting." He observed, "They had received their trench mortars and their infantry cannons only a short time before and did not know how to use them. Some regiments marched through the whole campaign without taking these indispensable weapons from their trains. They were, in consequence, badly in need of material with which to attack German machine gun nests at close range." McCormick's comments had some validity. For example, when the 145th Infantry attacked on the first day of the Meuse-Argonne Offensive, its Stokes mortars and 37 mm guns were at the trail of its formation, where it was nearly impossible for them to move forward to support the attacking infantrymen in any timely manner.[27]

Despite all of the shortcomings in training and leadership, in fairness to the infantrymen and support weapons crewmen it must be pointed out that some of the failure to more effectively use these tools simply came down to the physics of combat. All of the support weapons, and the ammunition they fired, were very heavy and cumbersome. One must keep in mind the difficulty that the crews had in keeping up with the relatively unencumbered riflemen over difficult, muddy, and shell-torn ground. A machine gun officer in the 79th Division stated that as he moved his unit forward to support the attacking infantry in the Argonne, "So deep were the shell holes and mine craters that the ammunition carriers were even unable to carry two small boxes of ammunition." The soldiers removed the ammunition belts from the boxes and wrapped them around their bodies to free their hands to crawl through the slippery obstacles. When or-

248

Machine gun crew from the machine gun company of the 113th Infantry, 29th Division. Although the M1917 Browning machine gun they carry was much lighter than the guns provided by the Allies, this photo still gives an indication of the weight and the awkwardness of the war's weapons.
(Photograph from the author's collection.)

dered to move his guns forward with the infantry during an attack in the Soissons Offensive, Lt. Malcolm Helms was dismayed when his overloaded machine gunners "soon fell behind the infantry" and could offer the doughboys no support because "we were floundering through the shell holes of our barrage with our heavier loads." Pvt. Ray Johnson, a machine gunner with the 37th Division, simply stated that keeping his guns up with the infantry led to "the exhaustion of our reserves of strength by the weight of our equipment."[28]

Although the preceding passages were from machine gunners, they were equally applicable to the crews of Stokes mortars and 37 mm guns. A number of sources point to the effectiveness of these weapons in destroying German machine gun positions but in the same breath complain about the difficulty of bringing them forward. For

Chapter 9

example, an officer with the 29th Division noted that "trench mortars by reason of their lack of mobility and difficulty of keeping them supplied with ammunition are not considered efficient weapons in mobile warfare." In a report issued three days after the Armistice, the commander of one of the AEF's most combat experienced units, the 2nd Division's 3rd Infantry Brigade, admitted that while his units had used their one-pounders and mortars before the infantry went over the top, "after that time" the weapons "could not keep up with the advancing troops."[29]

In addition to the problems of training and mobility, one of the other major drawbacks to the use of the support weapons was their logistical requirements. The ammunition for these weapons was heavy, cumbersome, and usually carried forward by hand. In any heavy engagement in which units encountered numerous enemy strong points or machine gun nests, the supply of ammunition for the infantry's support weapons would be exhausted quickly, and bringing up fresh ammunition was nearly always a chore.

The problem of bringing up ammunition was further exacerbated as the campaigns dragged on and the physical condition of the crews deteriorated. In some cases units tried to overcome the problem of keeping the support weapons supplied by providing their crews with carrying parties drawn from infantry companies. However, at times these carriers could be very unreliable. During the Aisne-Marne Campaign, American machine gunners complained that soon after the guns began to move forward, their infantry carrying parties "almost invariably abandoned the squad during the assault and took part in the fight with their own units." Without carriers, the crews were left to muddle through as best as they could. As his unit moved forward for the British army's attack on the Saint-Quentin Canal in late September 1918, Leslie Baker recalled that his company cracked from exhaustion while lugging fifty-pound ammunition boxes to the front lines. He credited his lieutenant for saving the day when the officer ordered his men to abandon the heavy boxes by the side of the road and move to the trenches without them. Although the officer's decision was perhaps correct, it dramatically reduced the effectiveness of his machine guns.[30]

The remaining nonorganic support weapons that were sometimes encountered at the battalion level or below, such as accompanying artillery guns and tanks, also require some discussion. Some of the AEF's units attempted to overcome the problem of unresponsive artillery by placing a 75 mm gun or other full-sized cannon under the control of the assaulting battalion commander. As for most things in the AEF, the variety of experiences with these weapons indicates that their use was a mixed bag. On one hand an officer from the AEF inspector general's office reported that the accompanying guns attached to an infantry battalion of the 82nd Division managed to destroy eight German pillboxes in one day. However, another division inspector found that in the infantry "the average battalion commander had not enough experience in handling a battalion of infantry to warrant his paying sufficient attention to his accompanying support battery," and many of the leaders forgot that they had the guns at their disposal. Some artillery officers also believed that the practice of providing the infantry with accompanying guns hindered their ability to mass artillery fires and usually led to the guns' "misuse or enforced idleness."[31]

251

The same ambivalence that characterized the use of the accompanying guns also typified the infantry's experience with tanks. Infantry-tank cooperation was seldom trained, and misperceptions about the weapon's capabilities and limitations were common among infantry leaders. One tank corps officer ruefully noted, "It's surprising what they asked us to do. Doughboys to Generals have set us up against places a battleship couldn't capture." While tanks often provided the infantry with the mobile firepower they needed to break into the German defenses, the weapon's mechanical unreliability and limited numbers, the lack of a solid system for tank-infantry cooperation, and the failure of infantry commanders to give the tanks definite and attainable tasks prevented tanks from playing a more significant role in the tactical world of most American infantry junior leaders.[32]

As infantry units became more experienced in the use of the support weapons discussed above, it seems that most became more comfortable with their use. This is illustrated in the experiences of the 89th Division's 353rd Infantry. At the close of the unit's first major combat action at Saint-Mihiel, the regimental commander reported, "The most weapons used were the Infantry rifle and the hand grenade, in connection with the machine guns attached to the 2nd or leading battalion. I saw very little use made of rifle grenades, very little of automatic rifles, and little or nothing accomplished by either the one-pounders or Stokes mortars. Each of the last mentioned special weapons fired a few shots, but usually there is reason to believe that some of the shots of the Stokes, as usual, fell short and injured our own men." However, by the time of the 353rd Infantry's attack on the Bois de Barricourt on 1 November 1918, the colonel had overcome his initial skepticism and fairly gushed at his unit's ability to coordinate the fire of rifles, Chauchats, 37 mm guns, and Stokes mortars to destroy German machine gun nests. However, there were some problems that time and experience still could not overcome. For example, he recalled that the effectiveness of his mortars was still hindered by "the impossibility of keeping of a supply of ammunition" for the weapons. Even with all this additional firepower, the officer still maintained that "as heretofore, the main work was done by the infantry rifle," and that it and automatic rifles were the weapons most utilized in overcoming the "points of resistance encountered."[33]

252   While the 353rd Infantry commander's comments may seem like the ranting of a disciple of the Pershing school of open warfare, he merely expressed the fact that at some time in the fight his infantry companies still had to close with and kill Germans aided only by the weapons they had at hand. The AEF's infantry companies and platoons were where "self-reliant infantry" truly existed. It was also at this level where the AEF's battles were fought and all of its shortcomings in training and leadership became most apparent. We will now turn to the organization of infantry companies to illustrate the challenges facing their junior leaders and what could be expected of them in combat.

At full strength, an American infantry company was composed of six officers, forty-eight NCOs, and 207 soldiers (see Appendix Table A-1). This was more than twice the size of the prewar infantry company. The commander could employ the firepower of sixteen automatic rifles, twenty-four rifle grenade launchers, and 192 rifles. This firepower was equally divided among each of the infantry company's four rifle platoons.

The full-strength rifle platoon consisted of one officer platoon leader, three buck sergeants, eight corporals, and forty-seven privates. The rifle platoon had four automatic rifles, six rifle grenade launchers, and forty-eight rifles. The AEF rifle platoon leaders generally employed their units either as one large mass, as half-platoons which divided the control of the unit between the platoon leader and platoon sergeant, or as four combat groups divided by weapons type. The last organization would have two combat groups of automatic riflemen, one of hand bombers and rifle grenadiers, and one that consisted solely of riflemen. (see Appendix Tables A-2 and A-3).

While the firepower of these companies and platoons was exponentially greater than that of these units of 1916, their ponderous size made them difficult for company commanders and platoon leaders to control. The AEF GHQ was quick to dismiss the concern of some senior officers that the increase in the size of infantry companies and platoons would overwhelm the already struggling junior leadership.

In May 1918 Col. Hugh Drum, the GHQ assistant chief of staff, responded to criticism of its large formations by arguing,

> The difficulty of infantry combat of today is due not so much to the inability to control men as it is the lack of men available to meet each new situation. Difficulties of control arise through lack of sufficient men and not through an excess. There is no doubt in my mind that the platoon leader that controls only 20 men in a task requiring 50 will make more tactical errors than if he had an excess of 30. The tendency to disperse 20 men so as to accomplish the task requiring 50 will often lead to disastrous results. Moreover it is false practice to organize an army on the assumed capabilities of the platoon leader.

Drum's convoluted rationale for retaining the bloated companies and platoons was not borne out in combat. The commander of the 7th Infantry Brigade, Brig. Gen. Benjamin Poore argued, "I believe that the infantry officers will agree that we have used too many men in our combat formations and the inevitable result was greater and probably avoidable losses. The companies were too large to be handled by officers of average ability and little experience. Even a highly trained and experienced officer found great difficulty in handling a company of 250 men." As will be seen, the difficulty that junior leaders had in controlling their massive units contributed to the tactical sluggishness of the AEF's units during the opening weeks of the Meuse-Argonne.[34]

As with the heavier support weapons, the armaments that were organic to the AEF's infantry companies also possessed inherent advantages and limitations. The most common infantry weapons were the M1903 Springfield and M1917 "Enfield" rifles. Although they were sighted for much greater ranges, their effective range was 350 to 600 meters, depending on the skill of the firer. The aimed rate of fire for the weapons was roughly fifteen to twenty rounds per minute. The basic combat load for the infantryman was the one hundred rounds that he carried in his cartridge belt, but this could be increased if the soldier carried one or two additional sixty-round bandoleers.

The most common automatic rifles used in the AEF were the French M1915 Chauchat, the M1918 Browning, and for US II Corps soldiers fighting with the British, the Lewis gun. The 8 mm Chauchat weighed nineteen pounds and had a maximum rate of fire of 240 rounds per minute. Its reputation for unreliability in the AEF led one American officer to characterize the weapon as "a villainous piece of unreliable makeshift." However, in the hands of a trained and careful gunner, it provided a fairly light and powerful addition to the firepower of the platoon. The chief source of the weapon's problems was its fragile 20-round magazines. The magazine had windows cut in one of its sides to allow the gunner to quickly check to see how many rounds it had remaining. Unfortunately, if the gunner was not careful, mud entered these windows and jammed the gun. The combat load for a Chauchat gunner and two assistant gunners/carriers was nineteen or twenty magazines. Since each magazine weighed two pounds, the choice of combat load could quickly overburden the crews.[35]

The divisions that arrived in France after June 1918 were armed with the Browning automatic rifle (BAR). The BAR weighed sixteen pounds and had a rate of fire of 550 rounds per minute using 20-round magazines. Although the BAR was lighter than the Chauchat, it was still nearly double the weight of the standard infantry rifle. The basic load for a BAR gunner and two assistant gunners/carriers was between twenty and twenty-four pounds depending on the cartridge belt and bandoleers used by the crew. The weapon was much more reliable than the Chauchat and was able to produce a large volume of killing or suppressive fire.[36]

Last, the company commander and platoon leaders could call on the firepower of their hand and rifle grenades. The most common rifle grenade used by the AEF was the French Viven-Bessiere (VB) grenade fired from a "trombone" grenade discharger mounted to the standard-issue M1903 or M1917 rifle. The grenade's range was up to two hundred meters. While the Ordnance Department's Bernard Crowell reported that "any man within 75 yards of an exploding rifle grenade is likely to be wounded or killed," the actual burst radius of the seventeen-ounce projectile was much less, and any enemy soldiers employing the barest of overhead cover had little to fear from the grenades.[37]

The combination of rifles, automatic rifles, and grenades theoretically gave the American small units the firepower they required to deal with most of the tactical situations they encountered. However, there was always a sharp divide between theory and reality. As with the heavier weapons at regimental or battalion level, the soldiers and leaders of companies and platoons often lacked a sound grasp of how to employ their organic weapons. As this was one of those variables on the margins of attritional war in which the leaders could gain a slight advantage in grinding away at the enemy, any inability to use these weapons constituted a major flaw in the combat leadership of the AEF's junior officers and NCOs.

These problems were not lost on some senior officers. In a 25 August 1918 review of the 1st Division's operations in the Aisne-Marne Offensive, Maj. Gen. Charles P. Summerall concluded that even in his veteran unit, infantry officers and NCOs were failing to get the best use from the weapons at their disposal. The report noted that when it came to the use of the Chauchat automatic rifle,

> these important weapons were virtually turned over to the individual carrier and gunners for such use as they were able to make of them. Many were not fired at all, others were fired at random. Automatic rifles were fired from the hip without need, and the ammunition was often needlessly expended. . . . Sometimes the guns were abandoned because the men were tired, or because the ammunition was exhausted. None but the automatic rifle teams had been taught to operate them, and some of these men had only a short period of service and did not know how to prevent or correct jams.[38]

These problems were not limited to the 1st Division. *Notes on Recent Operations, No. 1,* issued shortly after the campaign, recorded that "many organizations failed to use the fire of rifles and auto-rifles to assist in the advance." First Sgt. Harold C. Woehl, of the 32nd Division, reported that during the first day of action in the Oise-Aisne Offensive, his company "had no automatic rifles, no pistol ammunition, no hand or rifle grenades. Just our trusty Springfield rifles and plenty of guts." Although Woehl did not explain why these key weapons were missing, the end result of their absence was clear. As the company attacked toward Juvigny, it could not suppress the German machine guns that raked its ranks. On that day of fighting the company lost sixteen men killed and twenty-seven wounded.[39]

Part of the problem of properly using these weapons was the same one that plagued the larger supporting weapons: weight and ammunition consumption. Soon after the Soissons drive, a 1st Division officer noted, "It is evident . . . that the loads of the automatic rifle and machine gun men are very much greater than the loads of men carrying the service rifle." He went on to point out, "This difference of load, and consequent difference of mobility . . . at once make themselves felt when troops must move for long distances over difficult terrain. The men carrying the heavier weapons fail to keep up with those carrying lighter weapons and either fall behind or abandon the heavier weapon for a lighter one which enables them to follow."[40] It was not just that the weapons themselves were heavy, but also that the ammunition and accouterments that went with them added to their crews' burden.

Capt. Ashby Williams estimated that his riflemen marched into the Meuse-Argonne Offensive with sixty to sixty-five pounds of weapons and gear each. Although some units shed as much weight as possible from their soldiers' packs prior to going into action, the doughboys were still weighed down by loads of approximately forty to forty-five pounds.[41] With the extra weight of his weapon and ammunition, a Chauchat gunner would have gone into action with fifteen to eighteen more pounds of gear to carry than the average rifleman. This estimate would not have been much less for a gunner with a BAR. While these details may sound minor, that extra weight slowed down the responsiveness and agility of the weapons teams, made them more vulnerable to enemy fire, and encouraged the less disciplined of the gunners to ditch these valuable weapons at first opportunity.

General Summerall tried to address both the issue of weight and the proper employment of these weapons. In late August 1918 he directed that company commanders and platoon leaders place their automatic rifles under the control of a competent NCO trained in the tactical employment of the weapon. He ordered that all infantrymen would learn how to fire the Chauchat so anyone could crew the weapon in the event of casualties. To ease the load carried by the automatic rifle team, Summerall further stipulated that every soldier in an infantry company would carry an extra loaded Chauchat magazine into action.[42] Unfortunately, the standard M1910 American infantry pack had little enough space for the soldier's rations and clothing, let alone the rather bulky half-moon magazine for the Chauchat.

Although Summerall's actions reflected a proactive approach to the dilemma of combining firepower with maneuver, on the whole the AEF's small-unit leaders were often inept at employing their support and organic weapons in their attacks. As Col. George Marshall noted shortly after the war, company commanders were glacially slow "in learning how to combine fire action with maneuver."[43] When combined with the challenges of combat physics and the defenders' inherent advantages, the junior leaders' inexperience and lack of know-how made it exceedingly difficult for them to take advantage of the few opportunities they could wring from the margins of attritional war.

IO The School of Hard Knocks

THE FRENCH MARSHAL FERDINAND FOCH is said to have remarked that it took fifteen thousand casualties to train a major general. Although this assertion sounds rather cold-blooded to modern ears, Foch was offering an honest assessment of the grim internal logic of combat in the Great War. None of the major combatants were truly prepared for the bleak realities of the mass attritional warfare they encountered on the Western Front. Sadly, senior officers had to pass through the bitter schooling of experience, a schooling whose matriculation was paid for in time and the consumption of human life, before they could understand even the vaguest outlines of the battlefield challenges that they faced. If Foch was right about the number of casualties required to train a general in World War I, one wonders how many casualties did it take to train the war's company-level leaders?

Marshal Foch also made one other incisive observation. He once advised young officers, "There is no studying on the battlefield. It is simply a case of doing what is possible, to make use of what one knows, and, in order to make a little possible, one must know much."[1] Many of the problems faced by the war's senior leaders were the same as those faced by junior leaders; they differed only in scope and scale. Both had to find the enemy and divine their strength and intentions, employ fires to suppress or neutralize the enemy, coordinate fire with maneuver, match tactics and movement to the given terrain and enemy, plan and execute resupply efforts, and maintain command and control to keep their subordinates directed toward achieving the mission of the higher headquarters. On the margins of attritional war, the junior leader could only manipulate the variables of manpower, weapons, tactics, and the power of his leadership to squeeze every drop of advantage in combat. This was the "much" that Foch believed the leaders had to know "in order to make a little possible." As the captains, lieutenants, and NCOs learned how to accomplish these vital tasks, what were acceptable casualties of soldiers in the school of hard knocks?

Defining "acceptable losses" is problematic because they will always reside in the eye of the beholder. One could argue that the Allied armies' ability to grind the German forces down in 1918 meant that regardless of any wastage at the tactical level, the means of attrition ultimately achieved the strategic end. While accurate, this thought was far from comforting to American soldiers. During the Great War, American units remained in combat well after enduring casualties that would render modern US Army formations combat ineffective. However, the participant accounts suggest that the AEF GHQ and the American soldiers understood the definition of "acceptable losses." For example, the GHQ took steps to prevent such future bloodlettings after the shocking losses the Americans had taken during the Aisne-Marne Campaign. As there was no universal standard for acceptable losses or for defining the point at which a unit becomes combat ineffective, the author will leave it to readers to draw their own conclusions based on the evidence and statements made from the participants themselves.

One of the major variables on the margins of attritional warfare that the AEF's small-unit leaders could influence was their use of tactics. Here, small-unit tactics are defined as the art of applying or adapting formations, maneuver, and firepower to the specific challenges presented by an enemy on a given piece of terrain. The end result of tactics was simple: accomplishing the mission given to the unit by the higher headquarters and, in the process, inflicting as much damage on the enemy as possible while preserving as much of one's own manpower as feasible. To wring the best results from tactics, junior infantry officers and NCOs had to be well versed in the capabilities and limitations of the weapons and their soldiers and be able to adjust their formations on the fly to balance the needs of protection with firepower. Leaders also had to grasp how to use the terrain to minimize their own casualties while also placing the enemy at a disadvantage. Last, they had to be able to issue clear orders in the heat of combat that directed their subordinates to accomplish the leader's goals and intent.

It was in the area of tactics that all of the American shortcomings and shortcuts in training were laid bare. From the use of terrain and formations to the combining of fire and maneuver, American small-unit leaders generally showed themselves to be rather sluggish, doctrinaire, and ham-fisted in their application of tactics. This is not to cast blame on the officers and NCOs themselves for this tactical clumsiness. No infantry captain was willfully incompetent or clapped his hands in joy at the thought of leading his soldiers to destruction. The simple fact was that much of their training and experience had not prepared them for the complex task of leading 250 men across difficult terrain against a skillful enemy. As their training had been poor, their schooling at the hands of the Germans was painful. As one infantry battalion commander commented, since his junior leaders had failed to learn the required tactical skills and knowledge prior to combat, the "officers . . . must learn their business from day to day at the expense of their trade in human beings. The latter must bear the cost of this learning and pay the price of every experiment in the process."[2]

As early as the summer of 1918 the AEF GHQ noticed worrying signs that the tactical skills of the army's infantry officers and NCOs were somewhat lacking. The American losses during the Aisne-Marne Offensive were particularly heavy and unexpected. While insufficient artillery support and mistakes by senior French officers certainly contributed to the bloodletting, at the battalion and below it was much more a case of poor tactical skills rather than these operational missteps that added to the butcher bill.

The common mistake of American infantry leaders was to confuse tactical formations for tactics. Formations were merely a means of moving soldiers forward by balancing mobility with protection and the ability for the leader to deploy his soldiers rapidly to meet situations in battle using fire and maneuver. In a lecture given shortly after the Armistice, Brig. Gen. Frank Parker neatly encapsulated the tactical challenges that faced American small-unit leaders. He observed, "The object of the infantry attack is to come into physical contact with the enemy and to be alive at the same time. This war has demonstrated the fact firepower is practically independent of numbers; a small efficient, determined combat group consisting of very few men produces a most powerful and destructive fire. The problem then consists in pushing forward a maximum of fire with a minimum of personnel, this personnel so manipulated as to present a minimum target to hostile fire, artillery and small arms." Parker went on to state that the most important element of the division was the platoon, "as it is the substance out of which the Division is made—if this substance is not solid the division will be a weak structure, incapable of standing strain."[3]

In the summer operations, the AEF's junior leaders seem to have fallen into two broad camps: those whose formations and deployments were based on a hodgepodge of techniques, and those who rigidly kept to the tactical system illustrated in *Offensive Combat of Small Units.* In the *Notes on Recent Operations, No. 1,* which followed the Aisne-Marne Offensive, the GHQ staff reported that during the fighting "an endless variety of attack formations was used." It also provided an example in which a company commander had so failed to understand tactical deployments that he positioned a single platoon on a frontage of six hundred meters, negating any hope the platoon leader had of commanding his unit. An infantry NCO noted that during his company's fighting in August, his commander "quickly formed one 'skirmish line' of the old style" as his preferred method for attacking German positions. Both of these passages indicate that some small-unit leaders were assembling their tactical formations and techniques from a host of doctrinal and nondoctrinal sources. For example, the formations described by the NCO show the continuation of tactical concepts from the prewar *IDR*. That such a mélange of formations and tactics was still in use in the AEF in mid-1918 should come as no surprise. The fact that American units had been bombarded with rapidly changing tactical techniques from British, French, and American sources certainly encouraged an à la carte approach to formations and tactics.[4]

The other, more common extreme was those leaders who unbendingly followed the formations given in *Offensive Combat of Small Units*. Although the distances between soldiers and echelons in the manual were too short, there was nothing inherently wrong with the tactical concepts and intent of the manual's formations. The problem was in the leader's application of those formations to the terrain and enemy. In the AEF's *Combat Instructions,* Pershing decried the fact that

> attack formations of platoons, companies, and battalions are everywhere too dense and follow too rigidly the illustrations in the Offensive Combat of Small Units. Waves are too close together; individuals therein have too little interval. Lines are frequently seen with the men almost elbow to elbow and seldom at intervals greater than two to three paces. . . . All formations are habitually lacking in elasticity; there is almost never any attempt to maneuver, that is to throw supports and reserves to the flanks for envelopment. . . . Subordinate officers display little appreciation of the essential situation and how to best meet its requirements.

This point was also made by Capt. Manton Eddy of the 39th Infantry. His unit's push to the Vesle River on 3 August 1918 progressed in strict adherence to the formations pictured in *Offensive Combat of Small Units* without adjusting them in any way to account for the enemy's dispositions and the terrain the Americans encountered. Eddy later recalled that his unit's close formation "was a sight that must have made the German observers gasp in amazement, for before them lay an artilleryman's dream."[5]

Training units to move in formations and to change from one formation to another is relatively easy. In fact, Capt. B. A. Colonna, the commander of B Company, 311th Infantry, claimed that "the new formations were mastered remarkably quickly" in his unit.[6] The difficult part was for the leaders to understand the linkages between those formations and the tactical techniques of fire and maneuver that they had to employ to overcome the German defenders. These difficulties were magnified when the junior leaders experienced trouble with command and control due to casualties and the normal fog and friction of combat.

The Americans' tactical clumsiness was not lost on the Germans. A staff officer of the German 7th Army reported on 23 July 1918 that when the Americans attacked, "[Our] defense was too strong for the limited attack power of the enemy infantry. When the fire protection of the artillery ceased, when the tanks were lost, only seldom did it continue forward. It gave ground to every counter-attack even when made by inferior German numbers. As a result of its dense formations . . . it suffered heavy, bloody losses whenever it encountered our artillery fire." The end result of the dearth of tactical savvy in the summer of 1918 was high casualties as the Americans attempted to smother the German positions under the weight of the doughboys' mass formations. The consequence of some junior officer's tactical ineptitude left a lasting impression on Pvt. Horatio Rogers. He later remembered, "Turning back across the fields I passed between groups of dead Americans lying in short windrows as they had been mowed down by the machine guns from the woods. Wave after wave had evidently assaulted from the ditches along the road before the survivors had obtained a foothold in the woods."[7]

The dead Americans that Rogers viewed were far from rare. In the 1st Division alone, the losses were 234 officers and 7,083 men killed, wounded, missing, or captured. The 26th Infantry lost all of its field-grade officers, and one battalion of the 28th Infantry lost all save one of its officers on the first day of the battle. The remaining officer was wounded on the second day of the drive. The 28th Division's 110th Infantry suffered over 1,100 casualties in its repeated attempts to take Cierges by frontal attack on 28 and 29 July 1918. One of the unit's officers honestly admitted that after charge after charge "up the bare slope of the hill," the failure could be summed up as, "artillery was insufficient, team work lacking, and information regarding the terrain meager." Another observer noted, "The men moved with such precision that it looked more like a drill than a great battle."[8]

The veterans of Soissons and the other summer battles often profited from this school of hard knocks. Edward Johnston, an officer in the 1st Division, recalled that in the fighting of May to July, the unit's infantry regiments usually attacked in mass waves that moved "forward ponderously with heavy losses against hostile fire, with no apparent effort to utilize cover." The infantry's "attack formations were generally too thick," and the doughboys exhibited a "tendency to attempt to overcome resistance by shock rather than by fire and shock combined." He did note, however, that the survivors of these encounters gained a large degree of battlefield wisdom and experience. Johnston claims that during the Soissons battle French Moroccan soldiers taught his 28th Infantry how to advance by "moving at a run from shell-hole to shell-hole" in a style "utilized by the European veterans." Overall, however, Johnston had to admit "training in musketry, combat practice, employment of maneuver, and tactical use of its auxiliary weapons was so deficient [in the AEF] as to greatly increase its losses in the attack."[9]

In other units the process of internalizing the lessons of recent operations proceeded more slowly. Herevy Allen noted that the Soissons drive was a bleak coming of age for his soldiers. He emphatically stated, "It was the grim common sense of the 'doughboy' and not our obsolete and impossible tactics that won the ground. Oh! the precious time wasted in our elaborate, useless, murderous 'science' called 'musketry.' . . . Teaching it should be made a court-martial offense. It is murder in print. Battles were not fought in lines."[10] Although Allen's assessment was correct, there was still much to be learned by his unit. For all "the grim common sense of the 'doughboy,'" it did not stop his battalion from still using dense formations of infantry shortly afterwards in its disastrous attacks on Fismes and Fismette.

As the Americans moved into their first independent operation at Saint-Mihiel, the tactical acumen of the AEF's infantry captains, lieutenants, and NCOs was very much a mixed bag. Some of the AEF's more experienced divisions had made great strides in infantry-artillery cooperation and in instilling a degree of tactical flexibility in their small units. Some had learned the benefits of thoughtful preparations and in taking time to understand and logically responds to a given tactical situation rather than trying to bull forward. Col. H. P. Hobbs, the inspector of the 26th Division, observed, "Our troops had learned much during the Second Battle of the Marne. . . . The infantry worked with much more caution and much better team-work and control than they did during the Second Battle of the Marne. The advance was not delayed by this caution, in fact, much time was saved and our losses greatly reduced." Unfortunately, this coming of tactical wisdom was not universal.[11]

The Saint-Mihiel Offensive was the first real taste of combat for many of the AEF's divisions. Despite the efforts by the GHQ to use its *Notes on Recent Operations* and *Combat Instructions* to spare the new divisions from the mistakes of their predecessors, the actions of the green units at Saint-Mihiel highlight the fact that the AEF's methods of transmitting its hard-won lessons learned were largely ineffective. The new divisions repeated the errors of pushing infantry forward in mass formations and failing to match firepower with maneuver. The small-unit leadership exhibited in these units continued to illustrate the army's overarching problems with leader training and initiative.

The experience of the 82nd Division is illustrative of leadership problems in the new divisions and the AEF's failure to instill the lessons learned from its previous battles into the newcomers. While the division's role in the operation was small, Saint-Mihiel revealed that many of the division's infantry officers and NCOs were as deficient in the basic skills of map reading and small-unit tactics as had been those involved in the Soissons drive. The confusion caused by these deficiencies is best illustrated by comments made by the 326th Infantry's Lt. Justus Owens in a letter that he sent to his mother soon after the battle:

> We left our present positions about 9:00 p.m. . . . We headed for our objective . . . but hadn't gone far until we decided we were headed in the wrong direction. . . . It afterward turned out that we were headed in the right direction at first and lost out (and ourselves) by turning right. . . . We wandered around in the rain and slush and mire of no-mans land for several hours. . . . We finally located our woods about 2:15 a.m. It was still so dark that we could hardly see anything, so I placed my men in one corner of the woods and told them we'd hold tight until it got lighter.

Luckily for Owens and his soldiers, their objective had been abandoned by the Germans. However, his blundering attempts to find the objective and his failure to clear it while he still possessed the cover of darkness put his soldiers at great risk and gave his men grounds to question his leadership. Owens admitted that after their night of futile wanderings his soldiers were wet, tired, and in "bad humor."[12]

Not all of the 82nd Division's soldiers were as lucky as Owens's command. George Loukides, a private in H Company, 326th Infan-

263

try, noted that his officers "were not trained for combat and the privates paid for it." He recalled that at Saint-Mihiel his company lost "many killed" when their officers led an attack across a dangerously open field in broad daylight. Alvin York wrote that during the 328th Infantry's attack on Norroy the regiment's companies "got mussed up right smart," and his unit's inability to protect its flanks or adequately maneuver its units allowed the Germans to enfilade the American positions.[13]

The Saint-Mihiel operation also pointed to the continuing inability of infantry units to adapt their formations to the ground and the enemy and the Americans' penchant for resorting to frontal assaults when coming in contact with the Germans. Even experienced small-unit commanders continued to fixate on maintaining the dressing and alignment of their unit formations as they pressed forward. Pvt. Charles MacArthur was amazed that during one infantry attack by the veteran 42nd Division, "the doughboys were scrambling out of their trenches. . . . Their officers ran after yelling: 'Dress on the right you gosh dam lousy doughboys!' . . . The doughboys strung along like crowds following a golf match, slowly and deliberately, dressing on the right whenever they were told. Here and there a man stumbled and fell. The line moved on under a cataract of shrapnel and high explosive."[14] MacArthur's account was backed by Lt. Hugh Thompson, who remembered yelling for his soldiers to "keep your three-yard interval" and recalled the efforts he made to keep his soldiers in strict formation when they attacked in the first wave of the Saint-Mihiel Offensive. He soon admitted the futility of retaining the lock-step methods of his training under heavy enemy fire as "all thought of controlling the scattered line gave way to fearful self-preservation." Thompson realized that "we'd be killed if we lay still" and that "there was a gambling chance if we charged ahead," and he began to move his men forward by "a mad dash" from cover to cover.[15] Under the press of necessity and grim reality, Thompson had accidentally stumbled upon the proper use of movement and terrain.

Here again, the German defenders offered an honest assessment of the fighting skills of their American foes. The intelligence officer of the German Army Detachment C, facing the Americans at Saint-Mihiel, reported, "The American advance at the time of the infantry attack was entirety schematic, and betrayed a great lack of skill in the movement of the support waves following in dense formations over the terrain. . . . [They] gave an impression of awkwardness and indecision. Neither the officers nor men knew how to utilize the terrain. If they encountered resistance they did not try to seek shelter, but fell back walking upright. The Americans do not know how to move either forward or backward by crawling on the ground or by sudden rushes." The German noted that while the average American soldier "is doubtless[ly] brave," he "is ignorant of the proper behavior in the course of an attack," and "grenades put him to flight at once." Of the doughboys' officers, he wrote, "The command is extremely bad and without initiative. The enemy has obviously many officers, but they all lack an aptitude of command. Their embarrassment was obvious when they attained their initial success. They found themselves at a loss in the presence of a new situation, and were not capable of exploiting the success."[16] The Germans' pointed observations on the American performance at Saint-Mihiel provided sad evidence of how little the American doughboys had learned from the Aisne-Marne Offensive and how far the novices still had to go to be effective in combat.

For most of the new divisions, their participation in the Saint-Mihiel Offensive was not long or intense enough for them to gain the degree of experience required to season their officers and soldiers for the challenges that lay before them in the Meuse-Argonne. An infantry officer in the 89th Division wrote in his after-action report of the operation that "the formations adopted and the means at hand proved adequate for overcoming the resistance met." However, he was also circumspect enough to admit that "had the enemy chosen to occupy his works in stronger force and offered a stiffer resistance, it is believed that our lack of time for thorough consideration of orders and study of maps would have cost us severely." The false impression and illusions that the soldiers of the 5th Division had taken from Saint-Mihiel were later shattered by the realities they faced in attempting to take Cunel on 11 October 1918. As one officer stated, "The men still remembered the victorious rush at St. Mihiel and dashed forward impetuously. But it was a different enemy here, one who was sticking till the last and fighting for every foot of ground." Unfortunately, even if these units had been so inclined, the issuing of orders and the movement of troops to their staging areas for the Meuse-Argonne Offensive left no time to correct the leadership deficiencies brought to light by the Saint-Mihiel fight.[17]

The Meuse-Argonne Offensive was the AEF's largest and deadliest battle of the war. Pershing hoped that the campaign would vindicate his insistence on an independent American army trained in the fine art of open warfare. Instead, the Meuse-Argonne turned out to be a forty-seven-day ordeal that pushed the AEF to the breaking point. Gen. Hunter Liggett, the I Corps commander, noted that the region was "a natural fortress beside which the Virginia Wilderness in which Grant and Lee fought was a park."[18] The German defenders had occupied the area since 1914 and had placed much effort into preparing the fortifications of their Giselher, Freya, and Kriemhilde defensive lines.

Not only was the Meuse-Argonne crisscrossed by a vast array of natural and man-made obstacles, but the Germans had also made great strides in tying these obstructions into a system of mutually supporting machine gun and artillery positions. To add to these problems, since 1917 the Germans had attempted to offset their losses in manpower with increased numbers of machine guns in their battalions. For example, by 7 October the German 123rd Infantry Division was down to 89 officers and 1,705 men. Yet it could still wield 198 heavy and light machine guns, or roughly one gun for every eleven German defenders.[19]

Even with fully trained officers and soldiers, the Argonne Forest and the rolling hills of the Meuse region would have presented a for-

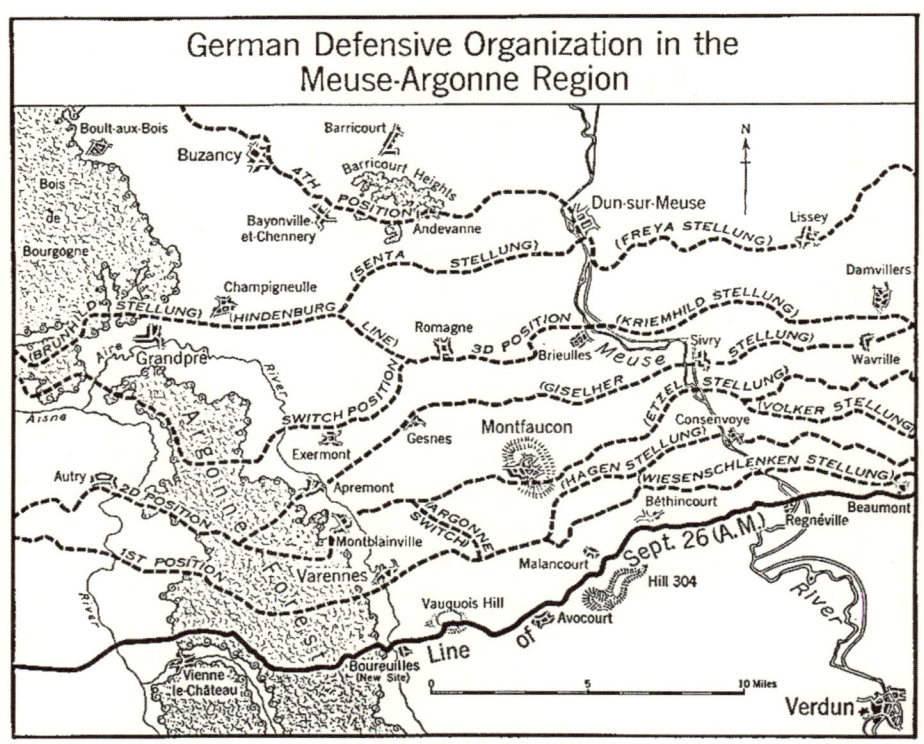

(*From American Battle Monuments Commission,* American Armies and Battlefields in Europe.*)

midable obstacle to any army. The complexities of the terrain and the German defense required that junior leaders have a "master's degree" in tactics, while most of the American junior officers were barely out of the grade school of the profession. The Meuse-Argonne Offensive was the litmus test of the AEF's junior leadership. The great losses and near disintegration of American units in the battle was ultimately the price the AEF paid for its failure to properly train and develop its company-level officers and NCOs.

It is evident from the reports of wartime German units and American staff officers of the AEF inspector general's office, postwar officer boards, and participant accounts that the AEF entered into the Meuse-Argonne with grave problems with its junior leaders. The continued inability of these leaders to combine formations, maneuver, and firepower to the terrain and enemy they encountered came at a great price to themselves and their units. The casualties caused by these missteps undercut the morale and effectiveness of the AEF and fueled a personnel crisis within its ranks that only deepened the army's other systemic problems.

Given the nature of the enemy defenses, any American attack that did not synchronize formations, maneuver, and firepower was likely to fail. In his *Combat Instructions,* Pershing fairly pleaded with his infantry platoon leaders and company commanders to suppress enemy machine gunners with American fire from the front or flank while maneuvering some portion of the unit to attack the German position from the flank. He urged that "where strong resistance is encountered, reinforcements must not be thrown in to make a frontal attack at this point, but must be pushed through gaps created by successful units, to attack these strong points in the flank or rear."[20] Pershing's admonitions were absolutely correct. But what seemed so easy on paper at Chaumont was much more difficult for junior leaders actually to carry out in the tangle of the Argonne.

To some extent the young leaders could be forgiven their tactical sins. Even well-planned and -executed attacks faced the reality that seeking the enemy's flanks was easier said than done. Soon after his battalion of the 79th Division started advancing on Montfaucon in the opening hours of the Meuse-Argonne Offensive, Maj. Charles DuPuy came to the disturbing realization that "the tactics which we had learned, proved to be of little value." He discovered, "We had always been taught to attack and take a machine gun by the flanks, but in trying to do so we simply ran into a frontal fire from a machine gun on one side or the other of the one we were trying to take, so that it was necessary a great many times to simply charge a gun from the front and both flanks, and take it regardless of our losses, which, per gun captured, averaged ten to twenty men." This dilemma was also noted by Brig. Gen. George C. Barnhart, who commanded the 1st Division's 2nd Infantry Brigade through some of the Argonne fighting. He noted that the tactics used by the AEF seldom enabled it to overcome enemy resistance without heavy losses and pointed out that when American units tried to flank one machine gun nest, the flankers merely stumbled into the fire of another well-sited gun.[21]

268

Col. E. G. Peyton, who commanded the 80th Division's 320th Infantry, echoed DuPuy and Barhart's observations in this description of the fighting in the Argonne: "Here in this irregular line troops gave battle to the German machine guns that could be seen only at rare intervals. The enemy intrenchments [*sic*] afforded every advantage in position, concealment and for enfilade fire. Time and again rushes were made from the front and flank against the nests only to be met by a curtain of lead that was absolutely impassable. . . . Here lives were needlessly lost in trying to rush through this curtain of lead." When faced with such tactical conundrums, the junior leaders either tended to push the attack to the limit and, as the 28th Infantry's commander, Col. Adolph Hugeut, put it, sacrifice "themselves and their men to put the Machine gun out of action," or became so cautious and hesitant that their units gained little from their losses.[22]

The AEF's junior leaders continued to make some of the same grave mistakes as had been made in the AEF's earlier campaigns. Soon after the Armistice, the AEF inspector general produced a summary of his inspectors' reports of American units in combat from 12 September to 11 November 1918. These observations were able to show the army's steep and costly learning curve during its heaviest period of action and a general assessment of the effectiveness of its combat leaders in overcoming the challenges thrown at them during the fighting.

The report noted that in the opening weeks of the Meuse-Argonne Offensive the "infantry would advance in their prescribed attack formation until they would run into machine gun fire[;] they would then halt and call for a barrage or for artillery preparation and would advance in frontal attack upon the machine gun nests, suffering heavy casualties" when the fires were not forthcoming. The inspectors pointed out that "serious losses at first were suffered through the fact that the infantry would be held up by some few machine gun nest [*sic*] long enough to loose [*sic*] the protection of their barrage." Toward the end of the operation, infantry junior officers had gotten better at combining suppressive fires to pin down machine guns while small parties of soldiers moved around the guns to attack them from the flanks. The artillery also became more adept at holding their creeping barrages when the attacking infantry became stalled. Sadly, the report noted a continuing problem: "Although orders were issued respecting formations in depth the reserve lines kept crowding forward whenever the attacking battalions were held up. In several instances this resulted in serious machine gun casualties in the support and even in the reserve battalions. The rear lines could not be made to see that their crowding forward did not help the attack but merely fed the men to machine guns."[23] Unfortunately, the inspector general failed to provide any analysis of why the AEF made progress in some tactical areas while it continued to struggle in others. A possible explanation is that heavy losses of junior leaders doomed the AEF's small units to a constant cycle of new groups of officers and NCOs endlessly having to learn the same lessons over and over again.

 The AEF inspector general noted that across the army leaders were making fundamental tactical errors or mistakes in judgment that were leading to ever greater numbers of casualties. Junior leaders all too often failed to make proper reconnaissance of the terrain over which they were to attack prior to the assault. Planning was based mostly on maps, with little to no effort made by the leaders to match the plan to the actual terrain and enemy in the area. The failure of reconnaissance meant that American attacks often blundered into German positions, causing unnecessary casualties and throwing off the timelines of the operation. A soldier in the 328th Infantry, for example, reported that when his company attacked near Sommerance on 14 October 1918, it lost thirty-eight men, including the platoon leader, in one brief encounter after the lieutenant led his unit across an open field without conducting even the briefest of reconnaissance.[24]

 The AEF inspector general, Maj. Gen. Andre W. Brewster, became so concerned about the lapses of sound combat leadership among the AEF's junior officers that he expressed his belief that drastic measures needed to be taken to correct the deficiencies. On 21 October 1918 he wrote to the AEF chief of staff:

> As soon as the present period of active operations ceases, a series of platoon leader schools should be established through which all platoon leaders should pass before the next period of active operations. Not only have platoon leaders shown lack of resource when confronted with situations on the battlefield but they seem to have forgotten the use of the various instruments of the offense in their platoons. When stopped, instead of developing all of their resources, they have frequently pushed forward blindly

269

and had their platoons suffer heavily. The platoon is the basis of the offensive and our platoons in this last offensive have not developed all of their offensive power nor have the platoon leaders kept their men together.[25]

Although the war ended before the inspector general's recommendations could be put into effect, the existence of the memorandum and the force of his statements highlight the fact that the GHQ was growing increasingly concerned with the abilities of leadership of its junior officers corps.

The inspector general and the GHQ had good reasons to be concerned. Time and time again, American officers were leading their men forward in mass formations against the frontal and flanking fire of machine guns. An artilleryman in the 32nd Division observed one of these attacks in the early days of the Meuse-Argonne drive. He was mesmerized by one attack in which "the infantry advanced with courage and seemed to be making good headway. We noted that all their dead lay in little circles. It was instantly clear to us that they must be entirely inexperienced, or they would never group together while enemy field-gunners had them in sight. . . . Occasionally their officers scattered them but it was difficult to keep them apart. They were evidently coming together for comfort and sympathy, forgetting that in their situation the best comfort lay in keeping far apart and offering enemy gunners the smallest possible target." The infantryman John Barkley saw the results of a similar one of these failed attacks launched by the 5th Division against Cunel in October 1918. He later wrote, "They had evidently tried to hold their formations as they burst from the woods, and had met with machine gun fire from the opposite hill. They were lying now in wave formations. Some of them must have rolled and twisted as they died, but not enough to disturb the outlines of that wave." However, Barkley's own leaders seemed little better in directing his company than had those of the dead men lying on the hillside. As his unit attacked through the Bois des Ogons, he recalled, "Suddenly a heavy rifle fire and automatic rifle fire opened directly ahead. I heard somebody yell, 'Lets Go!' and we ran straight forward. At the same time the Germans on the right end of the line opened up on us with dozens of machine guns." During the chaos of combat, Barkley's junior leaders simply limited their tactical options to massing their units against the closest source of fire and charging forward.[26]

The experience of trying to lead the oversized American units into combat in the Argonne sector was equally bewildering to many of the officers themselves. Lt. Joseph Lawrence admitted that during his unit's attack on the Bois des Chênes on 10 October 1918, "I had difficulty keeping my line straight and moving forward in the midst of the chaos, and the men gathered in groups in spite of all the sergeants and I could do." Lt. Maury Maverick, a new replacement officer with the 1st Division's 28th Infantry, was completely unprepared for his first experience in combat. When his veteran officers ordered a frontal attack near Exermont on 4 October, Maverick recalled,

> Most of us who were young American officers knew little of actual warfare—we had the daring but not the training of the old officer of the front. The Germans simply waited, and then laid a barrage of steel and fire. And the machine gunners poured it on us. Our company numbered two hundred men. Within a few minutes about half of them were either killed or wounded. . . . Everything happened that never happens in the storybooks of war. . . . There were no bugles, no flags, no drums, and as far as we knew, no heroes.

Under the circumstances encountered by Lawrence and Maverick, the leaders could only draw upon their meager training and experience to puzzle out a solution to their tactical problems.[27]

As Maverick stated, the young officers had all of "the daring but not the training" to cope with the German machine guns. In these situations, junior officers tended to pick the tactical paths of least resistance, in terms of the use of both terrain and tactics, to attack the enemy. Undertrained and inexperienced leaders and soldiers are predisposed to follow routes that offer the easiest way forward. They tend to follow paths and openings rather than try to push through areas that are thickly wooded or overgrown. This fact was not lost on the German defenders. A company commander in the 78th Division recalled that the Germans attempted to canalize the Americans by "barring all other approaches with barb wire" except for paths that led directly into the guns' field of fire. He noted that this tactic was very effective, because "after you've struggled in barb wire for a while you'll take a chance on machine gun bullets to get on a path." The officer claimed that after a few sorrowful experiences with this German technique, he and his men became more adept at avoiding them.[28]

Despite their growing awareness of these German machine gun tactics, American infantry units still continued to make many of the same mistakes in attacking them throughout the war. A veteran soldier in the 42nd Division, Martin Hogan, recalled that his officers and comrades came to use "all that they had learned about fighting against men in cover" in his unit's actions in the Argonne. What stayed with him, however, was that even after realizing that "the paths that led to the enemy machine-gun nests were almost unthinkably bad," leaders continued to rush through them "front-on, again, and again and again."[29]

It is interesting to note that these tactical failings were not limited to those American units fighting as part of the AEF. The 27th and 30th Divisions, attached to the British army during all of their active campaigning, exhibited similar lapses in leadership. In an after-action report of the Ypres-Lys Offensive, an officer from the 30th Division noted,

> The natural tendency of men seems to be to rush . . . [machine gun] nests in frontal attacks instead of using a flanking or enveloping movement. . . . Small unit leadership: next to lack of liaison the most glaring defect. Small unit leaders wait to be told how to do every little thing and use little or no initiative of their own, do not assume enough responsibility. In all training of men, they have been too dependent on officers telling them not only what to do but how to do it. . . . Platoon leaders have not had instruction and almost no practice in the actual use of maps and aerial photos.[30]

Thus, even with the benefit of close contact with experienced Allied troops and British schools, American junior leaders remained hobbled by their poor initial training.

American junior leaders also demonstrated shortcomings in their knowledge of gas warfare as well as in key leader tasks such as map reading. During the Great War the Americans lost over 1,400 men who were killed outright or later died in hospital wards due to gas poisoning. The AEF hospitalized over 70,552 men for gas exposure, and gas accounted for over 27 percent of the army's total combat-related casualties. An unknown number of these 70,552 men later died of complications brought about by gas exposure or limped on with poor health for the rest of their days.[31]

By 1918 the major European combatants had reached a point at which gas had, for the most part, become a weapon of harassment—one that degraded the effectiveness of their soldiers without causing an unmanageable number of casualties. This was not the case with the AEF. As the figures above show, the AEF lost considerable manpower to gas exposure. Most of the reason for the AEF's relatively high losses to gas was poor training and leadership. The reports of gas attacks against American forces compiled by E. W. Spencer in 1928 consistently noted that the AEF's officers, NCOs, and soldiers where woefully untrained in gas warfare. Additionally, failures in leadership, such as the lax enforcement of gas discipline and poor decisions by officers and NCOs, also caused unnecessary losses. For example, in a gas attack against the 26th Division on 10 October 1918, 111 casualties in the 101st and 102nd Field Artillery were caused "principally by premature removal of respirators," ordered by a junior officer. In another case, a battalion of the 386th Infantry had to hospitalize over two hundred men for gas exposure after it was hit by a German gas attack on 6 October 1918. An officer from one of the gassed companies "gave the order to remove masks in less than two hours after the bombardment." The division gas officer reported, "I cannot understand how any officer could be so ignorant of the effects of mustard gas to issue such an order and *this order was responsible for at least two-thirds of the casualties in this attack.*"[32]

273

Junior leaders also had some difficulty in mastering the art of land navigation. There is an old army joke that the most dangerous man on the battlefield is a second lieutenant with a map. Sadly, during World War I the second lieutenant in question was also joined by his captain and major when it came to map reading. In an examination given by the II Corps headquarters to a mixed group of field-grade officers immediately after the war, only five out of fifty-seven leaders tested could accurately locate map coordinates. The ability of leaders to use maps and compasses to report their positions, plan fire support, and find their way to tactical objectives was a vital combat skill that was not given enough emphasis during the training of the AEF's officers and leaders. Given the tangled terrain of the Argonne sector, leaders frequently had to follow compass directions just to move forward. Yet an officer in the 5th Division stated that training in the use of a magnetic compass was not given to his unit until a lull in the Argonne drive, after experience had demonstrated the unit's need for such equipment and training.[33]

Failures in land navigation could have critical implications for small-unit actions. Cpl. Berch Ford, a soldier in the 16th Infantry, recalled that during the Soissons drive, his platoon discovered that its lieutenant could not read a map and had taken the unit far beyond their objective. The unit was fortunate that the platoon sergeant was a long-service regular and reoriented that platoon after telling the officer that he was taking over command. During the Meuse-Argonne fighting Capt. Sam Woodfill got into an argument with a fellow company commander whose inability to read a map was jeopardizing his unit. Woodfill noted, "Instead of makin' for the woods I had just patrolled they were goin' straight toward another wood which we knew was chuck-full of Germans." While he managed to talk the other captain into changing his course, the mixed-up officer left Woodfill still declaring that he knew exactly where he was.[34]

While all of this paints a rather bleak portrait of American junior leadership, it must be stated that not all American leaders made these mistakes or failed to learn from their experiences. Some officers actively sought to adapt their tactics to the challenges that confronted them. The diary of Harold C. Woehl indicates that the officers of the 126th Infantry experimented with new tactics and methods of organizing and fighting after their heavy losses in the Aisne-Marne and Oise-Aisne Campaigns. On 20 September 1918, First Sergeant Woehl recorded that his company received orders from the regiment to practice fighting "in line of gangs." This was different from the combat groups listed in *Offensive Combat of Small Units*. Rather than having squads organized around a single weapon, as was done in combat groups, the combat gangs were all-arms squads built "with the Automatic Rifle as a group nucleus."[35]

First Lt. Fred Jankoska, the company commander for H Company, 126th Infantry, also displayed a marked ability to change his formations and tactics to suit the combat situations he encountered. For example, on 1 October 1918 he opted to send the company's first and second waves forward "a few men at a time filtering forward from shellhole to shellhole to advance our line" rather than attacking in the lines of platoons as directed by *Offensive Combat of Small Units*. Although the assault encountered heavy fire and slowly ground to a halt, the company had only three soldiers killed in the action. The next morning Lieutenant Jankoska again changed his method of attack. To cross an unavoidable hundred-yard-wide swath of open space he infiltrated the deadly ground by sending one man over at a time. The move forward proceeded at a snail's pace under constant enemy artillery and machine gun fire, but the company suffered the loss of only one wounded man.[36]

Why was Jankoska able to do this while other officers failed? Part of it was his longevity in the unit. He had been in the regiment since it had arrived in France and had been assigned with H Company since early August. Jankoska was a veteran of the school of hard knocks and had lived through the experience while many of his fellow company officers had fallen to death or wounds. If lessons are to be learned in small units, their leaders must live through the ordeal to tell the tale. Furthermore, Jankoska's superiors were willing to ignore the aspects of the doctrine coming out of Chaumont that did not match the realities of combat. This willingness to abridge or discard the "book" when it was not relevant supports Grotelueschen's contention that tactical adaptability in the AEF often came from below. Sadly, few of the divisions that arrived in the summer and fall of 1918 had the time, experience, or ability to advance to this level of tactical common sense.

It must also be noted that learning at the school of hard knocks could be very costly in terms of human lives. A 42nd Division infantry captain admitted to a division inspector in late October 1918 that the experience of combat had forced him to change his formations and tactics. He had originally placed all of his automatic rifles in his first wave of attack so they could immediately fire on any machine guns they encountered. Unfortunately, he found that in doing so, all of the automatic rifles were "put out of action early." He told the inspector that he intended to group all of his automatic rifles and rifle grenades in the second wave the next time his unit advanced. Although it was good that the captain was learning from his past mistakes, the automatic rifle squads decimated in his first attack probably did not share in the joy of their leader's tactical enlightenment.[37]

The available German sources also point to the lack of training, initiative, and skill of the American small-unit leaders. German officers were endlessly surprised by the density of the American formations and the penchant of their leaders to launch frontal attacks with little to no regard for the terrain and the German dispositions. The German commander facing the crack 1st Division on 4 October 1918 reported, "The mass concentration of men was so large that the one wave pushed the other forward in a certain sense." One afteraction report of the 31st Bavarian Infantry Regiment stated, "Even when deployed the enemy suffered bloody losses. The separate and isolated groups coming in carelessly at first, were at once subjected to the withering concentrated fire of light and heavy machine guns. . . . Gaping holes were torn in the lines of riflemen, entire columns being mowed down. . . . They were visible at great distances and offered excellent targets. . . . One could plainly observe that the unrest in the ranks grew every minute. Lone individuals and frequently entire detachments ran aimlessly about." Time and time again, German officers described the American attacks as being made in "dense masses" or "dense waves." As late as 20 October 1918, the commander of the German 170th Infantry Regiment reported, "In action the American appears very awkward. The hostile infantry attacks in large masses, at times in as many as 15 waves."[38]

Some German sources also made the same observations of the

American fixation for maintaining rigid formations regardless of the tactical situation as had been made by the AEF's inspector generals. Ernst Otto, a German officer facing the 2nd Division's attacks on Blanc Mont, praised the bravery and the tenacity of the Americans, but he later noted that his enemy's lock-step effort to keep formation and tendency to remain in dense "battalion-columns" resulted in many of the American losses. On 26 October 1918 the commander of the German 111th Infantry Regiment reported that the US Army "is effective solely on account of its mass action and its freshness." The Germans also observed, "A certain naïveté manifests," and that "the Americans conduct themselves rather boldly, indicating their inexperience." He went on to argue, "The advance of the hostile infantry was not consistent. On one hand it advanced in skirmish lines, another time in file, then again in light groups, even though the terrain was not favorable for such."[39]

Although it might be easy to discount the German views as sour grapes or wishful thinking, their observations were echoed in the comments made by officers sitting on the boards that collected the AEF's lessons learned from the war. During the Lewis Board, Brig. Gen. Malin Craig, one of the US Army's shining intellectual lights, argued that throughout the war "our men still inclined to go forward in close masses and to take shelter in masses, instead of reducing the size or the target by intervals." Col. Gordon Johnson pointed out that "the tendency to belt straight ahead within a given sector was the cause not only of many losses, but [also] of the failure to properly use all the means at hand for overcoming resistance." The members of the Superior Board concluded that "untrained leaders were too often found in the line" and that this "oftentimes jeopardized the chance of success and unquestionably increased the casualties within our ranks."[40]

Even some experienced American units could fall into the tactical traps described by the Germans and lamented by the postwar boards. George Cornish, an officer in the 1st Division's 26th Infantry, recalled that the normal attack formation used by his unit during both the Saint-Mihiel and Meuse-Argonne Offensives was to place two companies abreast in the lead echelon, with "each company with two platoons in the first wave and two in support [in the second wave]." Each platoon was to deploy "with one-half [its strength] in the front line and one-half in the second," with "50 meters between the lines, 100 meters between waves." The battalion's remaining two companies were to support the advance and remain 400 meters behind the assault companies. In this formation the battalion frontage was 600 meters, and its depth, with an attached machine gun company, was

1,200 meters. While these formations were nearly identical to those given in *Offensive Combat of Small Units,* their spacing was 100 meters wider than the frontage given in the manual.

Even with these changes, the formation proved too dense during the regiment's attack on the Exermont Ravine on 4 October 1918. The regiment's 1st Battalion was particularly hard hit and had "practically lost all of its officers and suffered about 50% casualties" during the advance. The 2nd Battalion maneuvered a bit more skillfully, breaking out of the "normal formation," moving forward "in squad columns widely deployed," and crossing "the exposed ground by infiltration." This suggests that even within the same regiment, one battalion commander might rigidly adhere to doctrinal formations and techniques while others more readily adapted to the situations they encountered. By the end of the day the 26th Infantry's attack had seized a mile of ground at a cost of 565 officers and men. Although it generally benefited from superior artillery support during the next seven days the regiment was in the line, and it advanced a total of five miles, the assaults had cost the regiment 41 of the 84 officers and 1,600 of the 3,300 men it had entered the fighting with on 3 October 1918.[41]

In addition to the poor use of tactics and formations by junior officers, a number of German officers noted the American tendency to lose any element of surprise by their predictable sequences of preparing for and executing attacks. The commander of the German 102nd Infantry Regiment reported on 1 November 1918 that the Americans generally provided their enemies indicators of their future actions long before going over the top. He observed,

> The Americans betray their offensive intentions repeatedly, by conspicuously orienting themselves on the terrain before the attack. They show themselves as individuals and groups, apparently officers, during the day, by walking around noisily with unfolded maps without any cover. The preparations for an attack are also repeatedly carried out in conspicuous fashion, so that our artillery and machine gun fire can disrupt them. . . .
>
> The [American] attack is carried out not in firing lines but in little groups in rows or packs, moving unskillfully over the terrain offering lucrative targets for machine guns and infantry. The American soldier is brave and bold, lacks the proper junior leadership and often shows himself to be improperly trained. If our artillery and machine gun fire comes into effect properly, the opponent is thrown into confusion and the attack comes to a halt.

The commander of the German 170th Infantry also noted that the doughboys always signaled the location and direction of their coming attacks with a very predictable morning assault preceded by a bombardment lasting several hours.[42]

The propensity of the Americans to telegraph their punch gave the Germans an obvious tactical advantage. As the Germans expected the Americans to make morning attacks preceded by artillery fires, it was not uncommon for the Germans to unleash their own artillery and machine gun fires on identified and possible American assembly areas for the coming attack. One German officer reported during the Argonne fight that "when annihilation fire is placed on the hostile assembly areas in a timely manner, a part of the force of the assault is taken from the enemy." The Germans were fully aware of the Americans' lack of basic training in gas warfare and used this knowledge to delay, disrupt, or halt the doughboys' attacks before they gained momentum. "The Americans are very much afraid of artillery fire and especially gas shells," a German officer noted, and "a few yellow cross shells [mustard gas] are sufficient to start the gas alarm and considerable confusion."[43]

The German officer's assertion was confirmed by the Americans themselves. During an attack in the Meuse-Argonne on Hill 378, an officer in the 2nd Battalion, 313th Infantry reported "the enemy fire and gas were so bad that the Battalion became somewhat disorganized and lost the barrage." Although the battalion gained its objective, it only did so with very heavy losses. Lt. Hervey Allen also admitted that when his battalion was hit with German gas, "the usual result was great trouble. Platoons and companies lost touch with each other, and there was great difficulty in giving orders or having them understood."[44]

One of the gravest criticisms that the Germans levied against the American junior leaders was their lack of initiative in combat. On 26 October 1918 the commander of the German 111th Infantry Regiment reported to his brigade commander, "All American attacks had a very limited objective. The successes were never exploited, as otherwise on many occasions this could have had disastrous consequences [for the Germans]." After his encounter with the Americans in September 1918, another German commander noted,

The [American] leadership was altogether clumsy. Evidently our opponent has many officers at his disposal. However, most of them do not possess the qualifications necessary of leadership. It was impossible to overlook the embarrassment displayed by the

Americans as soon as their initial aims were achieved. They help-lessly faced their new positions, unable to take any advantage of them. . . . Favorable opportunities to overtake and encircle us were allowed by them to go by. . . . As soon as the infantry, charging straight ahead, had achieved its goals, leaders, as well as the rank and file, were nonplused.

Last, a German regimental commander explained, "As soon as [the Americans] had a success [they] failed to exploit it but remained in position for hours without moving forward in the terrain."[45]

279

In addition to this lack of small-unit initiative, German officers frequently noted the benefit that their units gained from the Americans' failure to maintain adequate fire support for their assault units after the preliminary and rolling barrages. An officer of the German 169th Infantry maintained that "after the infantry attack is launched, in most cases there is no liaison between the artillery and the infantry; the American artillery always seems not to learn the location of the forward line until later and as a result either does not fire at all or rather [fires] far into the rear." Another German officer stated, "Liaison between [the American] infantry and artillery was poor. This manifested itself primarily during the [German] counter-attacks. Then the hostile artillery was silent in most cases."[46]

The failure of American infantry-artillery cooperation and of the AEF's junior leaders was a disastrous combination. In both cases the American inactivity allowed the Germans to launch their inevitable counterattacks to dislodge the Americans without fear of retaliation by the doughboys' artillery or much reaction from their infantry. A German infantry commander reported with great relish that in those situations "a single counter-attack impaired the whole activity of the enemy for days and cost less casualties and demand[ed] less on the nervous energy of the [German] troops than the wait under very heavy artillery fire without cover."[47]

Compounding this problem, American inspector generals had found that in the fighting at Saint-Mihiel and the Meuse-Argonne, "our infantry did not dig in during temporary halts. In many instances . . . the infantry did not dig in, and were quickly blown off the height by concentrations of enemy shell fire which invariably resulted in it becoming necessary to retake the position with loss of men." This fact was illustrated during an American attack against the German 151st Infantry on 11 October 1918. Although the American assault succeeded in collapsing the enemy's defenses, the failure of the Americans either to capitalize on their success by continuing their attacks against the fleeing Germans or to at least consolidate their gains allowed the German commander the time to halt the rout of his troops by deploying his regimental staff as skirmishers and collecting the men retreating from the lines. The counterattack launched by these hastily assembled scratch forces succeeded in pushing the Americans back to their start point.[48]

It was in instances such as these that the US Army's general failure to teach its junior leaders how to exercise intelligent initiative came back to haunt the AEF. Although junior officers were often cognizant of their own shortcomings and lack of experience, they generally resented the micromanagement of their superiors. The company-grade officers argued that the direct involvement of senior officers in the command and administration of their companies diminished their authority and leadership within their units. An infantry captain decried what he saw as "interference by officers higher than Company Commanders in those problems of responsibility and duty of the Company Commander, with the result that Company Commanders were often mere figureheads." One officer resented the fact that "in most places the junior officers, especially Reserve and National Guard officers, have been treated more as dishonorable and dishonest men . . . and not treated as officers should be treated." Another captain noted the tendency of senior leaders to treat their subordinates "as if they were irresponsible and had no idea of right and wrong."[49]

The senior officers seemed to justify their micromanagement on the grounds that junior leaders could not be trusted to carry out important assignments without close supervision. This perceived need to regulate junior officers further stressed already overburdened senior commanders and staffs and also created command climates in which initiative and independent decision making were not prized or encouraged. As a second lieutenant noted shortly after the war, "There are too many instances of Colonels doing Platoon Leader's work. There are too many lieutenants doing the work of sergeants. There is a tendency on the part of the General Staff itself to direct the simplest movements of small units in detail instead of giving a general outline in orders and leaving the details to be worked out by those who are commissioned for that purpose."[50]

There seems to be much truth in the accusations of both sides. In *Notes on Recent Operations, No. 3,* the AEF GHQ went so far as to admonish commanders for their constant "interference in the province of subordinates" when issuing orders. These admonishments did little to change the army's culture of micromanagement. Of his regiment's action in the Argonne, for example, the 42nd Division's

Lt. Col. William Donovan recalled, "There were green company commanders with the companies; liaison was not maintained; the barrage was not followed closely; there was not enough punch. There were times when I had to march at the head of the companies to get them forward. . . . [the] new men need[ed] some visible symbol of authority." While Donovan's micromanagement was perhaps justified, he seemed oblivious to the long-term implications of his actions.[51]

Shortly after the war, Maj. Gen. Robert Alexander stated that in his 77th Division there had been a general failure in "the development of individual initiative along proper lines" among his junior leaders. He noted, "It did not seem to me that the Junior officers and non-commissioned officers realized the importance of maneuvering as a means whereby successful results might be attained with a minimum of losses. This . . . was simply due to the fact that their instruction had either not been conducted along lines that would impress upon them the vital necessity for such maneuver, or the course of instruction through which they had been put had not made upon them the impression desired." The AEF inspector general agreed with Alexander's assessment and laid the fault for much of the army's problems on the lack of responsibility and initiative on the part of the junior officers. Unfortunately, the AEF's senior leaders failed to see their own culpability in their subordinates' failures. To a great extent the command climate that Pershing instilled in the AEF only served to exacerbate the problem of initiative in his commanders.[52]

While the senior officers often criticized and distrusted the abilities of their subordinates, the AEF's junior leaders frequently held equally low opinions of their superiors. Like their subordinates, the senior leadership had virtually no experience with massed artillery, machine guns, tanks, or the other technological changes affecting the battlefield. The junior leaders chafed under what they considered the field-grade officers' hypocritical criticisms of their competency. The junior officers countered contempt with contempt and lashed their superiors with keen accusations. One young officer blasted the "lack of experience and common sense in the handling of large bodies of troops by some higher officers," while another bluntly wrote, "The field officers and many general officers did not understand their work." An infantry lieutenant pointedly noted, "In battle, General and Field Officers remained far to the rear, but after the battle they came and bitterly criticize[d] the work of the combatants, when if the higher officers had been in their proper places they could have personally directed the fighting."[53]

The junior officers' most striking criticism was that their superiors often lacked basic command skills and were out of touch with the realities of modern warfare. In the Morale Branch survey a field artillery officer commented, "Many Commanding Officers were ignorant as to what their organizations were capable of doing in action. That is they expected the impossible at times and did not take advantage of things they could do at times." These charges had some merit. Soon after the war, field-grade officers in II Corps schools were found to be unable "to write a clear, concise message, and had small conception of the general tactical principles employed in offensive movements."[54]

The inability of senior officers to issue clear and timely orders often hobbled the operations of their subordinates. One 4th Division machine gun battalion commander complained, "Consideration is not given [by senior officers] to the absolute necessity to move and properly . . . [deploy] troops in combat. Battalion and company commanders were kept in the dark almost up to the last moment and then comes the order, the execution of which is to be practically immediate and with no chance to properly study the ground or maps. This was not because there was not time enough from the time the attack was planned but because the staff took too long and gave the line too little time."[55] Based on their study of the Great War, officers at the Infantry School later admitted that "instances abound in which attack orders were received after the hour specified by the order for the jump-off" of the attack. Given the minute planning required to make the American artillery effective, such last-minute issuing or changing of orders often ensured that the outwardly improvised attacks would fail or at least saddle junior leaders with a host of new complications. Incidents in which the foot dragging of senior headquarters in issuing orders directly caused unnecessary casualties among the attacking troops were rife in the AEF. A mix-up in orders for an attack by the 2nd Division's 23rd Infantry on 6 June 1918 so delayed the issuance of instructions to the unit's company commanders that they had to sprint back to their companies and push the men over the top with little guidance or direction. The attack commenced without artillery support, and the regiment's Stokes mortars and 37 mm guns were eliminated by enemy fire. The American attack was checked by German machine gun fire, and the regiment "suffered severe casualties."[56]

A similar series of events happened to units of the 36th Division when they were attached to the 2nd Division for an attack on Saint-Etienne on 8 October 1918. The 2nd Division's attack orders for a 6:15 a.m. attack were not briefed to the company commanders of the attached 141st Infantry until 6:11 a.m. on the morning of the assault. In fact, "the battalion commander of the front-line battalion of the 141st Infantry was just in the act of imparting detailed instructions to his company commanders when the rolling barrage commenced."[57] After losing the rolling barrage, the attacking battalions suffered heavy casualties, and the attack stalled shortly after it began. By the end of the day, the attacking 71st Infantry Brigade had lost 33 percent of its officers and 23 percent of its enlisted men.

The 36th Division's Capt. Ben Chastaine noted that prior to the abortive attack "only the smallest amount of information had been obtained of the position," and "no maps had been distributed to the companies and only the most general idea of the terrain was to be had." He denounced the fact that the combat instructions given to the officers were so rushed and sketchy that "they were useless as far as the commanders of the combat units were concerned." One company commander never even received these incomplete orders, for the runner sent to bring him to the battalion commander's hurried orders briefing was killed and his message went undelivered. When the absent commander noticed that the battalion was attacking, he belatedly set his own company into movement. The ill-planned attack was a disaster, and in the confusion the "troops of the supporting battalion coming up from the rear only served to make the line denser and increase the number of casualties."[58]

In instances such as these the fear of relief and disgrace led far too many senior officers to push attacks long after their chances of success were past. This reality was not lost on the enlisted men. On 11 October 1918 the 325th Infantry launched an attack across the Aire River near Saint-Juvin to unhinge the German defenses on a ridge to the north of the town. The frontal assault made grudging headway against the strong German positions. One of the regiment's soldiers, Sgt. Victor Vigorito, recalled that the order from higher headquarters was to "push steadily on, regardless of the cost." He noted that in trying to carry out those instructions "we lost 280 officers and men in a few minutes," and claimed, "it was the worst piece of wholesale murder I saw in the whole war."[59]

It is probable that the willingness of infantry commanders to attack without artillery support, or to follow orders that were based on faulty assumptions, stemmed from the fear of repercussions if their attacks did not go off as scheduled. When senior officers told their subordinates to take or hold ground "at all cost," too many officers seemed willing to follow those instructions without question. After the Germans dropped a lethal combinations of gas on the 82nd and 89th Divisions during a relief in place on 7 and 8 August 1918 near Seicheprey, the two divisions suffered a combined total of 47 men killed and another 759 evacuated for gas exposure. An investigation concluded that "the large number of casualties was in part due to the lack of knowledge, on the part of officers and men involved, concerning Mustard Gas." Furthermore, "When the men went into the line they were told *to hold their positions at all costs,* and apparently men and officers alike considered it almost a point of honor to remain despite the Mustard Gas."[60]

Another such incident occurred on 7 November 1918 when the 79th Division directed that each of its infantry company commanders send out a sixteen-man patrol to locate machine gun nests and bring back German prisoners. Capt. Arthur Joel, a company commander in the 314th Infantry, recalled, "One's first opinion naturally was that there must be some mistake in the orders. To send a patrol across the lines in broad daylight . . . seemed like suicide!" Despite great reservations and the gnawing feeling that the orders were a mistake, Joel philosophically shrugged that "orders were orders" and sent the men out. As he had predicted, the patrol was shot to pieces by machine gun and artillery fire. A similar event occurred on 10 November 1918 when 1st Lt. Glen Gardiner, of the 5th Division's 60th Infantry, delayed attacking Juvigny for approximately twelve hours. The officer's delay ensured that that objective was not captured prior to the time that the Armistice went into effect at 11:00 a.m. on 11 November. Gardiner claimed that the two companies given him for the attack were short of ammunition, lacked artillery support, and also had not been fed. For his decisions Gardiner's commander sent him to Blois for reclassification.[61]

Although AEF commanders stated that they wanted their junior leaders to exercise initiative, in choosing his men over the mission in the waning hours of the war, Lieutenant Gardiner had, of course, not exhibited the right kind of initiative. The failure of the AEF's senior leaders to reconcile their demands for absolute obedience to orders with the need for their small-unit leaders to exercise initiative left junior officers and NCOs unsure of what the right kind of initiative actually looked like. This failure led to some of the army's most ironic moments. When his unit was pulled out of the lines around Cantigny in late May 1918, Lt. Jeremiah Evarts was ordered to follow a given route to the rear area and not to deviate from the path. After finding that the route was under observation by German balloons and being shelled, Evarts quickly diverted his unit off the road and moved it across country by a safer path. Just as he reached his destination, the lieutenant was accosted by a division staff officer who questioned him about his failure to obey orders and stated that he would report his actions to the division commander. In the end, the young officer was reprimanded for his intelligent and common-sense use of initiative.[62]

The old army expression, "Shit rolls downhill," reflects the fact that when one's superiors are feeling the heat for some problem or are being goaded into action by their bosses, inevitably one (and consequently one's subordinates) will be spurred in a given direction. The same captains that complained of the micromanagement of their superiors seemed to have had little compunction about exhibiting the same behavior toward their own lieutenants and NCOs. In training and combat the American junior officers demonstrated a propensity for consolidating decision making and supervision into their own hands. On 17 October 1918 the 6th Division's inspector general reported,

> Experience and observation convinces me that in our Army sufficient responsibility is not placed upon *squad leaders*—and they are not impressed with that responsibility. The squad is *the* unit around and upon which our Army is [built], and can easily be supervised by one man—whereas it is absolutely impossible for one man (the Company Commander) to properly attend to the needs and wants of each man, by himself. We have yet to learn a proper sub-division of duties, and a distribution of responsibility, and only by making each squad a real "unit" is military efficiency to be obtained.[63]

The end results of these actions were predictable. Junior leaders who were separated from their commanders by accidents of action or casualties often found themselves unready to assume the mantle of command themselves. For example, during the fighting in the Argonne sector on 7 October 1918 the 82nd Division's inspector general complained, "The failure of some squad leaders to meet the demands of succession of command, and exercise aggressive control in platoons which had lost the lieutenants and sergeants" had exacerbated the unit's straggler crisis and caused much of the sluggishness that characterized the division's recent attacks.[64]

The failure to use their initiative when the tactical situation demanded often came with dire consequences for their units. During the Argonne drive, Lt. Joseph Lawrence waited impatiently for one of the trail companies to come up to support his unit's tenuous hold on the front lines. When the supporting company failed to appear, Lawrence sent runners to find the missing unit and lead them forward. The runners found the needed company firmly ensconced some seventy-five yards behind Lawrence's embattled unit. When Lawrence

285

pleaded for the commander to move his company forward, the officer replied that he had been placed there by their battalion commander, and he would not move unless ordered to do so by the major. When Lawrence found the major to inform him of the situation, the superior officer berated him for leaving his company, ordered the lieutenant out of his dugout, and promptly went back to sleep. Unfortunately, when the major discovered his mistake the next morning and ordered the support company forward, it was chewed up by German fire as it advanced in the daylight.[65]

286

It would be wrong to say that American junior leaders actively avoided taking the initiative. There are a number of examples in which their timely actions were critical to the accomplishment of missions or saving of American lives. For example, Stanley Herzog observed a particularly effective use of initiative by some junior officers of the 26th Division during the Aisne-Marne Offensive. After the doughboys of the 102nd Infantry had been held up by a German strongpoint, a lieutenant commanding an artillery battery, on his own, coordinated with the nearby machine gunners to bring the enemy positions under the synchronized and concentrated fire of their collective artillery and machine guns. This effective suppression allowed the infantry to advance and seize their objective. In another instance, when his regiment was held up by heavy German fire while trying to cross a Meuse River canal on 4 November, the 5th Division's Capt. Edward Allworth swam the river with the remainder of his company and dislodged the German defenders. However, these incidents transpired in spite of, rather than because of, the training the leaders had previously been given.[66]

In the final analysis Marshall Foch's assertion, that in combat "in order to make a little possible, one must know much," was borne out in the experiences of American junior leaders in the Great War. When it came to the stock-in-trade of combat leadership, the ability to use initiative and to match formations, maneuver, and firepower to challenges of the enemy and terrain, the Americans' training and experiences had left them woefully short on knowing much. What the junior officers and NCOs had failed to learn in training was later taught to them by the Germans in their unforgiving school of hard knocks. Although surviving infantry officers and NCOs learned from these ruthless schoolmasters, their tuition was all too often paid in the unnecessary losses of American soldiers and in the sluggishness and missed opportunities that characterized the AEF's operations during the war.

11 Combat Leadership and the Attritional Battlefield

ULTIMATELY, THE OUTCOME of the various and assorted lapses in small-unit leadership examined in the previous chapter was the line of dead and wounded they left in their wake. The AEF suffered over 256,000 battle casualties during the war. Of these, slightly less than 50,000 were killed in action or died of their wounds. While these statistics pale in comparison to the losses of the other powers, well over half of the American casualties occurred in the last seven weeks of the war. Of the AEF, Marshal Foch commented, "It can be stated that the percentage of its losses in relation to its effectives engaged and to the length of time it was in the field was found to be the highest of all the Allied armies in 1918."[1] These casualties had a telling effect upon morale, unit cohesion, and the combat effectiveness of the AEF's infantry companies, platoons, and squads. Although heavy losses were a reality on the Great War's battlefields, in the Americans' case many of them resulted from flaws in leadership at the battalion level and below.

Because the 82nd Division's casualty numbers make it an average AEF unit, its experiences are useful to illustrate the effects of poor tactics and leadership in the Meuse-Argonne. The division's first three days of combat in the Argonne shocked the doughboys with their ferocity and deadliness. In the six months that the 82nd Division had served in France prior to the Meuse-Argonne, the division had lost a total of 133 soldiers killed in action, 1,244 wounded or gassed, and 13 captured. From 7 to 10 October the 327th Infantry alone suffered the loss of 118 soldiers killed, 700 wounded, and 96 captured. When the 82nd Division was relieved from the lines on 30 October 1918 after twenty-three days of continuous fighting, it had lost 902 soldiers killed in action, 4,897 wounded, and 185 taken prisoner.[2] Nearly all of these soldiers were infantrymen. Other divisions suffered equally severe losses.

This level of casualties was devastating to the AEF's small units. A few examples illustrate this problem. After three days of impaling itself on the German defenses at Belleau Bois and the Bois d'Ormont in late October 1918, the 26th Division's 51st Infantry Brigade had been sadly depleted. The unit's frontal attacks had left many corporals commanding companies and had forced the consolidation of battalions and their subordinate units. Company A of the 102nd Machine Gun Battalion went into the lines with six officers and 172 men. When it came out of the line on 29 October, it was down to one officer, eight NCOs, and forty-seven privates. The unit's one officer was a lieutenant from the 101st Infantry who had been assigned to the company when it lost all its officers in combat. To top it off, by 4 November the company had lost three of its NCOs to direct commissions or officer training.[3]

In the 28th Division, Company E of the 107th Infantry had already lost a number of its officers to transfers and NCOs to the AEF candidates school before it had even entered its first combat in August 1918. In its attack on the Hindenburg Line on 28 September, the company strength stood at three officers and 170 men. Within twenty-four hours all of its officers were dead, and the unit had been whittled down to only 46 men. Although the company was left with only four sergeants and six corporals, on 2 October two of its remaining sergeants were ordered to the candidates school. Ten days later the unit lost its two remaining NCOs to officer training and one of its replacement officers to gas poisoning. When the depleted company reentered the fight on 18 October, it lost its newly assigned company commander when the novice officer led a frontal attack on a German machine gun nest.[4]

Between the times that it entered the lines around Saint-Mihiel on 22 September 1918 until it was relieved from their sector of the Meuse-Argonne on 5 November 1918, B Company of the 311th Infantry was decimated and rebuilt three times. A 26 September attack in the Saint-Mihiel sector did not go well for the company. German machine guns and artillery fire shredded its ranks, and every effort the commander made to flank the German guns merely landed it in another kill zone. In the confusion of the fight, the company commander lost all touch with one of his platoons and did not regain it until late in the afternoon, when he stumbled across its six unwounded survivors.[5]

On 12 October the company was brought up to strength after receiving 104 replacements. This levy contained a large number of men who "had never fired a rifle and were not familiar with the use of the gas mask." As there were only two officers in the company, all of the platoons were commanded by buck sergeants. The company reentered combat in the Meuse-Argonne three days later. In the midst of the fighting on 17 October, three of the company's NCOs, including one of the acting platoon leaders, were ordered to the Army Candidates School. When the company was relieved from the lines on 26 October, it was again down to a strength of less than two platoons. Furthermore, the regimental commander assigned one of the company's two officers to another unit to make up their shortage of leaders. Soon after leaving the front, the regiment ordered the company commander to attend the Army School of the Line for a six-week course. The company, now down to one officer and eighty men, was reorganized as a platoon.[6]

Before the company reentered the fight on 29 October, the regiment again replenished it with replacements. However, after only a single day of fighting the unit's manpower had shrunk to less than two platoons. Furthermore, the mortal wounding of the company's only officer and other losses of junior leaders meant that when the unit reorganized into two platoons, both were commanded by a corporal, and the company was led by a buck sergeant.

There are a number of things that the narratives of these three companies have in common. All highlight the underlying deadliness of the attritional war that the AEF faced. They also demonstrate the overall indifference of the AEF GHQ to the needs of its small units. Regardless of the condition of a unit coming out of action, or its need to retain NCOs and officers to rebuild its cohesion, the quotas for the AEF's bloated school system took precedence over all else. This was a key mistake, for the AEF never copied the British practice of leaving a portion of battalion officers and NCOs out of a given action to serve as a cadre to rebuild the unit in event of heavy losses. In the British army these cadres were expected to retain the unit's collected battlefield wisdom and know-how to pass on to the replacements that would soon fill its ranks. With the loss of junior leaders to casualties and schools, infantry companies all too often lost the continuity for passing on the unit's hard-won lessons learned.

The most deadly assignment in the AEF was that of a company-grade infantry officer. By far, infantry units contributed the greatest percentage of the AEF's fatal casualties. Approximately 52 enlisted infantrymen out of 1,000 were killed in action or later died of wounds. Infantry officers, though, suffered over 80 killed or died of wounds for every 1,000 officers from that branch. As a point of comparison, the AEF's casualties for artillerymen were 8 out of every 1,000 officers and only 6 out of every 1,000 enlisted men. On average, there were six men wounded for every one man killed.[7] Using this rough figure as a guide, on average over 560 infantry and machine gun officers out of 1,000 were killed or wounded during the Great War. Al-

2nd Lt. JAMES G. BROPHY

BORN APRIL 9, 1890
DIED SEPTEMBER 28, 1918

290

James Brophy was a lawyer before attending the first OTC at Camp Sheridan. He died of wounds he received during the Saint-Mihiel Offensive while serving with the 360th Infantry. His case was a prime example of the cost borne by junior officers during the war and illustrates the problem that the AEF faced in keeping its officer ranks filled. *(Photograph from Fort Sheridan Association*, The History and Achievements of the Fort Sheridan Officers' Training Camps.*)*

though there is no information on the casualty rates among infantry and machine gun NCOs, it is probable that their loss rates fell somewhere in between those of the officers and the privates.

These losses in leaders had significant ramifications for the combat effectiveness of the AEF. Officers tended to be the best-trained personnel in the new divisions. Although much of this training was rather poor, the loss of these struggling "subject matter experts" resulted in a relative decline of know-how in small units. As Darryl Henderson noted, the small-unit officers and NCOs played the vital role of setting and maintaining the behavioral norms of their units and served as the intermediaries between the higher headquarters and the individual soldiers. When these leaders were lost, cohesion and effectiveness declined.

In their seminal work "Cohesion and Disintegration in the Wehrmacht in World War II," Edward Shils and Morris Janowitz argued that German units continued to fight effectively as long as they remained part of a cohesive small-unit primary group that cared for its members' physical and emotional needs and possessed leadership that balanced the needs of the men with the demands of the mission. They argued that the effectiveness of the Wehrmacht began a steady decline in the winter of 1945 as soldiers became increasingly isolated physically from their comrades, when their familial ties to home were broken, when unit leaders became unable to meet their soldiers' physical survival needs, and when the close bonds between group members and leaders were shattered by casualties.

Shils and Janowitz maintain that in the German army of World War II the heavy loss of junior officers and NCOs "resulted in a reduction in the amount of face-to-face contact between officers and men and in reduced feeling of the officers' protective function." Furthermore, as the Germans were rebuilding by consolidating depleted outfits and filling them with replacements, "the top non-commissioned officers often did not have sufficient time to promote the growth of strong identifications between themselves and their men."[8]

In the fall of 1918 the AEF's small units suffered from many of **291** the same challenges that would bedevil those of the Wehrmacht in 1945. The leadership in the AEF's small units was constantly being rebuilt because of casualties or the loss of leaders to AEF schools and assignments. Lt. Joseph Lawrence recalled that of the eleven graduates of his AEF candidates school class that were posted with him to the 29th Division in September 1918, seven eventually made it into the fighting in the Meuse-Argonne. Of these seven, three were killed in action and two more were wounded and evacuated. Only Lawrence and one other officer survived the battle without a scratch. Likewise, Lt. Henry Thorn, of the 313th Infantry, reported that his regiment's four-day attack to seize Montfaucon had cost forty-five officers, twelve of whom were killed in action. To make matters worse, as soon as the regiment came out of the line, orders came down to send one officer per company to the II Corps schools. Their places were filled by a replacement captain and fifteen replacement lieutenants.[9]

The AEF's junior officers came and went in infantry companies with a bewildering rapidity. Company A, 1st Battalion, 308th Infantry, had seven different company commanders from July to November 1918. During the same period Company B had five commanders, Company C had six, and Company D had four. The battalion's turnover of lieutenants was just as great; Company C had twenty-one officers pass through its ranks in those same four months.[10] The 308th Infantry's experience was far from uncommon. In the 27th Division's 107th Infantry, on average each of the unit's line companies had more than sixteen captains and lieutenants assigned to them over the course of the war. The regiment's A Company suffered the most changes, with twenty-five officers passing through its ranks during the conflict.

The effect of this revolving-door changeover of leaders was the breakdown of the vital face-to-face relationships between leaders and their soldiers that resulted in a decline in the morale and cohesion in the AEF's infantry companies. As his unit entered the Meuse-Argonne, John Barkley noted that "officers were like passing shadows with us now. It hardly paid to try to get acquainted with them," for they generally and quickly became casualties. The casualties that his unit soon suffered in the Argonne later led him to note, "The regiment was in bad shape. We'd been cut to pieces a dozen times, and the remains reorganized so often that nobody knew what he belonged to." Following ten days of bloodletting in the Argonne, a soldier in the 312th Infantry observed,

> The previous days of fighting had depleted the numbers until there were left not more than an average of sixty men in each rifle company. No battalion could boast of more than five line officers, while the lack of non-commissioned officers was a serious handicap. A thorough reorganization was necessary, a division of rifle companies into two platoons in place of the customary four and a redistribution of officers to provide at least one to each company. . . . Hasty appointments of acting non-commissioned officers to lead the subordinate elements followed as a matter of course. No longer did the officer have an intimate personal knowledge of the individuals under his supervision.

A number of doughboys echoed these sentiments. For example, Pvt. Milton Sweningsen, an infantryman in the 35th Division, reported that his unit was so wracked by the loss and replacement of leaders that "I hardly knew the officers of my own company."[11]

These losses could have an immediate and negative effect on a unit's performance and cohesion. The 112th Infantry's Pvt. Charles Flacker recalled that his company suffered so many casualties among its leaders that low-ranking men from the ranks had to fill the positions. After that, he noted, it was "every man for himself" in the company. After the near disintegration of the 35th Division in the opening days of the Meuse-Argonne Campaign, Col. Robert McCormick remarked that "casualties among the officers were undoubtedly responsible for a great deal of the disorganization" and that "most of the straggling and confusion was caused by men getting lost and not having leaders, and not from any deliberate design to go to the rear in order to avoid further fighting." During the 5th Division's time in the Argonne, the division inspector reported that due to heavy losses, "in some organizations the officers had been on duty for a very short time, and did not know the men, nor did the men know the officers. Apparently a great many men did not know their officers by sight." He believed that this situation undermined the division's cohesion and contributed to its lackluster combat performance.[12]

The losses of junior officers and NCOs also created a nearly insurmountable personnel crisis for the AEF during the Meuse-Argonne Campaign. As previously discussed, the unexpectedly high loss of infantry officers in the summer and fall of 1918 had forced reductions in the time allotted to officer candidate training in the United States and in the AEF. As small units were deprived of leaders who had gained a modicum of experience in combat, the army filled their places with men whose lack of training led them to repeat many of the same tactical mistakes that had bedeviled or killed their predecessors. As Pvt. Herman Dacus laconically recorded, "Most of the replacement officers made good, if they were lucky enough to come through a couple of battles" unscathed.[13]

Here was one of the AEF's fundamental dilemmas. The AEF's junior leaders were caught in a vicious cycle in which inexperienced officers and NCOs were being killed and wounded only to be replaced by other inexperienced officers and NCOs. How could units learn from their previous experiences when their leaders, the ones tasked with disseminating this battlefield wisdom, were chewed up at such a rapid rate? Herein lay the reason why the AEF never truly experienced a marked improvement in its combat effectiveness during the war.

High casualties and the constant levying of NCOs to fill quotas for the AEF candidates school further undercut the ability of small infantry units to learn from their hard-won combat experiences. The records of Company H, 126th Infantry, reveal that the vast majority of the soldiers and officers it lost to wounds never returned to duty with the company. When the company came out of the lines from the Oise-Aisne drive on 31 August 1918, it had lost 26 men killed and two officers and 38 men wounded. Of the wounded, 23 never returned to the company, 8 returned to duty by the end of August, 6 more returned by the end of October, and 1 did not return until after the Armistice. While fighting in the Argonne campaign from 1 to 18 October, H Company lost two officers and 20 men killed in action, one officer and 107 men wounded, 3 missing, and 3 evacuated for illness. Of the wounded, 69 percent never returned to duty with the company, and 19 percent more only returned to duty after the company's final combat action. Although some of the wounded who did not return to the company for duty may have been assigned as replacements for other organizations, their experience was lost to H Company for the duration of the war. In addition to the loss of officers and soldiers to casualties, it must be noted that Company H also lost one first sergeant to wounds and two first sergeants to the AEF candidates school between 29 September and 26 October 1918.[14]

Company first sergeant (front row, center) with the NCOs of his infantry company. Based on their service chevrons, most of these veterans of the school of hard knocks had spent two years overseas, and at least three of them had been wounded in combat. The sergeant to the first sergeant's left has suffered two wounds. *(Photograph from the author's collection.)*

High casualties and the levying of NCOs for schools also exacerbated the AEF's already dire need for qualified NCOs. An officer with the 77th Division stated on 24 October 1918 that "the question of non-commissioned officers in Infantry Companies of this division is serious," due to heavy casualties among their ranks and the lack of men "with sufficient training" to replace them. After the Armistice the Lewis Board concluded that combat losses among infantry NCOs led to a drastic reduction in the quality and reliability of small-unit leaders in the last months of the war. The board maintained that "nearly every survivor who belonged to a rifle company, and who was not a complete mental failure, of necessity had to become a non commissioned officer in order to rebuild a cadre that could absorb the replacements." This fact, they maintained, led to the AEF's having to rely on a cadre of "poorly trained and rather dull non commissioned officers."[15]

The AEF's problem with junior leaders was not just that inexperienced second lieutenants were having difficulty commanding their platoons; it was also the fact that casualties were rapidly promoting those novices to company command. The loss of officers was so great in the 77th Division's 308th Infantry in the Meuse-Argonne that second lieutenants "just out of training school" were assigned as company commanders. Joseph Lawrence reported that the 113th Infantry suffered so many officer casualties that one of the newly commissioned officers who reported to the regiment with him on 6 October was a company commander by 10 October. When that officer was killed a few days later, he had held a commission for only twenty days.[16]

Often the result of these young officers' rapid ascent to their Peter Principle level was deadly. Horace Baker recalled that heavy officer casualties had placed the command of his battalion into the hands of a lieutenant in early November 1918. The officer was clearly out of his depth and decided that in the battalion's attack on Brandeville it was best to push the attack without putting out patrols. The unit quickly blundered into the German defenses, and the attack became a hasty retreat. Baker later confessed that given the strength of the German fire, he did not "deem it a disgrace to have fled from that fatal field." After regaining his composure, Baker again moved forward to try to rejoin his company. He discovered that "American soldiers, dead and wounded, but nearly all dead . . . actually lay in windrows, just as the lines had stood when the machine guns had mowed them down." The frontal assault had cost Baker's company nineteen dead and so many wounded that the company strength was reduced to 60 percent.[17]

The heavy losses in the AEF's infantry companies dealt a double blow to the cohesion of the units. One of the blows was the breakdown of the intense group identity that soldiers had tried to build with their comrades despite the endless levies of personnel that had wracked the divisions forming and training in the United States. Soldiers returning to their units after convalescence from wounds or sickness were often shocked by the changes in personnel that had occurred since their departure. When Pvt. Irving Abrahams returned to his company after being hospitalized for forty-five days for wounds he received during the Soissons drive, he noted, "In my old company I felt like a stranger, for so many of the old crowd had been killed at the time that I got mine." Pvt. Duncan Kemerer recalled that when he returned after a month in the hospital there were "very few of my old buddies left," and the company was now composed "of mostly drafted men, and all new officers." In the preceding month the company had lost all of its officers during the Oise-Aisne drive and had one captain and two lieutenants killed leading the company during the Meuse-Argonne.[18]

These rapid changes in leaders and the breakdown of buddy groups undercut the ability of small units to stand the shock of battle. When one fights with strangers, one never knows the other's abilities and reliability. As a soldier in the 111th Infantry stated, during the Argonne offensive his unit had learned how to spread out their formations and send scouts out "so we would not suffer heavy casualties by so many troops walking into a trap or German lines." But heavy losses in leaders and experienced soldiers meant that his regiment was less successful in using these techniques once it was moved to the Thiaucourt sector in mid-October. After being caught in the open during a 4 November 1918 attack, one doughboy recalled, "We were ordered to retreat which became . . . to resemble a route [*sic*] as everybody started to run" to escape the German fire.[19]

The other blow to unit cohesion was the steady decline of the soldiers' faith in their leaders. Shils and Janowitz argued that "the men must be sure that their officer is duly considerate of their lives: they must know that he does not squander his human resources, that the losses of life which occur under his command will be minimal and justified." Given the clumsiness of the American attacks and the casualties that resulted from them, some soldiers started to wonder if their leaders were "duly considerate of their lives." This problem was compounded when their leaders made decisions that seemed to the doughboys to lack common sense or purpose. On 11 October 1918, Pvt. Fred Takes wrote in his diary that he and his comrades were demoralized by their company commander's refusal to allow his platoon to pull back twenty-five yards from an exposed position being heavily shelled by the Germans. After suffering several losses, the men disobeyed the commander's orders and pulled back on their own. As a result of this shelling and a series of ill-planned attacks, Takes's company was down to only thirty-five men by 16 October. After his officers ordered the company to attack a German position that had withstood previous assaults, Takes wrote, "When we got the orders to go over the top at 5 a.m. we were disgusted, thinking they [his company and battalion commanders] wanted to kill us all off."[20]

Similarly, Wilbert F. Stambaugh noted that a replacement officer inspired little confidence in the men. He was disgusted by the fact that the "newly commissioned lieutenant did not understand action in war [and] tried to force us beyond our own artillery fire. In another instance, Sgt. William Triplett of the 35th Division witnessed the devastating failure of a frontal attack against the town of Cheppy on the first day of the Meuse-Argonne Offensive. Triplett noted that some adjoining companies from the 138th Infantry "tried to do a 'Charge of the Light Brigade.' . . . The stretcher men were gathering 'em in and lining 'em up for the rest of the day." The junior officers of the division continued to throw their soldiers against the German defenses in frontal attacks for the next three days. As one member of the division recalled, the units "simply had melted under machine gun fire." After four days of uninspired slugging the 35th Division had lost 8,023 men. The unit's morale was so shattered by the action that it was withdrawn from the line, never again to see significant combat.[21]

It did not take long for the soldiers to realize that their officers' lack of skill was hazardous to the average doughboys' health. They resented the rather nonchalant indifference that some officers held toward casualties. In his after-action report of the Argonne battle, Capt. John K. Taylor informed his regimental commander, "To hasten the movement of the men to the front line positions here, I told them not to mind the bullets, that most of them were from our own machine guns. Upon seeing two men fall dead and another wounded by my side, I overheard a man say 'our machine guns are sure hell.'"[22] In his cavalier bragging to his superior, Taylor seems to have missed the fact that his soldiers were being critical of his leadership and decisions.

Whether unskillfully charging into machine gun positions or carelessly and unnecessarily exposing their soldiers to fire, the leaders' poor combat leadership was breaking down their soldiers' trust and confidence. A sergeant in the 1st Division explained why some of the doughboys viewed their officers with a jaundiced eye by simply stating that the soldiers "could not understand why the officer was always the boss when often he did not know what he was talking about."[23] Some of the dissatisfaction expressed in the sergeant's statement was also evident in the remarks made by some of the wounded evacuated from the Argonne fighting. On 3 November 1918, an 83rd Division intelligence agent reported that "several of the men said that the American disaster at Argonne Wood was the result of bad officering. Companies were bunched together and were easy victims of Boche machine guns. Two of the men quoted a chaplain as saying 'someone will be held accountable for the officering here.'"[24]

The historian Leonard Smith argues that during the war, units of the French army developed an unofficial system whereby the soldiers regulated their aggressiveness based on a cost-benefit negotiation with their officers. If the soldiers believed that a given attack was of great importance to the nation, they willingly sacrificed themselves to win the contest. If, on the other hand, the benefits gained from the operation were not worth the cost in human lives, the soldiers held back their full support and made only half-hearted attacks. Smith notes that the breakdown of this social contract between the French leaders and their men, and the soldiers' own attempt to renegotiate the military agreement between themselves and the army's command, fueled

the mutinies of 1917.[25] The soldiers of the American army never came close to approaching the depths of disenchantment that helped push the French army into mutiny. However, there is evidence that some doughboys conducted similar cost-benefit negotiations with their officers or otherwise limited their leaders' command options. The AEF inspector general noted a troubling tendency among the army's junior officers not to follow orders or adequately supervise their execution. He stated, "When violations of such orders were brought to the attention of the junior line officers they would answer, 'well, I gave the order to stop it.'" This led to a condition in which "the soldiers do not obey or respect the orders of their officers, and the junior officers often appear timid about enforcing their orders."[26]

This polite refusal to follow certain orders seems to have been more common when inexperienced replacement officers were placed over veteran soldiers. Stanley Herzog recalled that in some instances officers did not know the limits of their soldiers' abilities and endurance. When soldiers reached these limits, they believed that they had met their end of the social contract and accordingly limited their aggressiveness. During the 26th Division's attack during the Aisne-Marne Offensive, a company of infantrymen near Herzog's artillery position "refused to go forward." He defended their actions by noting that it was "not that they were cowards, but they were so fatigued that there was no strength left in them."[27]

Where the bonds between the junior leader and his soldiers were strong, officers were sometimes forced to make agonizing choices between following the orders of their superiors or the desires of their enlisted men. One of the most noteworthy incidents of junior leaders choosing to place men above their mission occurred in early November 1918 and involved the 26th Division's 103rd Infantry. On 5 November 1918 an officer from the 78th Division discovered that soldiers in the regiment's outposts in the Bois d'Haumont had established an informal live-and-let-live truce with the Germans residing only twenty to seventy-five yards opposite them. The Germans had told the American soldiers in the area that they expected peace at any moment and saw no point in risking their lives. The Germans refused to surrender because they feared reprisals against their families if they were reported as deserters. Both the Americans and the Germans agreed that if forced to fire, they would aim high, give warning to the other of impending attacks or raids, and, in the event of artillery bombardment, would signal the guns to shift their fire away from the opposing front lines and outposts. An investigating officer also discovered that although "no officer would admit that he knew any condition of laxity existed," he believed that some of them were turning a blind eye to the practice. The incident ultimately resulted in the relief of the battalion, regiment, and brigade commanders of the units involved. It also gave Pershing an added pretext for relieving the 26th Division commander, Maj. Gen. Clarence Edwards, from command.[28]

Although the refusal of units to attack or efforts by doughboys to restrain the authority of their officers were rather rare, there were other avenues that soldiers could follow to circumvent the orders of their leaders. One path that some soldiers took to remove themselves from combat was to resort to self-inflicted wounds. As the AEF entered major combat in the summer of 1918, the number of these injuries began to rise. In July 1918 the 1st Division alone had 72 cases. The next month, the AEF inspector general received reports of 179 cases of self-inflicted wounds in the fourteen divisions he surveyed. Investigations discovered that 131 of these wounds were accidental and 13 were intentional, and in 25 cases there was no definite conclusion on the soldiers' intentions. On 17 October 1918 the 6th Division inspector general, V. M. Elmore, reported that between 6 September and 12 October 1918 his unit had 44 cases of self-inflicted wounds. Although 15 of the cases were accidental, the remaining 29 were "intentionally inflicted, with a desire to evade and avoid further active service." These figures point to both the poor level of weapons training in the AEF as well as the fact that some of its soldiers were so dissatisfied with their lot that they were willing to injure themselves to get out of their units. Luckily for the AEF, the number of self-inflicted wounds or efforts by soldiers to incapacitate themselves never reached the point that it impaired the army's effectiveness.[29]

The easiest and most common way for a soldier to remove himself from the control of his leaders was simply to straggle from the lines. The AEF's problems with straggling were evident from its earliest operations. Maj. Gen. Robert Bullard recalled that during the AEF's summer battles between the Marne and Aisne Rivers, "far back behind our lines and camps my provost marshal now began to gather large numbers of American soldiers" from various divisions. However, it was not until the Meuse-Argonne Offensive that the straggler problem reached crisis proportions and truly weakened the effectiveness of the AEF. After the war, Maj. Gen. Hunter Liggett estimated that approximately 100,000 soldiers had straggled from their units in the first month of the Argonne drive. Between 900,000 and 1.2 million American soldiers participated in the campaign. If Liggett's estimate is correct, approximately 10 percent of the army's manpower simply stopped fighting and straggled towards the rear.[30]

It is difficult to uncover how closely Liggett's figures match reality; however, there is enough evidence to give at least some indication. During the Argonne fighting the AEF inspector general stated, "One division reported that it had only 1,600 men in the front line including an Engineer battalion that had been sent forward. . . . This division was taken out of the line and upon arriving in its rest area it was found that the infantry regiments alone had in them 8,418 men." The inspector general rightly concluded that the 6,000 soldiers who appeared in the rest area were stragglers from the division's frontline units. When the 91st Division was pulled out of the Meuse-Argonne fighting on 4 October 1918, a V Corps inspector reported that in its ten days of combat the unit had lost 148 officers and 3,197 men killed or wounded. More alarmingly, the officer noted that 7 officers and 2,206 soldiers were missing, but he believed that most of them were stragglers. The lieutenant in charge of the town of Raucourt rounded up "between 600 and 700" stragglers from the 1st Division on 8 October 1918. Four days later, the 36th Division's military policemen claimed to have rounded up "500 men of the division classed as stragglers." The Second Army's military police arrested 439 men for being AWOL in October and another 370 men for the same offense in November. Although it is impossible to accurately establish the number of men absent from the AEF's combat units, these accounts make clear that the figure was substantial.[31]

What was equally clear was that the AEF's senior leaders were cognizant of the dangers that these stragglers presented to American operations. For example, the inspector attached to the 37th Division reported that on 2 October 1918 he had found so many stragglers in the unit's rear area that he estimated "that combat troops only had 80% in strength due to this fact." Given this drain on combat power, the AEF's senior officers took action to reign in these disappearing doughboys. The V Corps G-1, Col. A. W. Foreman, stated that by 18 October 1918 the number of stragglers had grown "to such an alarming proportion" in the First Army that the corps formed a forty-five-hundred-man "Hobo Barrage" to "systematically mop up and thoroughly search all dugouts, houses, hospitals, railheads, Y.M.C.A.'s, etc. in the area assigned to them." Additionally, his corps established three tribunals to interrogate all captured stragglers to determine if the soldiers were truly stragglers or had been unjustly arrested. In a short amount of time, the Hobo Barrage arrested 719 soldiers and returned over 150 "unauthorized stragglers" to their units.[32]

The commander of the 82nd Division, Maj. Gen. George Duncan, noted that after a spike in the number of stragglers from his unit, he was forced to order his subordinates to "post file closers behind each platoon, in addition to the usual straggler's posts" and to direct his MPs to search likely straggler hiding or congregation points in the unit's rear. He also required platoon leaders to carry a list of all their unit's members which was to be checked at halts or lulls in the battle to keep an accurate tally of their losses and quickly identify any men who had straggled from the lines. By these methods the division's strength rose by over five hundred fighting effectives between 25 and 29 October. In a similar move, on 30 October 1918 the commander of the 89th Division ordered his MPs to move their straggler line "forward to a point three hundred meters in the rear of the front line" and to move forward "in very close contact with the advancing infantry." Unfortunately, these steps were not enough to slow the number of stragglers. On 21 October 1918 the AEF inspector general, Maj. Gen. A. W. Brewster, reported that despite efforts to stem the tide of straggling with patrols and stationary posts, "any quick witted straggler can get through these lines, especially at night."[33]

The AEF was never able to resolve its straggler problem. As late as 9 November 1918 the Second Army provost marshal warned his subordinates that "straggling has been allowed to become a menace to the success of operations" and ordered them to "take such definite, immediate, and aggressive steps as will insure without question the immediate apprehension and return of these men to their proper places in [the] line." Between 28 October 1918 and 1 November 1918 the MP companies operating straggler posts in the First Army area of the Meuse-Argonne sector rounded up 613 stragglers. On 30 October alone, the MPs apprehended 193 stragglers. These stragglers came from twenty-two different AEF divisions, with the vast majority of them being infantrymen or machine gunners. These apprehensions were likely only a small fraction of the stragglers roaming the AEF's rear area. If Brewster was correct, and a "quick witted straggler" was able to avoid arrest, the number of absent soldiers probably continued to number in the thousands.[34]

Wendell Westover claimed that much of the problem with straggling and malingering stemmed from the impossibility of having any effective punishment for the reprobates. Even when commanders were successful in bringing charges, "the Court-Martial was so frequently overruled by soft, slab-sided desk hounds . . . that discipline was hard to enforce anyhow. What did they know about the added danger to an outfit going in, incomplete because some quitter had dropped out with ammunition? What did they know of the instant effect on morale by desertion of just one man at a critical time, to say nothing of the added losses if such spirit was allowed to extend, or the operation was hindered by lack of men?"[35] As senior commanders had limited the ability of their junior officers to punish wayward soldiers in any meaningful manner, the soldiers faced few repercussions for their straggling. In most cases the stragglers were merely returned to their units without further action.

While often depriving junior leaders of much of their coercive and legitimate power, senior officers were quick to blame these officers and NCOs for failing to maintain discipline within their units. An IG investigation of straggling in the First Army stated that the causes for the problem were a "lack of discipline among both the officers and soldiers," a "lack of personnel and supervision of the men by the battalion and company commanders," and a "lack of leadership by platoon leaders and sergeants."[36] The report maintained that one of the primary reasons for the crisis was that "platoon leaders do not know where their men are" and made little effort to track them down. The inspector general's belief that the failure of junior leadership led to straggling was accurate and legitimate. When examining the straggler crisis, one can see that junior officers and NCOs were unable to retain the control and discipline of their units. The reason for this failure was a host of leadership issues, some of which were beyond the control of the leaders involved.

The key question that must be resolved is, why did the doughboys straggle from the frontlines? Unfortunately, few stragglers gave any reasons for their absence other than that they were lost and had become separated from their units. For many soldiers this was an honest and accurate confession. The difficult terrain of the Argonne region, morning fog, and battlefield smoke resulted in a number of doughboys truly becoming detached from their commands. On 10 October 1918 the 82nd Division's inspector general reported that "an unestimated number of men, reported to be considerable, have returned to their regiments during the past 24 hours stating that they had become separated and temporarily lost in the woods or during darkness." The inspector remarked that "their present attitudes and desire to fight indicates the truth of most of these statements." Pvt. Ray Johnson, a machine gunner in the 37th Division, noted that during the Argonne drive some men, "being separated from their outfits by chance shell-fire or orders to spread out, wandered helplessly about or attached themselves to other advancing units."[37]

In moving through the jumbled terrain in the Meuse Argonne, the AEF's junior leaders' problems with commanding and control-ling their enormous companies and platoons became apparent. The experiences of Capt. Clarence Minick illustrate the difficulties that company commanders faced in maintaining control of their units. On 29 September 1918 his company participated in an attack to seize the high ground to the northwest of Montfaucon. After fight-ing through most of the morning, Minick's battalion halted to sort out some of the confusion and mix-up of units that had occurred ear-lier in the day. At 2:30 p.m. Minick's battalion was ordered forward to seize Gesnes. Shortly after leading his company forward, Minick discovered that he was missing most of it. The only elements under his control were one and a half platoons. Minick confessed that his company "was pretty badly disorganized." Despite these mix-ups, the captain still managed to take Gesnes but suffered heavy losses in the process. Minick did not find his wayward platoon and squads until 7:00 a.m. on 30 September.

Minick's battalion attacked again on 30 September 1918. The events of that day were as confused and tragic as those of the day before. The American lines were in such a state of disorder that Min-ick's battalion was filled with soldiers from various units of the 91st, 37th, and 35th Divisions. Officers simply corralled all the soldiers they came across and pushed them forward in the attack. The co-hesion of this pick-up team was sparse, and as the ad-hoc unit came under heavy German fire, soldiers began to melt away. By 1:00 p.m. the attack had ground to a halt, and the American troops returned to their jump-off line.[38] Although part of this debacle was due to cir-cumstances beyond Minick's control, the young officer still made little to no effort to ensure that his subordinates were briefed and ready for the attacks of 29 and 30 September.

This degree of confusion could even occur with experienced lead-ers and units. Despite nearly twenty years of soldiering, Samuel Woodfill was not immune from problems with command and con-trol. While his unit was attacking near Cunel on 12 October 1918, the terrain and vegetation of the area presented Woodfill a situation in which "the company was scattered through the wood; I had lost

303

control of them as a unit. It was every man for himself. . . . Those soldiers of mine, some of them so green that they'd hardly smelled powder before, were on their own now.[39] Woodfill was awarded the Medal of Honor for that day's action, but it was only after he basically shook off the mantle of company leadership and fought his way forward as an individual rifleman that he received that honor.

In trying to retain control of their units and direct them towards accomplishing the unit's missions, junior officers were often hobbled by their lack of trained and experienced NCOs. Given the size of the AEF's companies and platoons and the lack of effective tactical communications, junior officers were dependent upon their NCOs to aid them in leading their extended or scattered ranks. The inability of some NCOs to step into their leadership roles led to dire consequences. An infantry battalion commander remembered that after his companies were shelled, the unit lost all order and cohesion. He wrote, "Over the suddenly disorganized mass the mere handful of officers, without the slightest voluntary aid from the noncommissioned officers, are able to exercise but little control. . . . All semblance of organization has vanished."[40]

In another instance, Pvt. John Barkley recalled that when his unit's last officer was killed, the sergeant who now commanded them "completely lost his head" and started issuing nonsensical orders. A senior 82nd Division officer complained on 7 October 1918 that far too many of the unit's squad leaders had failed to "exercise aggressive control" in their units following the loss of their officers and NCOs.[41]

Given the little emphasis that the army had placed on selecting and developing its wartime NCO corps, it was little wonder that the NCOs often proved unable either to assist their officers or assume the mantle of leadership themselves. The 7th Division's Cpl. Frank Dillman went so far as to state that his officers had done so little to uphold the status and authority of their NCOs that "the boys virtually refused to work except when a commissioned officer was over them." Given the casualty rate among officers, this shortcoming was a fatal flaw. As an officer in the 119th Infantry lamented, his NCOs "have been too dependent on officers telling them not only what to do but how to do it." He criticized the failure to allow the NCOs "to make and correct their own mistakes with less interference from officers." It must be stated that these observations reflected a general systemic problem in the AEF. There were numerous occasions in which NCOs demonstrated superior leadership and tactical ability. However, in these cases the NCOs had been "thrown into the deep end" of the tactical pool and forced to swim out of grim necessity.[42]

Officers often compounded their problems with command and control, and further undermined the ability of their NCOs to operate on their own, by failing to brief their soldiers on the details and intent of their unit's missions. Pvt. John Nell, an infantryman in the 77th Division, remembered of his time in the Argonne: "We enlisted men never knew much about our movements, only what we were told and what we could see and hear." Without any clear conception of the details of their missions and cut off from the orders of their officers, some soldiers straggled because they simply had no clue of what else to do. While repairing roads behind the front on 1 October 1918, Capt. Thomas Barber found large groups of soldiers aimlessly wandering around the rear area. One party consisted of a sergeant and twenty-eight men, and another was made up of a corporal and thirty men. They claimed that their officers were all dead, and as they were without orders, they had simply moved to the rear to recover the packs that they had left when they went into action. When deprived of their leaders and uncertain of their missions, the bands of men Barber encountered simply chose the path of least resistance and straggled from the front. It is interesting to note that Barber made no effort to organize these stragglers.[43]

305

The available evidence suggests that the largest number of cases in which men straggled from the line were directly related to the failure of junior leaders and their superiors to live up to their end of the social contract. Masses of men simply left the lines because their officers had failed to provide for the soldiers' basic needs for food and water. Combat logistics; the forward push of rations, ammunition, and supplies; and the rearward movement of casualties had long been a sore spot in the AEF and was the cause of much straggling. For example, during the Aisne-Marne operation, the 2nd Division's MPs reported, "The difficulty of getting the food to the troops soon resulted in looting for the men were searching the whole country for deserted chickens, rabbits and scant food supplies left by the villagers. Looting and straggling went hand in hand for it was noticed that in nearly all cases where arrests were made the looter was found also to be absent without leave from his organization."[44] The problem with getting rations and supplies to the front lines became even worse when the AEF moved into the Meuse-Argonne region. The area had a limited road network, and four years of fighting and shelling had left large swaths of terrain nearly untrafficable for the army's heavy artillery and supply wagons and trucks.

Within days of the start of the offensive, soldiers were already complaining about their lack of rations. Officers in the 79th Division's 313th Infantry noted that during their attack to take Mont-faucon, "It had been nearly impossible to get rations and the food carried in the packs had been consumed . . . and together with the lack of food and rest, the troops were in a pretty exhausted state." Ultimately, the soldiers of the regiment went nearly four days without any food except for their reserve rations. Between 12 and 15 October 1918, the mess sergeant of Company H, 126th Infantry, was unable to bring rations up to the line. The units made do by having returning stretcher bearers bring up hard bread and cans of corned beef. When the company commander sent back rations-carrying parties on 16 October, the men were "too tired, wary, and weak to carry the marmite cans of hot food thru the back area brush and shell holes." As a result, the only rations the company received were hard bread. The artilleryman L. V. Jacks recalled that despite the best efforts of his unit's cooks during the Meuse-Argonne, the lack of food meant that they "tightened their belts, for downright starvation seemed imminent." If things were bad for artillery units behind the lines, they were even worse for the infantrymen battling forward of them.[45]

One of the fundamental tenets of the army's paternalistic leadership and the social contract between the leader and the led was that the officer would provide his soldiers the basic necessities to maintain life and health. Shils and Janowitz argued that one of the major factors in the decline of the Wehrmacht's cohesion and effectiveness was that its unit leaders became unable to meet their soldiers' physical survival needs.[46] Similarly, when leaders failed to live up to their end of the bargain in providing for their basic needs, doughboys felt justified in withholding their participation in military operations by straggling from the lines in search of food.

On 9 October 1918 the inspector general for the 82nd Division reported that more than one hundred soldiers from the 78th Division had straggled into his unit's rear area on the night of 8 October. He declared that "all of these men asked for food, stating that none of them had anything to eat since the night of October 7th," and "some men stated that they had had nothing to eat for a longer period than two days." All admitted that "no permission had been given to leave their camp," but their officers had made no effort to account for their men, nor given them any indication when food would arrive. He also noted that "the personal appearance of these men indicated a general disorganized condition, as evidenced by torn and shabby clothing, unbuttoned blouses and overcoats, failure to shave for several days."[47]

Pvt. Horace Baker, admitted that he "went on an exploring trip" from the frontline during the Meuse-Argonne fighting but stated in his defense that "the pangs of hunger were largely responsible for this." A soldier in the 82nd Division recalled that his unit was so short of food on 11 October 1918 that he was forced to rifle through the pack of a dead German to get the man's black bread. After two more days without food, he straggled from the lines to try to find some rations. Sometimes even officers were complicit in this form of straggling. Capt. Thomas Barber's company grew so short of food during the Argonne Offensive that he selected four men that he "judged good thieves" and sent them to the rear to beg, borrow, or steal whatever rations they could find. Finally, the captain himself left the front with eight men to forage the rear area for food.[48]

Field kitchens located in the rear of the firing lines attracted hungry soldiers like moths to a flame. The problem became so acute that one officer eventually placed a guard on his mess line and kitchen to keep stragglers "from sneaking in." The staff of V Corps noted, "It was found that permitting the Y.M.C.A. and other canteens to approach too close to the front lines induced straggling. Many men who did not intend to become stragglers slipped away to get a cup of hot chocolate or some cigarettes and were picked up as stragglers."[49]

Much of the problem with getting rations to the front was due to circumstances that were far beyond the control of the unit's junior leaders. Tangled supply routes and German actions frequently slowed the forward movement of supplies to a trickle. However, there were also instances in which the officers' failure to plan for the resupply of their units caused or exacerbated their soldiers' hardships. In a particularly egregious case, one officer took the idea that rank hath its privileges too far. Lt. Joseph Lawrence reported that even though his company had been without food, water, or sleep for three days, the company commander took the lion's share of the rations that managed to make it forward.[50]

Most cases involved less malice and forethought by officers, although they had the same effect on their soldiers. After bewailing the inability of his mess sergeant to bring up hot food, Capt. Thomas Barber later sheepishly admitted that he had ordered his company kitchen to remain in place until he ordered it forward and then had forgotten about it for over two days. A veteran of the 5th Division, Maj. Merritt Olmstead, blamed the division's disorderly withdrawal from combat on the failure of commanders to keep their superiors accurately apprised of the actual situation at the front, especially in the physical condition of their soldiers. He also noted that had "commanders been more interested in the welfare of their commands and given some personal attention to the supply of food, their men would not have gone hungry throughout 12–13 October."[51]

Shortly after the war, Col. E. V. Smith reported that "the greatest difficulty was met with by me in getting officers to properly handle supply and administration." He noted that his officers lacked the time and training to deal adequately with these areas, and, "since the war began most officers have been in a 'mental fog' due to the crowding and cramming process in vogue. They had small chance to learn company duties and, in consequence, discipline ran low, kitchens were neglected, equipment and clothing overlooked." As Smith realized, little had been done to prepare junior officers to deal with issues of logistics. As in Barber's case, this resulted in the inability of many officers to "do routine things," such as feeding their soldiers "routinely." For the individual doughboy it mattered not that his lack of food resulted from the failure of supply lines or of his leaders. He was tired and hungry, and the duly appointed representatives of the army, his officers and NCOs, seemed unable or unwilling to do anything about it.[52]

Shortages of food also worsened other problems that wore away the soldiers' health, stamina, and morale. During his unit's time in the Argonne, Pvt. Leslie Langille found that not only were his rations scarce, but also none of his superiors were ensuring that the food that did reach the troops was edible. He wrote, "We subsist on stuff called 'camouflage' by the men because it looks like and tastes worse than our camouflage nets. . . . It keeps one's bowels in a constant state of uproar and dysentery rages rampant in the outfit." An officer of the 36th Division recalled that the water situation in his infantry unit was more dire than shortages of rations. Unfortunately, the nearest water point was over a mile away from his unit, and "the line was too thinly held to send details to the rear for water." Another infantry officer stated that when the 35th Division's soldiers began their attack in the Meuse-Argonne, "Canteens were soon emptied, and they drank water wherever they could find it—in shell holes, crevices, and in fact any place that water was obtainable. The eating of cold rations out of unwashed mess kits, this drinking of foul water, and the exposure and strain, caused every man to suffer from dysentery." Dysentery and exhaustion, brought about by the physical and mental exertion of combat and the lack of food and clean water, became as great a scourge on the AEF's infantrymen in the Argonne as German shot and shell.[53]

On 19 October 1918, the First Army's inspector general reported that the 91st Division was in dire straights and needed at least seven days' rest to rebuild its strength. The division surgeon informed the inspector that after nineteen days of fighting and marching, "none of the men were fit for duty owing to dysentery, fatigue and stomach trouble." He also noted that "the 2,309 replacements recently received are all contacts with influenza, 40% now being sick with that disease." Furthermore, the inspector discovered that there were a "considerable" number of stragglers from the unit, and 955 men were still reported as missing.[54]

There were measures that junior leaders could have taken to lessen some of the physical discomforts endured by their soldiers. In the environment the Americans faced in the Meuse-Argonne, it was incumbent on junior leaders to see that their men were at least well clothed and equipped to deal with the cold and the damp. A 32nd Division infantryman reported that by 19 October 1918 the lack of basic necessities in his unit was causing great hardship. He was still wearing summer-weight underclothes, was suffering from dysentery, and recalled that the "lack of food caused me to be very weak." Another doughboy in the 82nd Division remembered that when the officers failed to supervise and discipline their soldiers, the men "threw away their raincoats and overcoats when they went over the top, so that later they had nothing at all to protect them from the cold and the wet. They went for days and days, sleeping in shell holes filled with ice-water, living on nothing but bully beef and water." This failure of officers and NCOs to maintain even this level of discipline meant that the health and combat efficiency of the units quickly flagged. During October the 82nd Division's medical staff reported an average of seven hundred soldiers per day in their hospitals suffering from influenza, diarrhea, and exhaustion. Oliver Q. Melton, commander of K Company, 325th Infantry, reported that between 16 and 30 October "everyone was sick and weak, many of the men were on the verge of a nervous breakdown." Although some of these problems were due to the inherent nature of combat, the failure of leaders to be more proactive in ensuring the comfort of their men certainly contributed to the predicament.[55]

The 82nd Division was not the only combat unit suffering from the combined effects of high casualties, loss of leaders, battlefield exhaustion, and shortages of supply in the Argonne. Inspector general reports from other divisions revealed the same poor physical conditions and morale in their units. The lack of strong junior leadership to provide for the soldiers' basic needs, build unit cohesion, and reinforce the soldiers' morale could have striking consequences. After only a week of the offensive, the First Army inspector general reported a disturbing conversation with the 3rd Division G1:

> Colonel Stone . . . stated that the 3rd Division relieved the 79th [the] day before yesterday. He says that the 79th Division was the most demoralized outfit that he had ever seen; that the men had thrown away a great deal of their equipment and that the 3rd Division has equipped a complete Machine Gun Company with the machine guns thrown away by the 79th; that the men are dejected and demoralized and apparently not the subject of any discipline. From his talk with different men of the 79th he was convinced that they were utterly unfit for any further operations.[56]

The 3rd Division itself suffered these problems as the campaign dragged onward. After a series of costly attacks, the 3rd Division inspector general reported on 15 October, "Although I am inexperienced in judging men under battle conditions, I wish to state that those officers and men whom I saw of the 38th Infantry appeared to me, to use a slang term, 'all in.'"[57] The day after this report was made, the military police rounded up over five hundred stragglers from the division loitering in the rear area. Weeks of frontal attacks, combined with the leaders' inability to care for their soldiers, had brought the AEF to exhaustion and the brink of dissolution.

While some doughboys straggled from the lines due to being lost, hungry, sick, or leaderless, other left the lines deliberately to avoid combat and the conditions described above. One officer described these stragglers as "shell-holers": men "who in an advance got into shell holes and then liked it so well that they stayed there while their comrades advanced unsupported by them."[58] The MPs simply termed these men "battle stragglers."

The Meuse-Argonne region was crisscrossed with dense woods and shelters, dugouts, and barracks that had been constructed by the French and Germans over the past four years of the war. The natural and man-made features provided a ready sanctuary for any doughboy seeking to escape combat. While soldiers straggling in the immediate rear of the front lines could offer the excuse that they were lost from their units, men hiding out in shelters and woods far behind the lines indicate that their straggling was a premeditated attempt to dodge the fighting.

It is impossible to determine how many of the stragglers left the front to avoid combat. There is some evidence to show that the numbers were relatively large. The 82nd Division's inspector general reported on 12 October that while most of the division's stragglers were simply lost, "a small minority, difficult to estimate, were undoubtedly, endeavoring to evade their duty and were collected from dugouts in Chatel-Chéhéry and elsewhere." Eight days later the division could still not account for 1,019 men, and the adjoining 78th Division reported that the woods in its area were "full of stragglers" from the 82nd Division.[59]

The commander of the 89th Division, Maj. Gen. William Wright, wrote that while moving through the rear area he "found a number of stragglers from the Eighty-ninth, Forty-second, and Second divisions. They were out in the woods and making themselves comfortable in the Boche dugouts and apparently with the intention of staying there." On one occasion in early October 1918 a detail from the 32nd Division's military police searching for stragglers in abandoned shelters found ninety men hiding in one large dugout. The reports indicate that a number of men were seeking to dodge combat by straggling and that they were doing so with relative ease.[60]

Another indication of the depth of the problem of combat avoidance was the number of men apprehended multiple times for straggling. On 14 October 1918 military policemen from the 32nd Division complained that they had apprehended a number of stragglers from the 5th Division and returned them to their units only to find the same men shortly afterwards again hiding out around Montfaucon. The commander of H Troop, 2nd Cavalry, assigned to straggler post duty with the First Army during the Meuse-Argonne Offensive, reported that the 103rd Infantry's Pvt. Raymond Wellman was a "professional straggler" who had been caught by H Troop's posts on at least two different occasions. Upon capture, Wellman stated that "he didn't want to go back to his outfit or any outfit." These incidents tend to support the point that the AEF had men who were so averse to fighting that they risked capture multiple times, that the fear of punishment in these men was rather small, and that despite repeated infractions, their small-unit leaders were unable or unwilling to do much about it.[61]

This last point needs further exploration. The acting first sergeant for Company K, 142nd Infantry, Archibald Hart, recalled finding a number of stragglers hiding in a German bunker while he was searching for water for his company. Hart noted that they had picked a spot near a supply route where they could steal food by night and then "return to their comfortable quarters near the water supply and, undisturbed, catch up on their sleep during the day." The sergeant opted not to report the men for two reasons. First, he believed that such activities were an officer's purview, and he did not think his new commander would make an effort to follow up on the matter because the soldiers were not from his company. Also, he concluded philosophically, "a cozy hideout, well to the rear and out of harm's way, was a proper place for a skulker," for "he definitely would be a liability in the front line, and his Company would function better if he kept himself out of the way."[62]

It is hard to argue with Hart's logic, but if his laissez-faire attitude toward straggling was indicative of the opinions of other AEF junior leaders, their inactivity only encouraged the practice. One straggling soldier marveled "at the freedom I had in the advanced regions." He admitted that he frequently left the front lines and at times strayed over two miles from his unit. The infantryman noted, "The peculiar thing is that usually I did not have permission to go and never once got into trouble for going."[63] In this situation, either the soldier's leaders never realized that he was gone or turned a blind eye to his actions. In either case, it was a lapse in basic combat leadership at the company level and below.

Unlike finding the reasons of those soldiers who left the lines because they were lost or hungry, discovering why battle stragglers sought to avoid combat is much more difficult to determine. Since few admitted their motives, any discussion in this area must be based on the observations of third parties or speculation. Some of the reasons certainly went back to issues with leadership in the AEF's small combat units. For example, Capt. Thomas Barber attributed much of the straggling to poor leadership and with men becoming fed up with the uncertainties and pettiness of everyday military life. In some cases the junior leaders set such a bad example for their soldiers by their own misconduct that the men were naturally bound to follow. Pvt. Ernesto Bisogno stated that at Chatel-Chéhéry "some officers ran like sheep" and abrogated their responsibilities by trying to save their own skins. Joseph Lawrence, an infantry officer in the 29th Division, reported that his company's first sergeant deserted the unit in the middle of the Argonne fight, taking with him "several other men of the company." He also noted the poor example set by a company commander nicknamed "Dugout Pete" for his refusal to leave the safety of his bunker during his unit's attacks.[64]

The effect of leaders setting a poor example could be devastating. When Col. J. A. Bauer inspected the 5th Division, he found that "the officers with the troops of this division appear 'jumpy'" and suggested that this fact explained many of the unit's twenty-five hundred stragglers. Bauer's assessment of the 5th Division was close to the mark. During the attack of the 3rd Battalion, 61st Infantry, on the Bois des Rappes on 15 October 1918, the unit's adjutant broke down after witnessing the death of the battalion commander and two company commanders. The adjutant then "became panicky and departed precipitately to the rear," and "the few men in his immediate vicinity naturally followed." The terror-stricken officer soon reported to the regimental commander that the unit "was all cut to pieces and what was left of it was retreating." This bogus report led to the entire regiment's being pulled back from the line, only to suffer heavy casualties over the next two days trying to recover the terrain it had previously taken.[65]

A similar case of officer straggling and poor leadership occurred in the 4th Machine Gun Battalion during one of the 2nd Division's October 1918 attacks on Mont Blanc. During a move to the front, Wendell Westover's first sergeant reported that one of the company's lieutenants had left the formation during the march, claiming to have been wounded, and was nowhere to be found. The sergeant offered his opinion that he hoped the officer would not come back and darkly hinted that "he won't live through the attack to-morrow if he does show up again."[66] The NCO's veiled threat indicates the hatred that soldiers often felt for leaders who failed to live up to the men's expectations and, perhaps, the lengths that some soldiers would go to rid themselves of a particularly odious or incompetent leader.

Sgt. Archibald Hart recalled that the 142nd Infantry contained one such officer, a lieutenant "known by sight to every man in the Regiment; known and in varying degrees hated." He noted that the officer "employed and flaunted a constant attitude of contempt for all enlisted personnel" and that no enlisted man "ever encountered him when he wasn't exhibiting by tone, manner and expression his utter distaste and aversion." In the Meuse-Argonne, one of Hart's soldiers straggled from the line rather than follow the unpopular officer, and only the sergeant's inadvertent intervention had kept another soldier who had already "drawn a bead on the Lieutenant's back" from killing the despised leader. Hart later recalled, "Sometimes in Camp Bowie one would hear vague predictions that this or that man, not necessarily an officer, would certainly get his once the outfit reached the front, but as a rule these ominous hints were ignored. . . . The Lieutenant, however, had planted his seeds of ill will over the entire regimental area, and no doubt on more than one occasion had stepped far enough beyond the bounds of his usual contumely to incur a bit of vindictive enmity."[67] It is impossible to determine how many unpopular officers or NCOs met their ends at the hands of their own soldiers, but Westover's and Hart's accounts illustrate that at least some disgruntled doughboys were pushed in that direction.

Some of the battle stragglers were simply men who had stayed at the front until they had reached the breaking point of physical and psychological collapse. One officer later wrote that after grueling weeks at the front under constant fire, men tried to slip to the rear for "a few minutes of relief from the hell on the line." He recalled that "this kept up all night, making it necessary for me to patrol the line. . . . I would drive one man back to his position and another would try to slip by." One infantryman blamed this type of straggling on the simple fact that commanders "had forgotten that there is a limit to human endurance."[68]

Other battle stragglers were perhaps motivated to avoid combat due to the realization that neither their own nor their leaders' training had prepared them for battle. The AEF's soldiers were "thinking bayonets" who could not have helped being cognizant of these shortcomings. Henderson, Shils, and Janowitz all argued that one of the major pillars of small-unit cohesion was the soldier's faith that their leaders will be "duly considerate of their lives" and that the inevitable casualties of war will still be "minimal and justified."[69] The fact that the AEF had large numbers of men actively seeking to avoid combat indicates that there were major problems with cohesiveness within the army's small units. In many of the AEF's small units, it was simply the case of the unwilling being led by the unready into the unknown. As soldiers weighed their chance of survival in combat and opted to vote with their feet, the quality of their leaders was undoubtedly one of the factors that influenced their decision.

It is interesting to note that many officers and NCOs blamed the straggler problem on replacements. The AEF inspector general noted that when the replacements consisted of "men who do not know the rudiments of soldiering [they] soon become either 'cannon fodder' or skulkers." A soldier in the 42nd Division corroborated this observation by noting that most stragglers from his unit "had been replacements newly arrived." Nine years after the Armistice, the war correspondent Thomas Johnson wrote in his aptly titled *Without Censor* that the war was hardest on those men, usually replacements, who had been shunted off to the front with very little training under their belt. He noted, "We could always recognize them on the roads of the battle area. They were paler, slighter, than the men who had had their proper hardening and had not just come from crowded transports, and they looked about nervously. Who could blame them?" Johnson recalled that "some of the youngest ones, scared boys, drifted to Y.M.C.A. hotels where they were fed and warmed and often got their nerve and went back to the front."[70]

If replacements did make up the majority of battle stragglers (and this point is far from certain), they had good reason to fly from battle. An infantry first sergeant in the 32nd Division mourned the fact that "replacements get the end of dirty things in the Army. They are shoved from pillar to post and back again. . . . They acquire buddies one day to have them leave the next day. . . . Their APO [Army Post Office] number is changed before they receive mail from the folks at home."[71] If these indignities were not enough, in an army not known for the quality and quantity of its training, replacements were often the worst trained of the lot.

The stories that circulated that some replacements had to be told how to load their rifles just before H-hour are far from apocryphal. In the summer and fall of 1918 the 2nd Depot Division's intelligence officer recorded the level of training of the replacements that arrived in France. These reports provide sad evidence of the breakdown of the stateside training of replacements in the last four months of the war. On 12 August 1918 one of the officer's agents reported that the 2,500 men who just arrived at the division from Camps Gordon and Hancock had "been in the service only a few weeks." A month later, another agent reported that the 597 draftees who had just arrived from Camps Pike, MacArthur, and Gordon "had all been in the army less than a month and have had little or no training." The men who reported on 29 October from Camp Pike had spent only one day on the rifle range and had no gas training before being shipped overseas. The men who arrived on the same day from Camp MacArthur were little better off. They had spent one or two days on the range and had been given six hours of gas training just before leaving for France. Lt. Hugh Thompson found that twelve of the replacements assigned to his company just before the Saint-Mihiel Offensive had never fired their rifles before, and that his other replacements "were not very sure of their rifles." The new men's only training before going into combat was allowing each man "to fire a clip (five rounds) into the soggy ground at his feet."[72]

The commander of the 77th Division's 307th Infantry, Lt. Col. Eugene Houghton, argued that the cohesion and morale of his unit suffered from the influx of new recruits. Of the 850–900 men he received just before going into the Argonne battle, "90% of them had never fired a rifle, nor thrown a grenade, nor had they the ordinary close order drill." He went on to note, "Since the action started it has been frequently reported to me by company and battalion commanders that it was practically impossible to handle these men over the present terrain. They had no idea what it meant to extend [formations] and would have to be led around from place to place. They were continually getting lost and straggling, and their officers and N.C.O.'s were practically strangers to them, it made them very difficult to handle them."[73] Sadly, there was little that a company's officers and NCOs could do to address the problems caused by this massive influx of ill-trained men. In the end, the problem with replacement training was merely only one last straw to the litany of problems that dogged the AEF's small-unit leaders throughout the war.

The end result of poor leadership, training, and unit cohesion was the AEF's butcher's bill in the Meuse-Argonne. On 19 October 1918 Gen. John Du Cane, of the British Mission to the Allied Armies, reported that the disjointed and ill-led American attacks in the Argonne did nothing but "suffer wastage out of all proportion to the results achieved." While this criticism seems hypocritical coming from an officer of an army that a year previously had suffered grievously as a result of equally disjointed attacks, Du Cane's observations hit very close to their mark. A more sympathetic, or at least tactful, French officer simply noted, "These young Americans lost a good many of their illusions in the depths of the Argonne." Even Brig. Gen. Harold Fiske, Pershing's chief of training, had to admit that in the final analysis, "it must be remembered that to the end most of our divisions were lacking in skill. Given plenty of time for preparation, they were capable of powerful blows; but their blows were delivered with an awkwardness and lack of resource that made them unduly costly and rendered it impracticable to reap the full fruits of victory." Fiske's sad confession was also an admission that the US Army had failed to properly train and develop its officers, NCOs, and soldiers to meet the challenges that confronted them on the battlefield without enduring soul-numbing casualties. Pershing had intended that the AEF fight like a master swordsman: a fighter able to dispatch his enemies with quick maneuvers and deadly thrusts. The AEF, however, was more like a blind giant: a creature groping to find its opponent, suffering wound after wound in doing so, but finally crushing the enemy with its superior weight when it finally found him.[74]

The changing nature of warfare demanded that the AEF's junior officers and NCOs be able to employ a host of new weapons and to use their initiative on the battlefield. Not only were the junior leaders unprepared for this challenge, but the AEF also seemed to work at times to keep the leaders from rising to the occasion. The AEF's huge combat formations were too ponderous for the half-trained leaders to command and control adequately. Improper and incomplete training, as well as the failure to instruct junior leaders to act independently, had not prepared the AEF's junior officers and NCOs for the enemy and the environment that they encountered in combat. Consequently, American operations throughout the war tended to be rather ham-handed, flat-footed, and tragically lethal affairs. These bloodlettings were deadly to unit cohesion, combat effectiveness, and the leaders themselves. As leaders sacrificed themselves while learning at this school of hard knocks, a deadly cycle was created in which leaders all too often killed themselves and their soldiers before gaining the critical battlefield wisdom required for survival. In the final analysis, American junior officers and NCOs could not grasp those opportunities on the margins of attritional war that would have allowed them a degree of battlefield success without the unnecessary dead that littered the AEF's battlefields.

12 Conclusions
A Tale of George and Henry

SERVING IN THE GREAT WAR had given the two men a shared experience that had helped to grease the wheels of their working relationship. Henry, the elder of the two, had risen to the rank of colonel and had commanded an artillery regiment during the war. George had been promoted to the rank of temporary colonel and had served as a corps chief of staff. Both men looked upon their military service during World War I as one of the great formative events of their lives. Although the two worked extremely well together, there was one subject that nearly ended their professional relationship.

In March 1941, Secretary of War Henry Stimson was confronted with a number of problems. Henry knew well that the United States was likely to be drawn into a war with Germany or Japan and that the nation was little better prepared for this eventuality than it had been in 1917. A few months before, Congress had approved the nation's first peacetime draft, and the some of the draftees were already in training. However, this growing force still needed officers. While ROTC units across the nation provided a pool of reserve officers with a better military education than had been given those who had attended land-grant colleges prior to World War I, there were still not enough men to lead the growing ranks of the army. To solve this problem Stimson proposed that the army reestablish the ninety-day Officer Training Camps used in the Great War.

This proposal was abhorrent to the army chief of staff, Gen. George C. Marshall. George was adamant that any future army raised for the approaching war be provided with competent junior leaders more realistically and thoroughly trained than the generation of 1917. For the first and only time in their long and fruitful relationship, George informed Henry that if the secretary went forward with his plan on resurrecting the OTC model, he would resign as chief of staff.[1]

This rare confrontation between Stimson and Marshall was an indication of how deeply some of the army's officers believed that the training and performance of the Great War's junior officers had been flawed. Marshall and other Regular Army AEF veterans understood the price that the American soldier paid for serving under ill-prepared small-unit leaders. The AEF's half-trained junior officers and NCOs usually fought bravely but seldom fought skillfully. At the "tip of the spear," these infantry leaders lacked the critical experience and the tactical and technical skills to take advantage of the slim opportunities available to the attacker on the margins of the attritional World War I battlefield. This was a major failing, because changes in the nature of combat brought about by improved weapons and the expanded breadth of the battlefield now required a decentralization of command and control that placed much greater responsibility upon small-unit leaders. As the war progressed, it was increasingly the junior leaders at the tip of the spear, and not the generals, that ultimately decided whether the senior commander's grand plans were properly executed. In the case of the AEF, this spear point was made of a brittle and untempered metal.

319

The failure of the leadership spear point was not a result of the poor human material that comprised the blade. In the majority of cases, the AEF's junior officers and NCOs were patriotic, adequately educated, dedicated to the cause, and brave to a fault. They were eager to learn and were well aware of the limitations of their training and experience. These leaders did the best they could under the conditions they faced, but far too often they sacrificed themselves and their soldiers in clumsy, ill-supported, frontal mass attacks. Their maladroit tactics generally showed an awkward inability to match formations, maneuver, and firepower to the terrain and the enemy they encountered. Furthermore, American officers and NCOs also tended to display a fatal lack of initiative that ceded hard-won and short-lived tactical gains to a more skillful and agile enemy.

Some of these problems were the result of the inherent realities of the Great War's battlefields that all the major combatants had to contend with during the conflict. Advances in communications and command and control had not kept pace with the changes in weaponry. Without reliable and responsive communications, it was exceedingly difficult for the attacker to gain the reliable and responsive artillery fire he desperately needed to level the playing field between himself and the defender. The lack of an effective system of tactical communications meant that it was also hard for the junior leader to direct his soldiers, inform his superiors of his progress, or change orders in the heat of combat.

Many of the difficulties faced by the AEF's small-unit leaders were due to systemic problems associated with the nation's lack of preparation to fight a modern war and to the inherent growing pains associated with building a mass army. The United States was the least prepared—materially, numerically, administratively, and intellectually—of the major combatants that entered the war on the Western Front in World War I. This meant that when the nation declared war on Germany, the pressing need for leaders ensured that quantity took precedence over quality.

The press of time, the prodding of the desperate Allies, and the rapidly changing military situation in France in 1918 ultimately forced the army to take shortcuts with the stateside training of its leaders and soldiers that later had detrimental effects on the battlefield. The American army's demand for officers in the spring of 1917 led the army to adopt the flawed OTC training and commissioning system. Training in the OTCs failed to impart the leadership, tactical, and technical skills needed by the fledgling officers in combat. The OTCs, and later COTSs, placed far too much emphasis on close-order drill, bayonet training, and other skills of questionable importance and too little stress on teaching their students critical subjects such as map reading, the military aspects of terrain, logistics and administration, and the tactics of combining firepower and maneuver in combat.

The army's process for selecting and developing NCOs was even more defective. The Regular Army's prewar ad hoc system in which company commanders selected NCOs from long-service privates was one that could not be readily adapted to the realities of a mass draftee army. Lacking the training and status to set them above the mass of doughboys, the AEF's NCOs remained privates with stripes. This institutional failure to more systematically train and develop NCOs placed additional burdens on the army's already overtaxed junior officers and greatly complicated small-unit command and control in combat. The overall lack of junior officer and leader know-how, combined with the systemic problems associated with mass mobilization, hobbled the army's efforts to build cohesive and effective combat units.

The army itself must also bear its share of blame for its problems with junior leaders. The tactical doctrine promulgated by the Regular Army from 1914 to 1917 continued to exalt the power of the rifleman and downplay the effectiveness of modern artillery and the machine gun despite all evidence to the contrary from the European fighting. The US Army's attempt to close the intellectual/doctrinal gap between itself and the European combatants in 1917 and 1918, as well as Pershing's effort to craft a uniquely American open warfare doctrine, led to such a mixing of American, French, and British methods that that the AEF never had a uniform doctrine during the war. Faced with a deluge of often conflicting doctrinal publications, new weapons, and evolving tactical techniques, the AEF's small units tended to freelance their own doctrines or slavishly adhere to the formations presented in *Offensive Combat of Small Units*. Both of these approaches had their own problems and contributed to the AEF's flat-footed and ham-fisted performance at the tactical level of war.

Through trial and error, some leaders and soldiers picked up workable tactical methods and techniques from the unsparing school of hard knocks. However, this method of learning through hard experience was also problematic. The AEF's small units experienced such a vast and rapid turnover of leaders and soldiers, as a result of casualties, schools, and other removals, that their ability to learn lessons from their combat experience was stunted. This massive loss in the ranks of junior leaders of infantry and machine guns led to an unbreakable cycle of ineffectiveness as half-trained leaders were supplanted by even less trained and less experienced replacement officers and NCOs. Although commanders and staffs at the regimental level and above were able to puzzle out some of the AEF's tactical problems, such as the effective employment of planned artillery fires, at the battalion level and below rapid changes in leaders often prevented these units from being effective learning organizations. Too many junior leaders simply did not survive their first tactical missteps, mistakes, or blunders to get second chances to improve.

321

The US Army further handicapped the ability of its junior leaders to succeed in combat by instituting short-sighted personnel policies that undercut unit cohesion and morale. The large-scale levying of personnel from American units training in the United States and France broke or disrupted the bonds that tied units together and wed the leader to the led. Pershing's efforts to overcome his army's inability to wage a modern war by establishing an expansive school system ironically hindered its combat effectiveness by further removing essential leaders from their units at key times in their unit's training and operations. When combined with other factors that undermined unit morale, such as heavy casualties, the loss of leaders, and the inability of junior officers to live up to their end of the social contract with their soldiers, these flawed policies led to small combat units with very fragile cohesion. The end result was an explosion of stragglers in the last months of the war that further dulled the AEF's combat effectiveness by depriving its units of large percentages of manpower at the front.

Furthermore, unexpected changes in the military situation in the spring and summer of 1918, and the high American casualties that resulted, threw the army's system for training officers and NCOs in such disarray that the quality and quantity of leader training actually decreased as the war went on. The press of events in 1918 added another burden on junior leaders by filling the ranks of their units with replacements who were often so poorly trained as to be nearly as great threat to themselves and their comrades as they were to the Germans.

Last, the inability of the Regular Army officer corps to reconcile its demands for obedience with the need for its junior leaders to exercise initiative in combat further hobbled the effectiveness of its units throughout the war. The regular officers' prewar assumptions about the discipline required of leaders and men to "cross the fire swept ground" led commanders at the regimental level and above to demand an exacting degree of compliance to their orders by their subordinates. This degree of obedience and micromanagement by senior officers reflected conceptions of combat that were increasingly obsolete on the modern battlefield.

The micromanagement by senior officers of their subordinates was also part and parcel of the command climate that Pershing instilled in the AEF. Pershing's willingness to sack senior officers who he believed lacked aggressiveness or ability created a climate of fear that permeated all levels of the AEF. Regimental and brigade commanders, fearful that the mistakes of their subordinates would reflect poorly upon their command, made little effort to develop their wayward junior officers and instead purged them from their units by sending the errant lieutenants and captains to the Reclassification Center at Blois. These actions sapped the initiative of junior officers who were already hindered by the knowledge of their own inadequate training and the weight of being responsible for the lives of their soldiers.

The inability of the AEF's junior leaders to gain tactical success without long casualty lists left an ambiguous legacy in the post–World War I American military. George Marshall, for one, understood the Americans' failings and worked during the interwar period to head off future leadership problems. While serving as the assistant commandant of the Infantry School, Marshall had tried to pass on some of the army's hard-won battlefield wisdom to new generations of its junior leaders.

In 1934 the Infantry School at Fort Benning published its classic *Infantry in Battle* to give "peace-trained officers something of the viewpoint of the veteran." The work contained vignettes on minor tactics and leadership, mostly drawn from American actions in the Great War, to better prepare a new generation of junior leaders for the realities of combat. Marshall, the driving force behind the publication, understood all too well the limitations of America's wartime leadership and its flawed training. In the work's introduction Marshall noted, "In our schools we generally assume that organizations are well-trained and at full strength, that subordinates are competent, that supply arrangements function, that communications work, that orders are carried out. In war many or all of these conditions may be absent."[2] In stating this realistic view of combat, Marshall wanted his students to understand the nature of fighting modern wars and the skills required by officers to meet its challenges.

323

Marshall also brought a darker legacy from World War I to his subsequent military endeavors. Historian Daniel Bolger has noted that the lessons of the war and the approach to tactics that Marshall attempted to convey in *Infantry in Battle* were too mired in the infantry-centric views of the Great War. He also argued that as the assistant commandant of the Infantry School, Marshall inculcated a generation of the school's students with his own rigid ideas of tactics, discipline, and leadership. When students such as Omar Bradley, Courtney Hodges, and J. Lawton Collins later rose to senior command in World War II, they placed Marshall's narrow doctrinal views into practice. As such, the operations of Bradley's First Army were characterized by a cautious set-piece approach to warfare and a tendency of Hodges and Collins to micromanage and dampen the initiative of their subordinates.[3]

Bolger maintained that the stern and forbidding Marshall also passed on to his Benning disciples a rather harsh view of how to deal with subordinates who failed to perform in combat. Those who failed had to be cut out of the unit like a cancer. As a commander in Western Europe in 1944 and 1945, Bradley placed this merciless vision into practice, and between June 1944 and May 1945 he relieved two corps commanders, eight division commanders, and numerous brigade and regimental commanders. The result of these sackings created a "zero defect" command climate in the First Army that left its subordinate commanders fearful and unwilling to question the decisions of their superiors.

What Bolger failed to discuss was that this willingness to relieve officers at the drop of a hat, and the poor command climate this practice created, was also a legacy of World War I. It is sometime said that children who are abused grow up to be abusers themselves as adults. As a senior AEF staff officer, Marshall saw first-hand Pershing's ruthlessness in ridding himself of those who failed to please him. This experience left an indelible impression on Marshall—one that he later passed on to his Benning students. One passage in *Infantry in Battle* even advises commanders to "relieve all unreliable junior officers."[4] As Bradley and his senior commanders sacked their subordinates with pitiless abandon, the ghost of Blois stalked in the shadows.

Not all of the AEF's legacy was negative or unproductive when it came to junior leadership. Many officers looked past the period's jingoism to admit that the army's failure to properly train and develop its junior leaders had blunted the combat effectiveness of the AEF. While officially adhering to the party line that superior American manpower and know-how had decisively contributed to the Allied victory, many of the AEF's senior leaders later admitted that the army had done a poor job of preparing for combat. In April 1919 the AEF GHQ convened an officer board headed by Maj. Gen. Edward M. Lewis to "consider the lessons to be gained from the experiences of the present war in so far as they affect the tactics and organization of the Infantry."[5] A number of the board's participants were critical of the leadership exhibited by the AEF's junior officers and NCOs during the war. Although the board's recommendations regarding leadership were of a general nature, the board's findings still indicated that the AEF's infantry officers were cognizant of the problem and searching for a solution.

The Great War experience also encouraged the army to expand its schools system and promoted professional development to counter flaws uncovered by the war in the regular officer corps. Although it was bitter medicine, the regulars had to admit that many of their ranks who rose to senior command and staff positions in the AEF were no better prepared for the shocks of modern war and the duties and requirements of their elevated positions than had been their ninety-day-wonder junior officers. Eisenhower, Bradley, and other future World War II commanders benefited from this renaissance of the cult of professionalism.

The army and Congress also took steps to correct the problems of junior leadership that became apparent during the war. As superintendent of the Military Academy from 1919 to 1922, Douglas MacArthur changed the curriculum for the Corps of Cadets to address some of the conspicuous deficiencies in junior officer leadership that he had observed in France. For the first time in the academy's history its curriculum included formal classes on military leadership. While opposition from the Academy's entrenched and conservative faculty meant that few of MacArthur's reforms survived his departure from West Point, his actions still forced an examination of the cadets' leadership and tactical training.

After the war, the army did not allow the ROTC program to languish from lack of attention and direction as it had done with its Land Grant College predecessor. The National Defense Act of 1920 strengthened the army's commitment to the ROTC and provided for a more rational and regulated system for maintaining a trained Officers' Reserve Corps. By 1922 even Pershing, the strongest guardian of the AEF's reputation, tacitly admitted that leadership had been lacking in his army and that reform was necessary. In a keynote address to the Reserve Officer Association, Black Jack stated, "A resolve has gone forth, embodied in the law of 1920, that never again shall our untrained boys be compelled to serve their country on the battle-field under the leadership of new officers with practically no conception of their duties and responsibilities."[6] Efforts to improve the quality of junior leaders in the 1920s and 1930s later bore fruit in World War II.

Ultimately, the World War II officer corps would be composed mainly of ROTC and Officer Candidate School (OCS) graduates, with the majority coming from the candidate schools. Unlike their World War I predecessors, these officers were commissioned after having proven themselves as competent enlisted men for a minimum of four to six months and after having demonstrated tactical and leadership abilities during their seventeen-week-long OCS course. Most of these OCS officers went on to attend an additional two- to three-month branch or company commander's course for more intensive technical, tactical, and leadership training. Although the World War II army also faced some of the same systemic problems with its mass mobilization (frequent levying of personnel from existing units for cadres or replacements, shortages of qualified instructors, and so on) as had the World War I generation, it did a much better job of developing competent officers than did the army of the Great War.[7]

Unlike their Great War predecessors, the officers and NCOs of World War II were generally given that most precious of wartime resources to hone their skills and leadership abilities: time. The peacetime draft of 1940, and the fact that most of the army's units did not see combat until after the late fall of 1942, allowed for a degree of training that would have been the envy of the doughboys. The Louisiana Maneuvers and other large-scale exercises allowed the army to work out some of the tactical and logistical bugs that had so plagued the army of 1918. Furthermore, developments in tactical communications and increases in infantry, armor, air, and artillery firepower also eased many of the problems that had faced junior combat leaders in World War I. Although the army's system for procuring and training leaders and units in World War II was far from perfect, on the whole it was a vast improvement over what had been done in the Great War.

Unfortunately, lessons learned are sometimes forgotten in the heat of a new crisis. The army in Vietnam faced problems with junior leadership that often resembled those of the Great War. Vietnam also showed that the army's problems with properly selecting and training junior leaders were not just limited to wars requiring mass mobilization. Lyndon Johnson's refusal to expand mobilization for the war and the army's own flawed individual rotation policy created a constant drain of junior leaders from American combat units. As an institution, the army was ill-prepared to fight a protracted attritional war without a call-out of the National Guard and Army Reserve. As such, it had to scramble to adapt its system for identifying, training, and developing junior combat leaders. The growing unpopularity of the war complicated this task and further hindered army efforts to recruit suitable men for officers and NCOs. As did their Great War predecessors, officers in Vietnam resorted to promoting privates and specialists to the NCO ranks rapidly. These so-called shake-and-bake sergeants lacked specialized training for their jobs and usually owed their positions to their length of time in-country.

The officer situation was equally bad. The widespread granting of educational deferments, the declining enrollment in ROTC programs, and the incessant demand for platoon leaders forced the army to turn to the OCS to obtain officers.[8] By 1967 over half of the army's lieutenants were the products of a four-month-long OCS course. Given the strains of supply and demand, the army could not afford to be very selective in the officers it commissioned. As was the case with the World War I OTCs, the press of time forced the OCS to commission officers who lacked the leadership ability and tactical training to lead soldiers in combat.

Lt. William Calley, the infamous leader of the My Lai massacre, was one such officer. Calley, a college dropout and unemployed misanthrope, left OCS for Vietnam in 1968 ill-trained and unfit for the position that he held. But in a larger sense it was the army's failure to screen, train, and develop its junior leaders properly that was one of the root causes of its morale and discipline problems from 1969 to the end of the war. As one colonel noted at the time, "We have at least two or three thousand Calleys in the army just waiting for the next calamity."[9] While the American armies of the Great War and Vietnam had difficulty fielding competent junior leaders for different reasons, the end result was the same. In both cases, ill-trained and unprepared leaders caused unnecessary casualties and eroded unit morale and cohesion.

For all of its problems, in the end the AEF accomplished its strategic goal. At a time when its allies were experiencing a flagging of their strength and morale, the promise of fresh, numerous, and young Americans gave heart and hope to the Allied cause. The Allies' combined weight of numbers and materiel slowly but surely ground the German army under the millstone of attritional war. The sad reality was that the AEF was an army of 1914 thrust into 1918. To paraphrase Abraham Lincoln, in the clashes of 1914 the European powers were all "green alike" and thus developed at a pace that gave none a marked or lasting advantage when it came to the evolution of tactical thought or weapons. From the time that the first American soldier stepped foot in France, the AEF suffered from having to play catch-up with armies that had been studying in the school of hard knocks for four years.

327

Although few American soldiers at the time would have agreed, the AEF was in fact fortunate that the German army it faced in the summer and fall of 1918 was not the German army of 1916, or the army of 1917, or even the army of March 1918. While the Germans remained tough schoolmasters for the Americans to the very end of the war, the doughboys' rather unskillful and costly attacks still wore down the strength and willpower of their Teutonic foes. Sadly, the difficulty that the AEF's junior infantry officers and NCOs had in learning how to control their units while attempting to combine firepower with maneuver had caused the Americans a degree of casualties that outweighed the tactical gains that resulted from the sacrifice. Despite the AEF's blunt and costly approach to war, the doughboys and their leaders still bled Woodrow Wilson to a seat at the peace table. Unfortunately, the AEF's success at the strategic level was cold consolation to those who slogged out the war in the army's small units.

Organization of AEF Infantry Rifle Companies and Platoons

Table A-1. Infantry Rifle Company, 26 June 1918 AEF
Table of Organization and Equipment

HQ PLT	RIFLE PLT	RIFLE PLT	RIFLE PLT	RIFLE PLT
CPT	LT PLT LDR	LT PLT LDR	LT PLT LDR	LT PLT LDR
1LT XO	PLT SGT	PLT SGT	PLT SGT	PLT SGT
1SGT	2 x SGT	2 x SGT	2 x SGT	2 x SGT
Supply SGT	8 x CPL	8 x CPL	8 x CPL	8 x CPL
Mess SGT	15 x PFC	15 x PFC	15 x PFC	15 x PFC
4 x Cooks	32 x PVT	32 x PVT	32 x PVT	32 x PVT
Co. clerk				
4 x Mechanics				
2 x Buglers				
4 x PFC runners/ signalmen				
5 x Wagoners (From regt. supply co.)				

TOTAL STRENGTH & EQUIPMENT
   6 Officers, 48 NCOs, 207 Soldiers
      219 x Rifles
      16 x Automatic Rifles
      30 x Rifle Grenade Dischargers*
* Includes 6 spares held in company supply

Table A-2. Infantry Rifle Platoon, 26 June 1918 AEF
Table of Organization and Equipment, Formed as Half-Platoons

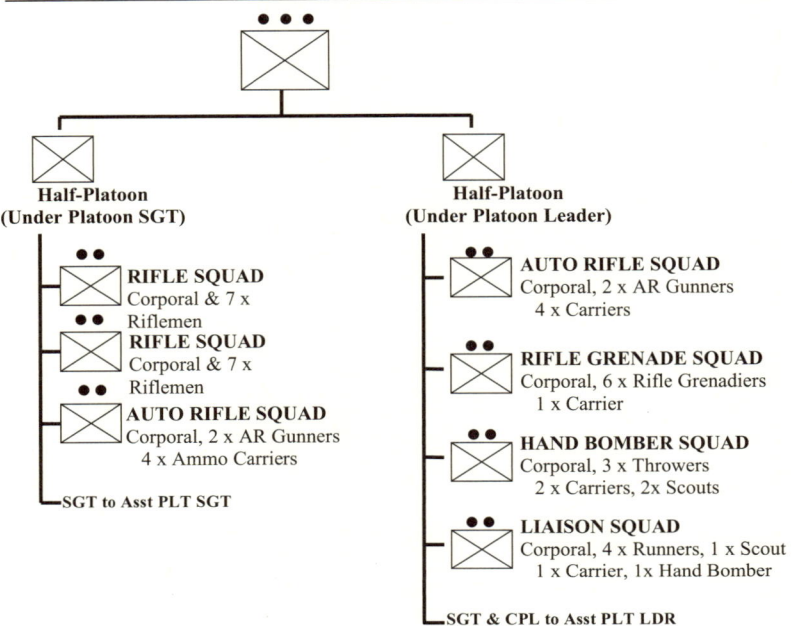

330

TOTAL STRENGTH & EQUIPMENT
1x Lieutenant , 3x Sergeants, 8x Corporals, 47x Privates 48 x Rifles 4 x Automatic Rifles 6 x Rifle Grenade Dischargers

Table A-3. Infantry Rifle Platoon, 26 June 1918 AEF
Table of Organization and Equipment, Formed as Combat Groups

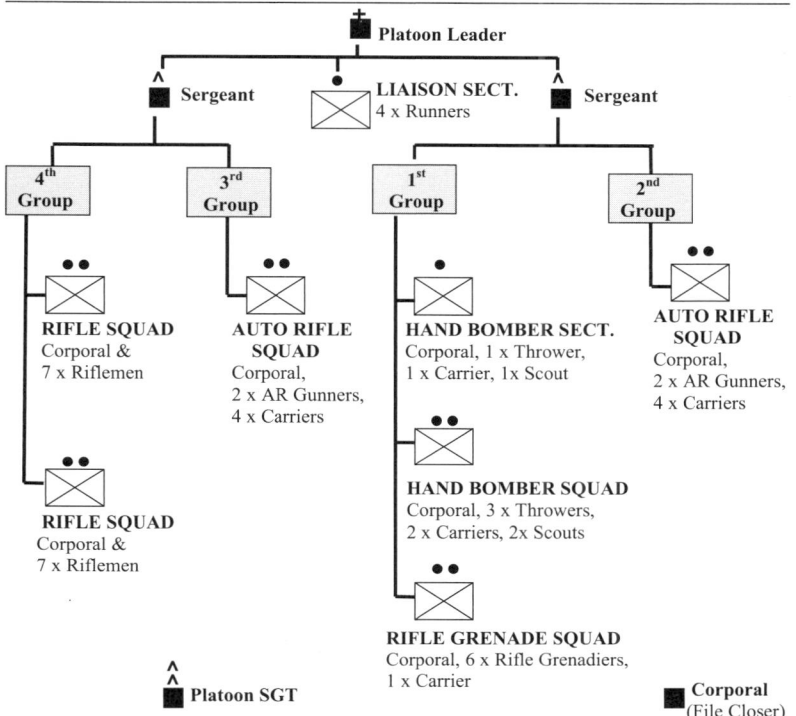

Platoon Leader

Sergeant

LIAISON SECT.
4 x Runners

Sergeant

4th Group

3rd Group

1st Group

2nd Group

RIFLE SQUAD
Corporal &
7 x Riflemen

AUTO RIFLE
SQUAD
Corporal,
2 x AR Gunners,
4 x Carriers

HAND BOMBER SECT.
Corporal, 1 x Thrower,
1 x Carrier, 1x Scout

AUTO RIFLE
SQUAD
Corporal,
2 x AR Gunners,
4 x Carriers

HAND BOMBER SQUAD
Corporal, 3 x Throwers,
2 x Carriers, 2x Scouts

RIFLE SQUAD
Corporal &
7 x Riflemen

RIFLE GRENADE SQUAD
Corporal, 6 x Rifle Grenadiers,
1 x Carrier

Platoon SGT

Corporal
(File Closer)

TOTAL STRENGTH & EQUIPMENT

1x Lieutenant, 3x Sergeants, 8x Corporals, 47x Privates
48 x Rifles
4 x Automatic Rifles
6 x Rifle Grenade Dischargers

Notes

Abbreviations

AWCHS	Army War College Historical Section/Branch, Army General Staff
CARL	Combined Arms Research Library, Fort Leavenworth, KS
CGSS SIRS	Command and General Staff School Student Individual Research Study
INJ	*Infantry Journal*
JMSIUS	*Journal of the Military Service Institution of the United States*
JUSCA	*Journal of the United States Cavalry Association*
NARA	National Archives and Records Administration
USACMH	US Army Center of Military History, Washington, DC
USAMHI	US Army Military History Institute, Carlisle Barracks, PA
WPD	War Plans Division

Chapter 1

1. Alvin C. York, *Sergeant York: His Own Life Story and War Diary,* ed. Tom Skeyhill, 240–55.
2. Ernesto Bisogno, "The Life and Death of Charles Clement," *American Legion Magazine,* March 1938, 50–51.
3. "Record of the Trial by General Courts-Martial of Captain Charles G. Clement, 328th Infantry," 15 July 1918, Box 5977, Docket no. 120515, RG 153, NARA, 17.
4. Ibid., 5.
5. Ibid., 18.
6. Bisogno, "Life and Death of Charles Clement," 51–52.
7. York, *Sergeant York,* 222, 226–27, 241.
8. US War Department, *War Department Annual Report, 1916,* 242; Leonard Ayres, *The War with Germany: A Statistical Summary,* 16–21.
9. Ayres, *War with Germany,* 22, 30.
10. Arthur F. Barbeau and Florette Henri, *The Unknown Soldiers: Black American Troops in World War I,* 58.

Chapter 2

1. Rudyard Kipling, "Only a Subaltern," *Soldiers Three: A Collection of Stories,* 184.
2. US War Department, *Field Service Regulations of the United States Army, with Corrections to May 21, 1913,* 51, 62, 161.
3. Col. Robert L. Bullard, "The Military Study of Men," *INJ* 8, no. 3 (November–December 1911), 327; Maj. Gen. David C. Shanks, *Management of the American Soldier,* 4–5. The material in the booklet was compiled from a set of articles that Shanks had written in the *Infantry Journal* from November 1916 through March 1917.
4. Maj. Frank R. McCoy, *Principles of Training,* Vol. 3 of *The National Service*

Library, ed. Maj. Charles E. Kilbourne, 144–45; Capt. George Haltzell, "The Proper Training of an Infantry Company," *INJ* 5, no. 5 (March 1909), 640.

5. Shanks, *Management,* 20; Edward M. Coffman, *The Regulars: The American Army, 1898–1941,* 151.

6. John M. Schofield, *Forty-Six Years in the Army,* 535–36 (emphasis in the original); Brig. Gen. J. P. Farley, "Military Service for College Men," *JMSI* 50 (1912), 114.

7. Edward M. Coffman, *The Old Army: A Portrait of the American Army in Peacetime, 1784–1898,* 281. Also see Adjutant General's Office, General Order 116, 1890; General Order 57, 1892; and General Order 17, 20 February 1903.

8. US War Department, *War Department Annual Report, 1910,* 1:160; *1905,* 1:17.

9. James A. Moss, *The Noncommissioned Officer's Manual,* 17.

10. Capt. Frank Tebbetts, "Leadership," *JUSCA* 27, no. 111 (July 1916), 19.

11. Ernest Fisher Jr., *Guardians of the Republic: A History of the Noncommissioned Officer Corps of the U.S. Army,* 178; US War Department, *Regulations for the Army of the United States, 1913, Corrected to April 15, 1917,* 71; Moss, *Noncommissioned Officer's Manual,* 18; 1st Lt. William Edwards, "The Squad Leader and His Squad," *JUSCA* 28, no. 4 (January 1913), 776.

12. *War Department Annual Report, 1908,* 3:178.

13. Ibid., *1909,* 3:143.

14. Moss, *Noncommissioned Officer's Manual,* 34–35.

15. Ibid., 20 (emphasis in the original).

16. Capt. Francis Greene, "Important Improvements in the Art of War during the Past Twenty Years and Their Probable Effect on Future Military Operations," *JMSIUS* 4, no. 13 (1883), 24; Capt. George Baltzell, "The Proper Training of an Infantry Company," *INJ* 5, no. 5 (March 1909), 645–46.

17. US War Department, *Infantry Drill Regulations (Provisional),* 1919, 99–100.

18. J. Franklin Bell, "Reflections and Suggestions: An Address by General J. Franklin Bell," 17 March 1906, CARL, 3; O. O. Ellis and E. B. Garey, *The Plattsburg Manual: A Handbook for Military Training,* 17.

19. Capt. Dana Merrill, "Infantry Training," *INJ* 9, no. 1 (July–August 1912), 69–70; "Reveries of an Old Field Officer," *JUSCA* 27, no. 113 (January 1917), 15–18.

20. "A School for Noncommissioned Officers," *INJ* 12, no. 8 (April 1916), 955; Maj. James Chester, "Comment and Criticism on Moral Preparation of the Soldier," *JMSIUS* 32 (1913), 111.

21. Capt. Charles Crawford, *Weapons and Munitions of War,* Part I, *Infantry Weapons,* 8; Lt. Col. R. K. Evans, "Infantry Fire in Battle," *INJ* 5, no. 6 (May 1909), 819.

22. War Department, *Infantry Drill Regulations,* 68–69, 104–9.

23. Henry E. Eames, *The Rifle in War,* 57.

24. Baltzell, "Proper Training," 646.

25. Lincoln C. Andrews, *Fundamentals of Military Service,* 42 (emphasis in the original).

26. Baltzell, "Proper Training," 658; McCoy, *Principles of Training,* 203 (emphasis in the original).

27. Maj. Spencer Cosby, report, 4 November 1914, and "French Casualties," report, 25 August 1915, in "Reports by U.S. Military Attaché in France, 1914–1917," Microfilm File 8698, Roll 219, RG 165, NARA.

28. Capt. J. W. Barker, report, 11 November 1915, "A Study on the Attack in the Present Period of the War," in ibid., and André Laffargue, "A Study on the Attack in the Present Period of the War," *INJ* 13, no. 2 (September–October 1916).

29. US War Department, *Instructions on the Offensive Conduct of Small Units;* Report, 22 March 1916, in "Reports by US Military Attaché in France, 1914–1917."

Chapter 3

1. "Reserve Officers' Training Camps," *INJ* 14, no. 6 (December 1917), 467.

2. US War Department, *War Department Annual Report, 1913,* 1:151–52.

3. Morrill quoted in Gene M. Lyons and John W. Massland, "The Origins of the ROTC," *Military Affairs* 23, no. 1 (Spring 1959), 3; Ira L. Reeves, *Military Education in the United States,* 68–69; *War Department Annual Report, 1913,* 189; Capt. Richard Stockton Jr., "Military Schools and the Nation," *INJ* 11, no. 1 (July–August 1914), 25.

4. US War Department, *Study on Educational Institutions Giving Military Training as a Source for a Supply of Officers for a National Army,* 4–5.

5. "An Act to Promote the Efficiency of the Militia and for Other Purposes," 21 January 1903, 32 Stat. L. 778; US Congress, House, Committee on Military Affairs, *To Increase the Efficiency of the Military Establishment of the United States,* 64th Cong., 1st sess., 31 January 1916, 7.

6. *War Department Annual Report, 1918,* 1:1109; Timothy K. Nenninger, "The Army Enters the Twentieth Century," in *Against All Enemies,* ed. Kenneth J. Hagen and William R. Roberts, 219–22.

7. *War Department Annual Report, 1919,* 300.

8. "Schedule of Training and Instruction of a Class of 338 Newly Commissioned Second Lieutenants of Infantry, Cavalry, and Field Artillery at the Army Service Schools, Fort Leavenworth, Kansas, January 3, 1917, to April 1, 1917," Correspondence of the War College Division, Microfilm M1024, Roll 142, File 7541–17, RG 165, NARA; Maj. Frederick Lafferty, *Roster of Officers of the First Provisional Class, 1916,* 3.

9. Lafferty, *Roster of Officers,* 4.

10. Brig. Gen. Joseph Kuhn, "Training Camps for Candidates for the Officers' Reserve Corps," memorandum, 13 April 1917, "Letters, Memorandums, Reports, etc., of the Citizens Training Camps, Officers' Training Camps, Central Officers' Training Schools, and Student Army Training Corps," Correspondence of the War College Division, Microfilm File 9226, Roll 261, RG 165, NARA (hereafter cited as "Letters, etc., OTCs").

11. US War Department, *Special Regulations No. 49: Training Camps for Reserve Officers and Candidates for Appointment As Such, May 15–August 11, 1917,* 11–13, 22–23.

12. Brig. Gen. H. P. McCain, memorandum, 23 April 1917, "Letters, etc., OTCs."

13. Emmett J. Scott, *Scott's Official History of the American Negro in the World War,* 564–65; Brig. Gen. Joseph Kuhn, "Appointment of Provisional Second Lieutenants in the Regular Army," memorandum for the Chief of Staff, 23 April 1917, "Letters, etc., OTCs."

14. Leonard Ayres, *The War with Germany: A Statistical Summary,* 21–22.

15. McCain, memorandum, 23 April 1917.

16. John G. Clifford, *The Citizen Soldiers: The Plattsburg Training Camp Movement, 1913–1920,* 228–34.

17. Fort Sheridan Association, *The History and Achievements of the Fort Sheridan Officers' Training Camps,* 40–172.

18. Ralph B. Perry, *The Plattsburg Movement: A Chapter of America's Participation in the World War,* 190, 202; *War Department Annual Report, 1917,* 1:23.

19. Brig. Gen. Lytle Brown, "Enlisted Candidates for the 4th Officers' Training Camps," memorandum, 5 August 1918, "Letters, etc., OTCs"; Commander, COTS, Camp Pike, Arkansas, report ending August 24, 1918, 26 August 1918, War Department General Staff, AWCHS, G5 Schools, 7–51.3, Box 186, Entry 310, "Camp Pike, Ark, Infantry COTS," RG 165, NARA.

20. 1st Lt. William McKinley Briggs, Camp Zachary Taylor, Kentucky, 159th Depot Brigade, File WWI 2655, World War I Veteran Survey, USAMHI; Transcript of interview of Gen. John E. Hull by Lt. Col. James W. Wurman on 22 October 1973, Washington, DC, Oral History Collection, USAMHI.

21. Carol S. Gruber, *Mars and Minerva: World War I and the Uses of Higher Learning in America,* 25–27, 52–56, 62–64; Letter from Reggie Bradley to Adelaide Bowen, 1 June 1918, Entry 435, Box 1, Adelaide Bowen Papers, Special Collections, Robert W. Woodruff Library, Emory University; Laurence Stallings, *The Doughboys: The Story of the AEF, 1917–1918,* 117–18; Henry Berry, *Make the Kaiser Dance: Living Memories of a Forgotten War: The American Experience in World War I,* 128, 133–35, 263n; H. W. Brands, *TR: The Last Romantic,* 752–53, 812–15.

22. War Department, *Special Regulations No. 49,* 32–55.

23. Ibid., 10.

24. Col. Henry C. Cabell, "Report of Line Officers' Training Schools from the Declaration of War to the Discontinuance of Schools," 28 February 1919, Army General Staff, AWCHS, G5 Schools, 7–52.8–52.9, Box 201, NM-84, Entry 310, RG 165, NARA (hereafter cited as Cabell Report), 3–5. Of the 27,341 officers commissioned from the first OTCs, 2 were commissioned as colonels, 1 as a lieutenant colonel, 235 as majors, 722 as captains, 4,452 as first lieutenants, and 18,929 as second lieutenants. The breakdown by branch was: Infantry, 14,484; Field Artillery, 4,565; Quartermaster Corps, 3,067; Engineers, 1,966; Cavalry, 1,660; Coast Artillery, 1,062; Ordnance, 385; and Statistical, 152.

25. "The Products of the Training Camps," *INJ* 14, no. 1 (July 1917), 468; "Reserve Officers' Training Camps," *INJ* 14, no. 6 (December 1917), 467.

26. Fort Sheridan Association, *History and Achievements,* 208.

27. Gus Dittmar, *They Were First,* 88.

28. "Advance Extract Copy of Program of Training for Training Camps for Candidates for Commission in the Army of the United States, May 15 to August 31, 1918," "Letters, etc., OTCs."

29. Francis Spears et al., *Damitall: Twentieth Company, Central Officers' Training Camp, Camp Gordon, Georgia,* 9.

30. Dittmar, *They Were First,* 75; Charles Sorust to Adelaide Bowen, 20 November 1917, Entry 435, Box 1, Special Collections, Robert W. Woodruff Library, Emory University; Milton E. Bernet, "The World War As I Saw It," manuscript, 89th Div., WWI 2340, World War I Veteran Survey, USAMHI, 21.

31. Transcript, Hull interview by Wurman, 22 October 1973, Washington, DC, USAMHI Oral History Collection.

32. *War Department Annual Report, 1919,* 313–14.

33. Dittmar, *They Were First,* 74.

34. Dwight D. Eisenhower, *At Ease: Stories I Tell to Friends,* 132; Dittmar, *They Were First,* 105.

35. "Advance Extract Copy of Program," "Letters, etc., OTCs."

36. Dittmar, *They Were First,* 112–14; Fort Sheridan Association, *History and Achievements,* 208–13.

37. Morale Branch of the War College and WPD to the Chief of Staff, "Replies to Officers' Questionnaires," 5 November 1919, NM-84, Entry 378, Box 6, RG 165, NARA, 52 (I would like to thank James "Ty" Seidule for providing me a copy of this intriguing report.); F. L. Miller, "The War to End All Wars," manuscript, Special Collections, Robert W. Woodruff Library, Emory University, 3.

38. Fort Sheridan Association, *History and Achievements,* 353; Dittmar, *They Were First,* 187. Also see Fort Sheridan Association, *History and Achievements,* 208–9.

39. Dittmar, *They Were First,* 193–94.

40. Milton E. Bernet to "Sue," 26 July 1917, 89th Div., file WWI 2340, World War I Veteran Survey, USAMHI.

41. Brig. Gen. H. P. McCain to Chief of Staff, "Special Training for Officers," memorandum, 1 August 1917, "Letters, etc., OTCs"; Snow quoted in Byron Farwell, *Over There: The United States in the Great War, 1917–1918,* 67.

42. Paul Azan, *The War of Positions,* 14, 23–25.

43. Maj. N. K. Averill to the Chief of Staff, "Defects in Our Training, Especially Candidates for Officers," 10 August 1917, "Letters, etc., OTCs."

44. AEF G-5 to AEF Chief of Staff, 4 July 1918, memorandum, Timberman-Fiske Papers, USAMHI.

45. Brig. Gen. H. P. McCain, "Foreign Officers as Instructors at Officers' Training Camps," memorandum, 16 April 1917, "Letters, etc., OTCs"; Brig. Gen. Joseph Kuhn, "Foreign Officers as Instructors at Officers' Training Camps," memorandum, 27 April 1917, "Letters, etc., OTCs."

46. Col. William Johnston, "Foreign Officers as Instructors at Officers' Training Camps, Dissent of Colonel W. H. Johnston, General Staff," memorandum, 27 April 1917, "Letters, etc., OTCs" (emphasis in the original).

47. Commander, COTS, Camp Lee, Virginia, "Weekly Report," 24 August 1918, AWCHS, G5 Combat Training, 7–56.2–56.4, Box 207, Entry 310, "COTS Lee," RG 165, NARA.

48. US War Department, *Notes on Infantry, Cavalry, and Field Artillery,* Part 3, 23.

49. McCain, memorandum, 23 April 1917, "Letters, etc., OTCs"; Brig. Gen. Joseph Kuhn, "Text-books and Maps Required for Use at Officers' Training Camps," memorandum, 25 April 1917, "Letters, etc., OTCs."

50. Maj. Gen. H. L. Scott to Department Commanders, "Composition of Officers' Training Camps Cadre," undated memorandum (ca. August 1917), "Letters, etc., OTCs."

51. Fort Sheridan Association, *History and Achievements,* 186.

52. Lucian K. Truscott Jr., *The Twilight of the Cavalry: Life in the Old Army, 1917–1942,* 4; 1st Lt. William McKinley Briggs, Camp Zachary Taylor, Kentucky, 159th Depot Brigade, File WWI 2655, World War I Veteran Survey, USAMHI.

53. All quotes are from Morale Branch of the War College and WPD to the Chief of Staff, "Replies to Officers' Questionnaires," 5 November 1919, 35 and 52. This problem was also noted by March in his *Annual Report, 1919,* 306–7.

54. Officer Candidate Raymond Phelan to Secretary of War Newton Baker, 31 July 1917, "Letters, etc., OTCs."

55. Fort Sheridan Association, *History and Achievements,* 344–46.

56. Gen. John J. Pershing to the Adjutant General, "Training," 11 July 1917, Correspondence of the War College Division, Microfilm M1024, Roll 142, File 7541–17, RG 165, NARA.

57. Fort Sheridan Association, *History and Achievements,* 212–13.

58. Photos of training in ibid., 350, 381; *The Plattsburger,* 14; Bernet, "World War As I Saw It," 69.

59. O. O. Ellis and E. B. Garey, *The Plattsburg Manual: A Handbook for Military Training,* 144–51.

60. Cabell Report, 6–7. Of the 17,237 officers commissioned from the second OTCs, 59 were commissioned as majors, 1,557 as captains, 7,469 as first lieutenants, and 8,125 as second lieutenants. The breakdown by branch was: Infantry, 10,857; Field Artillery, 3,642; Signal Corps, 1,262; Coast Artillery, 1,001; Ordnance, 382; and Cavalry, 93.

61. "Memorandum of Information, Officers' Training Camps, August 27–November 26, 1917," 4 June 1917, "Letters, etc., OTCs."

62. Hugh L. Scott, *Some Memories of a Soldier,* 556.

63. General Service Schools, *Studies in Minor Tactics;* Timothy K. Nenninger, *The Leavenworth Schools and the Old Army,* 90–92; Fiske quoted in Maj. Frank R. McCoy, *Principles of Training,* Vol. 3 of *The National Service Library,* ed. Maj. Charles E. Kilbourne, 198.

64. Capt. Dawson Warrington to Brig. Gen. Mark L. Hersey, Head of the American Mission to France, "The New Method of Training Aspirants and Officers at St. Cyr," report, 17 July 1917, Records of the AEF, G5 Schools, Army Candidate School, Box 1637, File 350, "Information in Regards to Schools and Courses," RG 165, NARA, 1, 5–6.

65. Ibid., 2, 13.

66. Ibid., 17.

67. Adjutant General's Office, *The Personnel System of the United States Army,* Vol. 1, *History of the Personnel System,* 47.

68. Ibid., 217–22; Vol. 2, *The Personnel Manual,* 261.

69. Dittmar, *They Were First,* 71.

70. Fort Devens OTC Yearbook Committee, *The Pick: 3rd O.T.C., Camp Devens, Mass.,* 62.

71. Dittmar, *They Were First,* 120, 152.

72. Egmont W. Ruschke, ed., *Lieuie VI: Being the Chronicle of the Battle of Camp Lee as Fought by the Deathless Sixth Battalion, Central Officers' Training School, Camp Lee, Virginia,* 68.

73. Lt. Col. Jennings C. Wise, "The Soldier's Life in Battle," *INJ* 16, no. 11 (May 1920), 929; Machine Gun Officers' Training School, *Four Months of Sand,* 54; Ruschke, *Lieuie VI,* 28.

74. Fort Sheridan Association, *History and Achievements,* 235.

75. Maj. William K. Gunn, "Summary of Instruction of Candidates," memorandum, 17 August 1918, AWCHS, G5 Schools, 7–51.3, NM-84, Box 186, Entry 310, "Camp Sherman, Ohio, Fourth Officers' Training School Summary of Instruction of Candidates," RG 165, NARA.

76. Robert M. Yerkes, *Psychological Examining in the United States Army,* Memoirs of the National Academy of Sciences 15:7–12.

77. Ibid., 21–23; quote from Clarence Yoakum and Robert Yerkes, *Army Mental Tests,* 13.

78. Yoakum and Yerkes, *Army Mental Tests,* 14, 15.

79. General Order 74, 14 August 1918; quotes from Yerkes, *Psychological Examining,* 108, 115.

80. Yerkes, "Psychological Examining," 779.

Chapter 4

1. US War Department, *War Department Annual Report, 1918,* 268.

2. Col. Henry C. Cabell, "Report of Line Officers' Training Schools from the Declaration of War to the Discontinuance of Schools," 28 February 1919, AWCHS, G5 Schools, 7–52.8–52.9, Box 201, NM-84, Entry 310, RG 165, NARA (hereafter cited as Cabell Report), 9.

3. Brig. Gen. H. P. McCain, "Proposed Changes in Plans for the Third Training Camps," memorandum, 20 December 1917, "Letters, Memorandums, Reports, etc., of the Citizens Training Camps, Officers' Training Camps, Central Officers' Training Schools, and Student Army Training Corps," Correspondence of the War College Division, Microfilm File 9226, Roll 261, RG 165, NARA (hereafter cited as "Letters, etc., OTCs"); Cabell Report, 9.

4. Statistical information drawn from Fort Devens OTC Yearbook Committee, *The Pick: 3rd O.T.C., Camp Devens, Mass.,* 10, 12, 16, 22, 27.

5. Col. P. D. Lochridge, "Admission of Civilians to the Third Series of Training Camps for Officers," memorandum, 21 November 1917," "Letters, etc., OTCs."

6. Dwight D. Eisenhower, *At Ease: Stories I Tell to Friends,* 141; Brig. Gen. Lytle Brown, "Suggestions Concerning the Use of Well Educated Enlisted Men," memorandum, 17 July 1918, "Letters, etc., OTCs."

7. Egmont W. Ruschke, ed., *Lieuie VI: Being the Chronicle of the Battle of Camp Lee as Fought by the Deathless Sixth Battalion, Central Officers' Training School, Camp Lee, Virginia,* 28.

8. Adjutant General to Commanding General of All Departments and Staff Corps and Departments, telegram, 1 May 1917, in *Special Regulations No. 49: Training Camps for Reserve Officers and Candidates for Appointment As Such, May 15– August 11, 1917,* 29–30.

9. Lowell Thomas, *Woodfill of the Regulars,* 219.

10. Gus Dittmar, *They Were First,* 77.

11. Machine Gun Officers' Training School, *Four Months of Sand,* 32.

12. Commander, COTS, Camp MacArthur, Texas, report, 15 November 1918, "Letters, etc., OTCs."

13. Lochridge, "Admission of Civilians," 21 November 1917.

14. Fort Devens OTC Yearbook Committee, *The Pick,* 12, 16, 21, 26, 30.

15. Cabell Report, 9–10.

16. "Advance Extract Copy of Program of Training for Training Camps for Candidates for Commission in the Army of the United States, May 15 to August 31, 1918," "Letters, etc., OTCs."

17. Carl Wilhelm et al., *Pass in Review: The Book of the Fourth Officers' Training School, Camp Dodge, Iowa, 1918,* 47.

18. Fort Devens OTC Yearbook Committee, *The Pick,* 86; Wilhelm, *Pass in Review,* 56.

19. Brig. Gen. Lytle Brown, "Enlisted Candidates for the 4th Officers' Training Camps," memorandum, 5 August 1918, "Letters, etc., OTCs."

20. Statistic drawn from Maj. William K. Gunn, "Summary of Instruction of Candidates," memorandum, 17 August 1918, AWCHS, G5 Schools, 7–51.3, NM-84, Box 186, Entry 310, "Camp Sherman, Ohio, Fourth Officers' Training School Summary of Instruction of Candidates," RG 165, NARA.

21. Robert M. Yerkes, *Psychological Examining in the United States Army,* Memoirs of the National Academy of Sciences 15:863–65

22. Ibid.

23. Wilhelm, *Pass in Review,* 9, 11–12, 15, 17, 19.

24. Gunn, "Summary of Instruction of Candidates," 17 August 1918; Wilhelm, *Pass in Review,* 39.

25. Gunn, "Summary of Instruction of Candidates," 17 August 1918.

26. Wilhelm, *Pass in Review,* 22.

27. Brig. Gen. H. P. McCain, memorandum, 2 August 1918, and Brig. Gen. Lytle Brown, "Graduation of Infantry Candidates, 4th Officers' Training Schools," memorandum, 8 August 1918, "Letters, etc OTCs"; Wilhelm, *Pass in Review,* 22.

28. Brig. Gen. H. P. McCain, memorandum, 18 May 1918, "Letters, etc., OTCs."

29. Col. D. W. Ketcham, "Future Officers' Training Schools for Infantry, Field Artillery and Cavalry," memorandum, 14 March 1918, "Letters, etc., OTCs."

30. Maj. Gen. Leonard Wood to Adjutant General, "Training Camps for Officers," 18 April 1918, "Letters, etc., OTCs"; McCain, memorandum, 18 May 1918.

31. Adjutant General's Office, "Central Officers' Training Schools for Candidates for Commission in the Infantry, Field Artillery, and Machine Gun Units," undated memorandum, but staffed by the War Plans Division on 16 June 1918, and "Cen-

tral Officers' Training Schools for Candidates for Commission in the Infantry, Field Artillery, and Machine Gun Units," draft special regulations, AWCHS, G-5 Schools, 7–52.8–52.9, Box 201, Entry 310, "Training Schools for Officers," RG 165, NARA.

32. "Central Officers' Training Schools for Candidates for Commission in the Infantry, Field Artillery, and Machine Gun Units," draft special regulations.

33. War Department General Staff, AWCHS, G5 Schools, 7–52.2–52.9, Box 201, Entry 310, "Camp Gordon COT School," RG 165, NARA; Headquarters Central Infantry Officers' Training School, "General Orders, Special Orders, Training Memorandums, Examinations, Miscellaneous," AWCHS, G5 Schools, 7–50.4, Box 178, NM-84, Entry 310, "Orders & etc., ICOTS, Camp MacArthur, Texas," RG 165, NARA.

34. Maj. Gen. Leonard Wood to Adjutant General, "Officers," 4 September 1918, "Letters, etc., OTCs."

35. Candidate Henry P. Fry to Maj. John R. M. Taylor, COTS, Camp Gordon, Georgia, 26 July 1918, AWCHS, G5 Schools, 7–52.2–52.9, Box 201, Entry 310, "Camp Gordon COT School," RG 165, NARA; Robert O'Hair, Student Army Training Corps, Indiana University, World War I Veteran Survey, USAMHI.

36. Capt. V. S. Hebbert to Group Commander, British Military Mission, Camp Gordon, Georgia, "Progress Report for Period Ending July 25, 1918," report, AWCHS, G5 Combat Training, Supervision by Allied Instructors, 7–56.5–56.9, Box 208, NM-84, Entry 310, File "Camp Gordon, Atlanta, Georgia—Progress Reports—July 1918, Central Officers Tn School," RG 165, NARA; Maj. Elvid Hunt, "Liaison Visit to Infantry Training Centers at Camps MacArthur, Pike, and Gordon," memorandum, July 1918, G5 Schools, 7–51.3, Box 185-A, NM-84, Entry 310, "Camp MacArthur, Texas, Infantry Training Center," RG 165, NARA.

37. Hebbert to Group Commander, "Progress Report."

38. "Capt. Hebbert's Report on C.O.T.C., Camp Gordon," memorandum for Col. Fleming, 30 July 1918, War Department General Staff, AWCHS, G5 Combat Training, Supervision by Allied Instructors, 7–56.5–56.9, Box 208, NM-84, Entry 310, File "Camp Gordon, Atlanta, Georgia—Progress Reports—July 1918, Central Officers Tn School," RG 165, NARA.

39. Commander, COTS, Camp Pike, Arkansas, "Report Ending August 24, 1918," 26 August 1918, AWCHS, G5 Schools, 7–51.3, Box 186, Entry 310, "Camp Pike, Ark, Infantry COTS," RG 165, NARA; Commander, COTS, Camp Pike, "Report Ending September 7, 1918," 9 September 1918, ibid.; Memorandum for Director, WPD, 2 October 1918, "Letters, etc., OTCs."

40. Commander, COTS, Camp Lee, Virginia, "Weekly Report," 14 September 1918, AWCHS, G5 Combat Training, 7–56.2–56.4, Box 207, Entry 310, "COTS Lee," RG 165, NARA.

41. Commander, COTS, Camp Lee, Virginia, "Report on Central Officers' Training School, Camp Lee, Va.," 17 August 1918, ibid.; Commander, COTS, Camp Gordon, Georgia, "Weekly Report," 19 October 1918, AWCHS, G5 Schools, 7–52.2–52.9, Box 201, Entry 310, "Camp Gordon COT School," RG 165, NARA.

42. Chief, Training and Instruction Branch, WPD, "Three Months' Course of Instruction," memorandum, 6 November 1918, AWCHS, G5 Schools, 7–50.4, Box 180, NM-84, Entry 310, "Reports of Camp Grant Infantry COTS," RG 165, NARA.

43. Eighth Company, Central Officers' Training School, Camp Lee, Virginia, *Take His Name*, 5–9.

44. Ruschke, *Lieuie VI*, 11.

45. Machine Gun Officers' Training School, *Four Months of Sand*, 29, 36–38, 71, 84.

46. Commander, COTS, Camp MacArthur, Texas, "Weekly Report," 15 September 1918, "Letters, etc., OTCs"; ibid., 19 October 1918 (emphasis in the original);

"Inspection Visit to Various Camps," memorandum for the Chief, Training and Instruction Branch, 2 October 1918, AWCHS, G5 Schools, 7–51.3, Box 185-A NM-84, Entry 310, "Camp MacArthur, Texas, Infantry Training Center," RG 165, NARA.

47. Commander, Camp Grant ICOTS, "Weekly Report, Week Ending November 2nd, 1918," memorandum, 4 November 1918, AWCHS, G5 Schools, 7–50.4, Box 180, NM-84, Entry 310, "Reports of Camp Grant Infantry COTS," RG 165, NARA; Col. Robert Getty, memorandum, 4 November 1918, AWCHS, G5 Schools, 7–51.3, Box 184, NM-84, Entry 310, File "Infantry Replacement Training, Camp Gordon, Ga.," RG 165, NARA.

48. Commander, COTS, Camp Lee, Virginia, "Report on Central Officers' Training School, Camp Lee, Va.," 17 August 1918, AWCHS, G5 Combat Training, 7–56.2–56.4, Box 207, Entry 310, "COTS Lee," RG 165, NARA; Commander, COTS, Camp Lee, Virginia, "Weekly Report," 7 September 1918, ibid.; Commander, COTS, Camp Pike, Arkansas, "Report Ending August 24, 1918," 26 August 1918, AWCHS, G5 Schools, 7–51.3, Box 186, Entry 310, "Camp Pike, Ark., Infantry COTS," RG 165, NARA.

49. Commander, COTS, Camp MacArthur, Texas, "Weekly Report," 26 September 1918, and "Weekly Report," 19 October 1918; Office of the Third Assistant Secretary of War, "Procurement of Officer Material," memorandum, October 1918, all in "Letters, etc., OTCs."

50. Commander, COTS, Camp MacArthur, Texas, "Weekly Report," 26 September 1918, "Letters, etc., OTCs" (emphasis in the original).

51. Cabell Report, 23; Brig. Gen. Henry Jervey, "Graduating Class Central Officers Training Schools of Infantry," memorandum, 27 September 1918, "Letters, etc., OTCs."

52. Memorandum for Director, WPD, 2 October 1918, "Letters, etc., OTCs"; Cabell Report, 13.

53. Cabell Report, 14, 15.

54. Commander, COTS, Camp Lee, Virginia, "Weekly Report," 7 September 1918.

55. Commander, COTS, Camp MacArthur, Texas, "Report of the Commander," 15 November 1918, "Letters, etc., OTCs."

56. *War Department Annual Report, 1919,* 299–304, 320–21. The SATC eventually encompassed 518 colleges and more than 135,000 students.

57. Memorandum from the Office of the Third Assistant Secretary of War for dissemination to the field, undated (ca. October 1918), "Letters, etc., OTCs." No subject is given, but the memo explains the status of the SATC.

58. Cabell Report, 24; Memorandum for Director, WPD, 2 October 1918, "Letters, etc., OTCs."

59. Cabell Report, 15, 16.

60. Chief, Training and Instruction Branch, WPD, "Three Months' Course of Instruction," memorandum, 6 November 1918, AWCHS, G5 Schools, 7–50.4, Box 180, NM-84, Entry 310, "Reports of Camp Grant Infantry COTS," RG 165, NARA.

61. Statistical information drawn from Machine Gun Officers' Training School, *Four Months of Sand,* 45–46, 54, 71–72.

62. Statistical information drawn from Eighth Company, COTS, Camp Lee, Virginia, *Take His Name,* 15–26.

63. Statistical information drawn from Ruschke, *Lieuie VI,* 30–51.

64. Commander, COTS, Camp MacArthur, Texas, report, 15 November 1918, "Letters, etc., OTCs."

65. Commander, COTS, Camp MacArthur, Texas, "Weekly Report," 19 October 1918, "Letters, etc., OTCs."

66. *War Department Annual Report, 1919,* 300.

Chapter 5

1. Lt. Col. Jennings C. Wise, "The Soldier's Life in Battle," *INJ* 16, no. 11 (May 1920), 929.

2. Milton E. Bernet, "The World War As I Saw It," manuscript, 89th Div., WWI 2340, World War I Veteran Survey, USAMHI, 35.

3. Anonymous, *Wine, Women and War: A Diary of Disillusionment,* 3, 9; Ben Chastaine, *Story of the 36th,* 17.

4. William E. Hocking, *Morale and Its Enemies,* 131–32.

5. Lt. Col. J. W. McAndrews, *Address to the Second Class of Provisional Second Lieutenants, April 17, 1917,* 7; Fort Devens OTC Yearbook Committee, *The Pick: 3rd O.T.C., Camp Devens, Mass.,* 15; Maj. C. A. Bach, "Leadership," *INJ* 14, no. 8 (February 1918), 607.

6. US War Department, *War Department Annual Report, 1918,* 1:672.

7. Adjutant General's Office, General Order 143, 14 November 1917; *War Department Annual Report, 1918,* 1:672.

8. Dwight D. Eisenhower, *At Ease: Stories I Tell to Friends,* 143–44.

9. Charles Miller, *The Customs of the Service: Also Some Suggestions and Advice,* 28.

10. Myron Adams, ed., *The Officer's Responsibility for His Men,* viii, 46.

11. Miller, *Customs,* 27 (emphasis in the original).

12. McAndrews, *Address to the Second Class,* 6; C. H. Hitchcock, "A Letter to a Training Camp Student," *INJ* 14, no. 1 (July 1917), 49–52.

13. US War Department, *Infantry Drill Regulations, 1913, corrected to December 31, 1917,* 99.

14. "Why Is It?" *Cavalry Journal* 28, no. 116 (October 1917), 268–72.

15. H. R. H., "Why Is It? An Answer," *Cavalry Journal* 28, no. 117 (January 1918), 221–23.

16. One of Them [pseud.], "Why Is It? Another Answer," *Cavalry Journal* 28, no. 118 (April 1918), 556–58.

17. Hitchcock, "Letter to a Training Camp Student"; Bach, "Leadership," 607–8.

18. US War Department, *Infantry Training,* 5, 7, 11–12, 20–29.

19. Capt. Charles Dienst et al., *They're from Kansas: History of the 353rd Infantry Regiment, 89th Division, National Army,* 10–11; B. A. Colonna, ed., *The History of Company B, 311th Infantry, in the World War,* 10.

20. 328th Infantry Historical Committee, *History of the Three Hundred and Twenty-eighth Infantry Regiment,* 7–8; G. Edward Buxton Jr., ed., *Official History of 82nd Division, American Expeditionary Forces: "All American" Division, 1917–1919,* 4.

21. Lonnie J. White, *Panthers to Arrowheads: The 36th Division in World War I,* 55–56; Christian A. Bach and Henry N. Hall, *The Fourth Division: Its Services and Achievements in the World War,* 21, 27.

22. Summary of the Reports of the French Advisory Mission, 16 February 1918, AWCHS, G5 Combat Training, Supervision by Allied Instructors, 7–56.5–56.9, Box 208, NM-84, Entry 310, File "Reports of Activities of Advisory Mission," RG 165, NARA; Bach and Hall, *Fourth Division,* 19; Frank H. Ward, ed., *Camp Sherman Souvenir,* 59; Lonnie J. White, *The 90th Division in World War I,* 40.

23. George H. English Jr., *History of the 89th Division, U.S.A.,* 28–29; Lucian K. Truscott Jr., *The Twilight of the Cavalry: Life in the Old Army, 1917–1942,* 19; Transcript of interview of Gen. John E. Hull by Lt. Col. James W. Wurman, 22 October 1973, Washington, DC, USAMHI Oral History Collection.

24. Robert S. Sutliffe, *Seventy-First New York in the World War,* 41.

25. Acting Intelligence Officer, 83rd Division, "General Information," report, 28 February 1918, Records of the General Staff, Entry 377, Correspondence Related to

Morale at Army Installations, RG 165, NARA (hereafter cited as Morale at Army Installations), Box 14, Camp Sherman file.

26. Bach and Hall, *Fourth Division,* 23–24.

27. P. Benson Oakley to Helen Oakley, postmarked 26 March 1918, from Camp Hancock, GA, in author's collection.

28. Carol R. Byerly, *The Fever of War: The Influenza Epidemic in the U.S. Army during World War I,* 8–9.

29. Dienst, *They're from Kansas,* 3–4.

30. English, *History of the 89th Division,* 21.

31. Charles W. Camp, *History of the 305th Field Artillery,* 13–14; Kerr Rainsford, *From Upton to the Meuse with the Three Hundred and Seventh Infantry,* 2–3.

32. Maj. J. C. Wise, "Organization and Initial Training of a Company," *INJ* 14, no. 3 (September 1917), 201–10; Maj. Charles R. Tips, "Selecting and Training Military Leaders," *INJ* 15, no. 7 (January 1919), 541.

33. Dienst, *They're from Kansas,* 7–8.

34. John W. Nell, *The Lost Battalion: A Private's Story,* 5.

35. Maj. George F. Arps, "Science as Applied to the Selection of Noncommissioned Officers," *INJ* 15, no. 7 (January 1919), 577–78; D. B. Gallagher, *The Battle of Bolts and Nuts in the Sector of Cognac Hill,* 12; Charles G. Campbell, Evacuation Ambulance Company No. 1, World War I Veteran Survey, USAMHI.

36. Capt. John S. Stringfellow, *Hell! No!* 35.

37. Col. H. O. Williams, "Report on Inspection of Troops at Camp Sherman, Ohio," memorandum, 31 July 1918, AWCHS, G5 Schools, 7–51.3, Box 185-A, NM-84, Entry 310, "Camp MacArthur, Texas, Infantry Training Center," RG 165, NARA.

38. Dienst, *They're from Kansas,* 13; Morale Branch of the War College and WPD to the Chief of Staff, "Replies to Officers' Questionnaires," 5 November 1919, NM-84, Entry 378, Box 6, RG 165, NARA (hereafter cited as Morale Branch Officers' Survey), 69.

39. Manuscript in the Julian L. Schley Papers, Box 1, United States Military Academy Library Special Collections, 11–12.

40. Maj. Thomas Swann, "The Top-Sergeant," *INJ* 15, no. 12 (June 1919), 952; Frank B. Tiebout, *A History of the 305th Infantry,* 19; Gallagher, *Battle of Bolts and Nuts,* 13.

41. Diary of Capt. Clarence J. Minick, Clarence J. Minick Papers, Liberty Memorial Archives, Kansas City, MO.

42. Wise, "Soldier's Life in Battle," 930.

43. US War Department, *Manual for Noncommissioned Officers and Privates of Infantry,* 9–10.

44. French Military Mission, "Improvements in the Condition and Instruction of Non-Commissioned Officers," 10 January 1918, AWCHS, G5 Combat Training, Supervision by Allied Instructors, 7–56.5–56.9, Box 208, NM-84, Entry 310, File "Reports of Activities of Advisory Mission," RG 165, NARA; "Report of General Claudon on his visit to Fort Oglethorpe, Georgia," 4–6 March 1918, ibid.

45. Report of Inspection of the Commander, British Military Mission, 24 May 1918, AWCHS, G5 Combat Training, Supervision by Allied Instructors, 7–56.5–56.9, Box 208, NM-84, Entry 310, File "Inspections, British Groups Stationed at Camps," RG 165, NARA.

46. Morale Branch Officers' Survey, 68–69.

47. Charles R. Tips, "Selecting and Training Military Leaders," *INJ* 15, no. 7 (January 1919), 543.

48. "Scheme for Proposed Non-commissioned Officers' Training School, Camp Devens, Massachusetts," 30 July 1918, AWCHS, G5 Schools, Box 202, NM-84, Entry 310, RG 165, NARA.

49. Training Bulletin for Noncommissioned Officers, Camp Lee, Virginia, Infantry Replacement and Training Camp, 3 September 1918, AWCHS, G5 Schools, 7–51.3, Box 185, NM-84, Entry 310, RG 165, NARA; Intelligence Officer, Camp Gordon, GA, "General Information Report, Week Ending September 2nd, 1918," and "Singing School," report, 17 September 1918, Morale at Army Installations, Box 5, Camp Gordon file.

50. Tips, "Selecting and Training Military Leaders," 543; Brig. Gen. Lytle Brown, "Appointment of Noncommissioned Officers of Those Students at Central Officers' Training Schools Who Fail to Receive Commissions," memorandum, 5 August 1918, "Letters, etc., OTCs."

51. General Headquarters, AEF, *Report of Officers Convened by Special Orders No. 98, GHQ AEF, 09 April 1919,* 9–10, Army War College Library, USAMHI.

52. Rainsford, *From Upton to the Meuse,* 16.

53. Quoted in James H. Hallas, *Doughboy War: The American Expeditionary Force in World War I,* 26.

54. W. A. Sirmon, *That's War: An Authentic Diary,* 21; Maury Maverick, *A Maverick American,* 113; Bolte quoted in Edward M. Coffman, *The War to End All Wars,* 57.

55. Transcript of interview of John G. Oechsner by CSM Erwin H. Koehler, 19 January 1982, Noncommissioned Officer Oral History Program, USAMHI; P. Benson Oakley to Helen Oakley, postmarked 26 April 1918, Camp Hancock, GA, from the author's collection; William F. Clarke, *Over There with O'Ryan's Roughnecks,* 25.

56. Milton E. Bernet, "The World War As I Saw It," manuscript, 89th Div., WWI 2340, World War I Veteran Survey, USAMHI, 132.

57. Truscott, *Twilight of the Cavalry,* 6–10.

58. *War Department Annual Report, 1919,* 299.

59. Morale Branch Officers' Survey, 77; Sirmon, *That's War,* 57.

60. Alvin C. York, *Sergeant York: His Own Life Story and War Diary,* ed. Tom Skeyhill, 46, 210; Col. H. O. Williams, "Report on Inspection of Troops at Camp Sherman, Ohio," memorandum, 31 July 1918, AWCHS, G5 Schools, 7–51.3, Box 185-A NM-84, Entry 310, "Camp MacArthur, Texas, Infantry Training Center," RG 165, NARA.

61. "Memorandum on Infantry Instruction," *INJ* 16, no. 7 (January 1920), 509 (emphasis in the original).

62. Williams, "Report on Inspection of Troops at Camp Sherman, Ohio."

63. English, *History of the 89th Division,* 34–35.

64. Hugh L. Scott, *Some Memories of a Soldier,* 613–14 (emphasis added).

65. "Extracts from the Report of Major De Reviers, Chief of the Atlanta Sub-District (French Military Mission) for the Two Weeks from March 1st to March 15th," AWCHS, G5 Combat Training, Supervision by Allied Instructors, 7–56.5–56.9, Box 208, NM-84, Entry 310, File "Reports of Activities of Advisory Mission," RG 165, NARA.

66. Col. James Martin, Acting Chief of French Advisory Mission, "Report of Instruction," 18 April 1918, ibid.

67. Maj. N. K. Averill, "Defects in Our Training, Especially Candidates for Officers," letter, 10 August 1917, "Letters, etc., OTCs" "Extracts from the Report of Major De Reviers"; "Report of General Claudon on his visit to Fort Oglethorpe, Georgia, March 4–6, 1918," G5 Combat Training, Supervision by Allied Instructors, 7–56.5–56.9, Box 208, NM-84, Entry 310, File "Reports of Activities of Advisory Mission," RG 165, NARA.

68. French Military Mission, "Summary of the Reports of the French Advisory Mission," 16 February 1918, AWCHS, G5 Combat Training, Supervision by Allied

Instructors, 7–56.5–56.9, Box 208, NM-84, Entry 310, File "Reports of Activities of Advisory Mission," RG 165, NARA.

69. "Extracts from the Report of Major De Reviers."

70. Sutliffe, *Seventy-First New York,* 47; Leslie Baker, *The Company History: The Story of Company B, 106th Machine Gun Battalion,* 10–11.

71. Rainsford, *From Upton to the Meuse,* 1.

72. Ibid., 11; L. Wardlaw Miles, *History of the 308th Infantry,* 18.

73. Gerald F. Jacobson, *History of the 107th Infantry U.S.A.,* 11–12.

74. Miles, *History of the 308th,* 18; Capt. Ernest F. McKeighan to "Dearest Girl," 2 September 1917, Capt. Ernest F. McKeighan Papers, Liberty Memorial Archives, Kansas City, MO (emphasis in the original).

75. Alan D. Graff, *Blood in the Argonne,* 41; Dienst, *They're from Kansas,* 15; P. Benson Oakley to Helen Oakley, postmarked 26 April 1918, Camp Hancock, GA, from the author's collection.

76. Lincoln C. Andrews, *Military Manpower,* 74–75; William E. Hocking, "Fundamentals of Military Psychology," *INJ* 14, no. 10 (April 1918), 719, 721.

77. Buxton, *Official History of 82nd,* 1–2; US Army Center of Military History, *Order of Battle of the United States Land Forces in the World War: American Expeditionary Forces,* Vol. 2, *Divisions,* 349; Manuscript in the Julian L. Schley Papers, Box 1, United States Military Academy Library Special Collections, 7, 14.

78. Manuscript in the Schley Papers, 15–16; Buxton, *Official History of 82nd,* 2; David M. Kennedy, *Over Here,* 24, 157.

79. Buxton, *Official History of 82nd,* 3; Kennedy, *Over Here,* 24–25, 63–69; Frank A. Holden, *War Memories,* 27–28.

80. US Army Center of Military History, *Order of Battle,* Vol. 2. For examples, see "Record of Events" for 39th, 40th, 77th, 81st, and 91st Divisions.

81. Intelligence Officer, 86th Division, "Summary of General Information, March 11th, 1918," report, Morale at Army Installations, Box 6, Camp Grant file.

82. Intelligence Officer, 89th Division, report, 25 March 1918, ibid., Box 5, Camp Funston file; Intelligence Officer, 31st Division, report, 22 February 1918, ibid., Box 17, Camp Wheeler file; Miles, *History of the 308th Infantry,* 14.

83. Rainsford, *From Upton to the Meuse,* 6–7; Manuscript in the Schley Papers, 16.

84. Tiebout, *History of the 305th,* 26.

85. Assistant Intelligence Officer, 86th Division, "Conditions in 91st Division and Auxiliary Organizations," report, 4 March 1918, Morale at Army Installations, Box 9, Camp Lewis file.

86. Col. H. O. Williams, "Report on Inspection of Troops at Camp Sherman, Ohio," 31 July 1918, AWCHS, G5 Schools, 7–51.3, Box 185-A NM-84, Entry 310, "Camp MacArthur, Texas, Infantry Training Center," RG 165, NARA.

87. Dienst, *They're from Kansas,* 2–3.

88. Intelligence Officer, Camp Funston, KS, "Morale at Camp Funston, Kansas," report, 19 July 1918, Morale at Army Installations, Box 5, Camp Funston file.

89. Rainsford, *From Upton to the Meuse,* 16.

90. Deinst, *They're from Kansas,* 14.

Chapter 6

1. Robert S. Sutliffe, *Seventy-First New York in the World War,* 67.

2. James W. Rainey, "Ambivalent Warfare: The Tactical Doctrine of the AEF in World War I," *Parameters* 13, no. 3 (September 1983), 34–46; James Rainey, "The Questionable Training of the AEF in World War I," *Parameters* 22, no. 4 (Winter 1992–93), 89–103; Timothy K. Nenninger, "Tactical Dysfunction in the AEF,

1917–1918," *Military Affairs* 51, no. 4 (October 1987), 177–81. Nenninger further expanded his argument in *Military Effectiveness,* Vol. 1, *The First World War,* ed. Allan R. Millett and Williamson Murray, 142–49.

3.　　Mark E. Grotelueschen, *The AEF Way of War,* 343–52.

4.　　US War Department, *Instructions for the Training of Platoons for Offensive Action,* 5.

5.　　John J. Pershing, *My Experiences in the World War* 1:154.

6.　　US Army, 1st Division, "Program of Training for the 1st Division, A.E.F.," 6 October 1917, in *World War Records, First Division,* Vol. 20, *Training First Division,* unpaginated.

7.　　General Headquarters, AEF, "Program of Training for the 2nd Division," G5 Schools, Army Candidates School, Box 1637, File 350, "Information in Regard to Schools and Courses," RG 120, NARA.

8.　　Ibid.

9.　　US Army Center of Military History, *The United States Army in the World War, 1917–1919* (hereafter cited as *US Army in the World War*), Vol. 14, *Reports,* 306.

10.　　Ibid., Vol. 3, *Training and Use of American Units with the British and French,* 36–37, 59–61.

11.　　Ibid., 63–64.

12.　　Ibid., 71, 83–84.

13.　　Ibid., 91–92.

14.　　"Final Report of the Assistant Chief of Staff, G-5," ibid., Vol. 14, *Reports,* 300–301.

15.　　Maj. Gen. Charles M. Clement, ed., *Pennsylvania in the World War: An Illustrated History of the Twenty-Eighth Division* 1:334–35.

16.　　Joseph Sanborn, *The 131st Infantry in the World War,* 29–35; *US Army in the World War* 3:162.

17.　　Grotelueschen, *AEF Way of War,* 60–63, 203–5. Not everyone agrees with Grotelueschen's conclusions. Douglas V. Johnson argues that the 1st Division's early training was characterized more by show than substance. He concludes that the unit's training emphasis was "on soldierly bearing and appearance" and that "fighting matters appear to be decidedly in second place" (Douglas V. Johnson, "Training the First Division for World War I," in *Cantigny at Seventy-Five: A Professional Discussion,* ed. Steven Wiengartner, 69–71).

18.　　Robert L. Bullard, *Personalities and Reminiscences of the War,* 103.

19.　　George C. Marshall, *Memories of My Service in the World War: 1917–1918,* 15.

20.　　Robert Bruce, *A Fraternity in Arms: America and France in the Great War,* 105; Col. Robert R. McCormick, *The Army of 1918,* 64, 65.

21.　　Edward M. Coffman, *The Regulars: The American Army, 1898–1941,* 209; George Marshall, *The Papers of George Catlett Marshall,* Vol. I, *The Soldierly Spirit, December 1880–June 1939,* ed. Larry I. Bland and Sharon Ritenour, 120.

22.　　Coffman, *The Regulars,* 209; Bullard, *Personalities,* 106.

23.　　McCormick, *Army of 1918,* 72–73, 113.

24.　　Ibid., 104.

25.　　W. A. Sirmon, *That's War: An Authentic Diary,* 92.

26.　　Quoted in Henry Berry, *Make the Kaiser Dance: Living Memories of a Forgotten War: The American Experience in World War I,* 363.

27.　　Chester E. Baker, *Doughboy's Diary,* 49.

28.　　Clement, *Pennsylvania in the World War* 1:287, 313; 2:467.

29.　　Charles Minder, *This Man's War,* 90–94, 125.

30.　　Baker, *Doughboy's Diary,* 49; H. G. Proctor, *The Iron Division in the World War,* 26; Clement, *Pennsylvania in the World War* 2:570; Kerr Rainsford, *From Upton to the Meuse with the Three Hundred and Seventh Infantry,* xix.

31. Peyton March, *The Nation at War*, 258; Sirmon, *That's War*, 96.

32. James J. Cooke, *Pershing and His Generals*, 82; Society of the Fifth Division, *The Official History of the Fifth Division*, 54; Clement, *Pennsylvania in the World War* 2:634, 666, 689.

33. Anonymous, *Wine, Women and War: A Diary of Disillusionment*, 14; US Army Center of Military History, *Order of Battle of the United States Land Forces in the World War: American Expeditionary Forces*, Vol. 2, *Divisions*, 347. Also see Capt. Edgar Fell, *History of the Seventh Division, 1917–1919*, 50, 56.

34. *US Army in the World War* 3:294–95.

35. Frank Sibley, *With the Yankee Division in France*, 48–49.

36. Christian A. Bach and Henry N. Hall, *The Fourth Division: Its Services and Achievements in the World War*, 49–50, 58–59.

37. James E. Pollard, *The Forty-Seventh Infantry: A History 1917–1918–1919*, 28–33.

38. Diary of Capt. Clarence J. Minick, Clarence J. Minick Papers, Liberty Memorial Archives, Kansas City, MO.

39. McCormick, *Army of 1918*, 71; *US Army in the World War* 14:301.

40. Sirmon, *That's War*, 163, 176–80; Tony Ashworth, *Trench Warfare, 1914–1918: The Live and Let Live System*, 14–20, 129–42; William S. Triplet, *A Youth in the Meuse-Argonne*, 95.

41. Clair Kenamore, *From Vauquois Hill to Exermont*, 45; Charles B. Holt, *Heroes of the Argonne*, 41; Joint War History Commissions of Michigan and Wisconsin, *The 32nd Division in the World War*, 44.

42. Elmer F. Straub, *A Sergeant's Diary in the World War*, Indiana Historical Collections 10:43, 118; Sgt. Richard McBride, 325th Inf., 82nd Div., manuscript, WWI 2178, World War I Veteran Survey, USAMHI; Kenamore, *From Vauquois Hill*, 65.

43. Robert Alexander, *Memories of the World War, 1917–1918*, 16–18, 111–12.

44. Evan A. Edwards, *From Doniphan to Verdun: The Official History of the 140th Infantry*, 42; Emil B. Gansser, *History of the 126th Infantry in the War with Germany*, 87; Joint War History Commissions, *32nd Division*, 44; John Cutchins and George Stewart, *History of the 29th Division, 1917–1919*, 91.

45. Hoyt, *Heroes of the Argonne*, 51.

46. Frank B. Tiebout, *A History of the 305th Infantry*, 74.

47. Sirmon, *That's War*, 166; Cutchins and Stewart, *History of the 29th Division*, 120.

48. G. Edward Buxton Jr., ed., *Official History of 82nd Division, American Expeditionary Forces: "All American" Division, 1917–1919*, 15–16; US War Department, *Field Service Pocket Book*, 117.

49. Charles Heller, *Chemical Warfare in World War I: The American Experience, 1917–1918*, 91–94; Archibald Hart, *Company K of Yesterday*, 37; E. W. Spencer, *History of Gas Attacks upon the American Expeditionary Forces during the World War*, 2.

50. Leslie Baker, *The Company History: The Story of Company B, 106th Machine Gun Battalion*, 87–88; Anonymous, *Wine, Women, and War*, 113.

51. Reports quoted in *US Army in the World War* 3:162, 208–9, 221–22.

52. Ashby Williams, *Experiences of the Great War*, 60.

53. B. A. Colonna, ed., *The History of Company B, 311th Infantry, in the World War*, 26.

54. Final Report of Gen. John J. Pershing, *War Department Annual Report, 1918*, 1:561.

55. Hunter Liggett, *Commanding an American Army: Recollections of the World War*, 18; AEF G-5 to AEF Chief of Staff, "Report of Training in the American Expeditionary Forces," 4 July 1918, 29, Timberman-Fiske Papers, USAMHI, Carlisle Barracks, PA.

347

56. L. Wardlaw Miles, *History of the 308th Infantry,* 33; Capt. John S. Stringfellow, *Hell! No!* 146; Capt. Paul Schmidt, *Co. C, 127th Infantry, in the World War,* 17; Society of the Fifth Division, *Official History,* 61.

57. Colonna, *History of Company B,* 19.

58. Major Beaugier, French Military Mission, to General McAndrew, Commandant, American Army Schools, 10 December 1917, G5 Schools, Army Candidates School, Box 1639, File 325.16, "Instructors French," RG 120, NARA.

59. Kenneth E. Hamburger, *Learning Lessons in the American Expeditionary Forces.*

60. *US Army in the World War* 14:329–33.

61. Cutchins and Stewart, *History of the 29th Division,* 49; Charles Dienst et al., *They're from Kansas: History of the 353rd Infantry Regiment, 89th Division, National Army,* 31.

62. Bullard, *Personalities,* 179.

63. Colonna, *History of Company B,* 22–23.

64. War Department, *Instruction for the Offensive Combat of Small Units,* May 1918, 10–11, 20, 24–27.

65. Ibid., 12–14, 27, 32–34.

66. The "half-platoon" was just that. The fifty-nine-man platoon was cut in half, with one of the half-platoons under the control of a "Sergeant, Assistant to Commander" and the other half under the direct control of the platoon leader. The platoon leader, however, was still responsible for the overall direction and control of the whole element. The "combat group" was the breaking down of the platoon into specialties.

67. War Department, *Offensive Combat,* May 1918, 20.

68. Minder, *This Man's War,* 154, 162–64, 199.

69. Bullard, *Personalities,* 64.

70. Report of Infantry Specialist School, Langres, France, 11 November 1918, in *US Army in the World War* 14:347.

71. Morale Branch of the War College and WPD to the Chief of Staff, "Replies to Officers' Questionnaires," 5 November 1919, NM-84, Entry 378, Box 6, RG165, NARA (hereafter cited as Morale Branch Officers' Survey), 44, 54; Gansser, *History of the 126th,* 52–53.

72. *US Army in the World War* 14:303–4.

73. Alexander, *Memories,* 16–18.

74. *US Army in the World War* 14:354.

75. Cpl. Fred Takes, 325th Inf., 82nd Div., manuscript, WWI 1760, World War I Veteran Survey, USAMHI.

76. Morale Branch Officers' Survey, 54.

77. Sirmon, *That's War,* 104, 119.

78. Fell, *History of the Seventh Division,* 22; Sibley, *With the Yankee Division,* 46 Marshall quoted in Millett and Murray, *Military Effectiveness,* Vol. 1, *The First World War,* 147.

79. C. L. Crane, "The Great War: 1917–1918–1919," original unpublished diary in the possession of Conrad Crane, Army War College. A copy of the diary is in the possession of the author.

80. William D. Henderson, *Cohesion: The Human Element in Combat,* 108–9.

81. Walter L. Wolf, 129 Inf., 33rd Div., World War I Veteran Survey, USAMHI; Connell Albertine, *The Yankee Doughboy,* 73; Dienst, *They're from Kansas,* 35; Madden quoted in Berry, *Make the Kaiser Dance,* 383–84.

82. *US Army in the World War* 3:213.

83. General Headquarters, AEF, *Report of Officers Convened By Special Orders No. 98, GHQ AEF, 09 April 1919,* Army War College Library, USAMHI, Annex R, 19.

84. Chief of Training Section, "Organization of the Army Candidates School at Langres," memorandum, 17 November 1917, G5 Schools, Army Candidates School, Box 1637, File 350, "Information in Regard to Schools and Courses," RG 120, NARA.

85. AEF Adjutant General, "Course of Instruction in Minor Tactics," memorandum, 21 December 1917, ibid.; Final Report of Gen. John J. Pershing, *War Department Annual Report, 1918,* 1:560.

86. Headquarters, Infantry Candidates School, "Outline of Instruction for a Three Month's Course, Infantry Candidates School," memorandum, 29 October 1918, Entry 424, G5 Schools, La Valbonne Infantry Candidates School, Box 1521, File "Training Schedules," RG 120, NARA.

87. *US Army in the World War* 14:298; Joseph D. Lawrence, *Fighting Soldier: The AEF in 1918,* 55.

88. Director, Army Candidates School, memorandums nos. 10 and 21, 20 October and 22 October, respectively, Entry 424, G5 Schools, La Valbonne Infantry Candidates School, Box 1521, File "Training Schedules," RG 120, NARA.

89. *US Army in the World War* 14:340; G5 Schools, Army Candidates School, Box 1639, File 352.16, "Instructors," RG 120, NARA.

90. "Nominal Roll of French Officers Detailed for Duty at the Army Candidates School, for the Course Commencing on April 1, 1918," 31 March 1918, G5 Schools, Army Candidates School, Box 1639, File 325.16, "Instructors French," RG 120, NARA.

91. Commandant, Army Schools, "French Personnel," memorandum, 26 January 1918, ibid.

92. Headquarters, Army Schools, American E.F., to Directors, Army School of the Line, Army Infantry Specialists' School, and Army Candidates' School, memorandum, 30 August 1918, G5 Schools, Army Candidates School, Box 1639, File 325.16 "British Instructors," RG 120, NARA.

93. Assistant Director, Army Candidates School, "Recommendation for Conduct of the Next Course," memorandum, 2 June 1918, G5 Schools, Army Candidates School, Box 1637, File 350, "Information in Regard to Schools and Courses," RG 120, NARA.

94. *US Army in the World War* 14:340.

95. "List of Candidates of the 3rd Battalion, Army Candidates School, Sent Here against Their Wishes," G5 Schools, Army Candidates School, Box 1637, File 352 "Candidates Miscellaneous"; and Commanding Officer, 3rd Battalion, Army Candidates School, "Illiterate Candidates," memorandum, 10 October 1918, G5 Schools, Army Candidates School, Box 1639, File 352 "Candidates Rejected," RG 120, NARA.

96. *US Army in the World War* 14:298; Headquarters Army Schools, Statistical Section, "Memorandum for Monthly Returns Section, Statistical Division, AGO, AEF," 11 January 1919, G5 Schools, Army Candidates School, Box 1619, File 065, RG 120, NARA; AEF GHQ to Commanding General, 29th Division, telegram, 18 September 1918, G5 Schools, Army Candidates School, Box 1625, File 300.4, RG 120, NARA.

97. Gerald F. Jacobson, *History of the 107th Infantry, U.S.A.,* 40; AEF GHQ Special Orders No. 304, 31 October 1918, author's collection; Miles, *History of the 308th Infantry,* 67–68.

98. Joint War History Commissions, *32nd Division,* 35–36; Gansser, History of the 126th, 51–52, 57.

99. Fell, History of the Seventh Division, 43; Ben Chastaine, *Story of the 36th,* 47–48.

100. *US Army in the World War* 3:213; Sgt. Richard McBride, 325th Inf., 82nd Div., manuscript, File WWI 2178, World War I Veteran Survey, USAMHI; Diary of

Sgt. William R. Phillips, William R. Phillips Papers, Liberty Memorial Archives, Kansas City, MO.

Chapter 7

1. Robert L. Bullard, *Personalities and Reminiscences of the War,* 267; Robert Alexander, *Memories of the World War,* 105.

2. Deputy Chief of Staff, SOS, "The Reclassification System of the A.E.F. (Blois)," report, 15 May 1919, Entry 465, Box 2257, "Reclassification System Combat Officers' Depot," 1–5, RG 120, NARA; Lt. John J. McInerny, "A Brief Summary of the Activities of the Casual Officers' Depot at Blois," in bound report summary, "Physical Classification System of the AEF at Blois," Entry 465, Box 2254, RG 120, NARA.

3. McInerny, "Brief Summary," 21–22.

4. These records are found in Entry 465, Box 2254, "Reclassification System Combat Officers' Depot," RG 120, NARA." The "proposed eliminations for inefficiency" listed the officer's name, rank, unit, and source of commission, as well as a brief summary of the findings of the board or commanding officer who sent the officer for reclassification. The date, location, and circumstances of these boards are a mystery. Internal evidence in the files in the form of memorandums and marginal notations dealing with the cases shows that boards met from as early as March 1918 to as late as March 1919. Most of the boards seemed to have been held in the summer and fall of 1918.

5. Commanding General, Combat Officers' Depot, "Re-classification of Officers," report, 22 May 1919, Entry 465, Box 2254, File "The Reclassification System at Gondrecourt," 1, RG 120, NARA.

6. Deputy Chief of Staff, SOS, "Reclassification System of the A.E.F. (Blois)," 22.

7. James G. Harbord, *The American Army in France, 1917–1919,* 426; Commanding General, Combat Officers' Depot, "Re-classification of Officers," 6; Harvey L. Harris, *The War as I Saw It,* 57.

8. James G. Harbord, *Leaves From A War Diary,* 343–4.

9. John W. Castles, manuscript, Castles Papers, United States Military Academy Library Archives, West Point, NY, 22; Henry Berry, *Make the Kaiser Dance: Living Memories of a Forgotten War: The American Experience in World War I,* 60; Colby L. McIntyre, *The Old Man of the 103rd: The Biography of Frank M. Hume,* 53.

10. This analysis was based on comparing the individual case reports for artillery officers in the grade of major through second lieutenant in Entry 541, Boxes 2286–2319, "Reclassified Officers National Army and National Guard, Blois," RG 120, NARA (hereafter cited as Blois case reports).

11. Deputy Chief of Staff, SOS, "Reclassification System of the A.E.F. (Blois)," 6.

12. Ibid., Table 2.

13. US War Department, *War Department Annual Report, 1919,* 676.

14. Hunter Liggett, *A.E.F.: Ten Years Ago in France,* 259, 261.

15. Col. M. G. Spinks, "Major Problems of the Inspector General, AEF, and Their Solution," lecture given to the Army War College, 9 October 1933, File AWC 401-A-5, USAMHI, 9; US Army Center of Military History, *The United States Army in the World War, 1917–1919,* Vol. 14, *Reports,* 401.

16. Maj. Gen. David C. Shanks, *Management of the American Soldier,* 4, 5.

17. Capt. James Archer, Blois case reports, Box 2286. Also see Maj. Birdsall P. Briscoe, Box 2289; Maj. Walter A. Marden, Box 2305; and Capt Charles H. Ingram, Box 2302, all in ibid.

18. Capt. Wilbur N. Farson, ibid., Box 2296.

19. 1st Lt. Albert C. Pate, ibid., Box 2310. Also see Capt. Frank L. Irwin, ibid., Box 2302.

20. Capt. Joseph E. Mirandon, ibid., Box 2307.

21. Capt. Raymond E. Copeland, ibid., Box 2293.

22. Ben Chastaine, *Story of the 36th,* 47; 2nd Lt. Mancel Coghlan, Blois case reports, Box 2292.

23. Capt. Charles M. Harrington, Blois case reports, Box 2299.

24. Anonymous, *Wine, Women and War: A Diary of Disillusionment,* 108.

Chapter 8

1. US Army Center of Military History, *The United States Army in the World War, 1917–1919,* Vol. 16, *General Orders,* 56.

2. Cpl. Alonzo M. LaVenture, 111th Inf., 28th Div., World War I Veteran Survey, USAMHI; Sgt. Maj. Mervyn F. Burke, Headquarters Troops, 1st Div., ibid.

3. Capt. B. A. Adams, Base Censor, "Examination of Mail of 78th, 79th, 80th, and 81st Divisions," memorandum, 28 January 1919, and Maj. Albert T. Rich, Assistant Inspector General, to the Inspector General, First Army, "Report of Investigation Concerning Morale in the 81st Division," 1 March 1919, both in Entry 588, Box 129, RG 120.

4. Howard Supple, 137th Inf., 35th Div.; Pfc. Elmer Stovall, 1st Ammunition Train, 1st Div.; Pfc. Henry P. King, 23rd Inf., 2nd Div., all in World War I Veteran Survey, USAMHI.

5. Carroll J. Swan, *My Company,* 32–33; Hervey Allen, *Towards the Flame,* 7.

6. George Marshall, *The Papers of George Catlett Marshall,* Vol. 1, *The Soldierly Spirit, December 1880–June 1939,* ed. Larry Bland and Sharon Ritenour Stevens, 127; Albert M. Ettinger, *A Doughboy with the Fighting 69th,* 147, 160–65.

7. US Army Center of Military History, *The United States Army in the World War, 1917–1919,* Vol. 3, *Training and Use of American Units with the British and French,* 208–9, 221–22.

8. US National Guard Bureau, *Report on Mobilization of the Organized Militia and National Guard of the United States, 1916,* 145.

9. William D. Henderson, *Cohesion: The Human Element in Combat,* 108.

10. Ibid., 112–15.

11. Charles Minder, *This Man's War,* 108; Ray N. Johnson, *Heaven, Hell, or Hoboken,* 114–15.

12. L. V. Jacks, *Service Record by an Artilleryman,* 211–12; Evan A. Edwards, *From Doniphan to Verdun: The Official History of the 140th Infantry,* 83.

13. Chester E. Baker, *Doughboy's Diary,* 59; Stanley J. Herzog, *Helmets: Second Battle of the Marne,* 57–70, 220–24.

14. Thomas H. Barber, *Along the Road,* 102–3; Charles MacArthur, *War Bugs,* 216.

15. Barber, *Along the Road,* 103.

16. Herzog, *Helmets,* 23–29; Elmer F. Straub, *A Sergeant's Diary in the World War,* Indiana Historical Collections 10:178.

17. William F. Clarke, *Over There with O'Ryan's Roughnecks,* 31.

18. US War Department, *War Department Annual Report, 1919,* 1:192, 674–75.

19. AEF General Order 56, 13 April 1918; AEF General Order 78, 25 May 1918. The author has found no evidence that such detachments were ever formed.

20. For an example, see Wendell Westover, *Suicide Battalions,* 182–87.

21. Capt. Edward L. Tanner, Entry 541, Box 2315, "Reclassified Officers National Army and National Guard, Blois," RG 120, NARA (hereafter cited as Blois case reports).

22. Intelligence Officer, 38th Division, Camp Shelby, Mississippi, "Report," 4 March 1918, in Records of the General Staff, Entry 377, Correspondence Related to

Morale at Army Installations, RG 165, NARA (hereafter cited as Morale at Army Installations), Box 13, Camp Shelby file.

23.   Cpl. Paul E. Maxwell, Camp Lee, 314th Field Artillery and Veterinary Training School, manuscript, File WWI-6694, World War I Veteran Survey, USAMHI, 25–26.

24.   John L. Barkley, *No Hard Feelings!* 232–33.

25.   MacArthur, *War Bugs,* 119; Leslie Langille, *Men of the Rainbow,* 134–35.

26.   Raymond B. Fosdick to Secretary of War Newton Baker, "Report to the Secretary of War on the Relation of Officers and Men in the A.E.F.," 17 April 1919, Entry 376, Box 18, RG 165, NARA.

27.   General Headquarters American Expeditionary Force, *Report of Officers Convened by Special Orders No. 98, GHQ AEF, 09 April 1919,* 9–10, Army War College Library, USAMHI, Annex S, 9; Morale Branch of the War College and WPD to the Chief of Staff, "Replies to Officers' Questionnaires," 5 November 1919, NM-84, Entry 378, Box 6, RG 165, NARA (hereafter cited as Morale Branch Officers' Survey), 34.

28.   Leslie Baker, *The Company History: The Story of Company B, 106th Machine Gun Battalion,* 60, 63; Ettinger, *Doughboy,* 147.

29.   Horace Baker, *Argonne Days in World War I,* 73, 78; MacArthur, *War Bugs,* 203–4.

30.   W. A. Sirmon, *That's War: An Authentic Diary,* 166; Westover, *Suicide Battalions,* 60; John S. Stringfellow, *Hell! No!* 143–44.

31.   Westover, *Suicide Battalions,* v; Minder, *This Man's War,* 190.

32.   Allen, *Towards the Flame,* 90; B. A. Colonna, ed., *The History of Company B, 311th Infantry, in the World War,* 12–13.

33.   Baker, *Doughboy's Diary,* 61.

34.   Jeremiah M. Evarts, *Cantigny: A Corner of the War,* 46–49.

35.   Fosdick, "Report to the Secretary of War," 17 April 1919.

36.   Sgt. Charles Strikell, 5th Field Art., 1st Div., World War I Veteran Survey, USAMHI; Morale Branch Officers' Survey, 24–28, 34.

37.   Harvey L. Harris, *The War as I Saw It: 1918 Letters of a Tank Corps Lieutenant,* 7; Raymond S. Tompkins, *The Story of the Rainbow Division,* 18.

38.   Colonna, *History of Company B,* 14.

39.   Allen, *Towards the Flame,* 14; Morale Branch Officers' Survey, 25.

40.   Joseph D. Lawrence, *Fighting Soldier: The AEF in 1918,* 20–21, 31; Sgt. Richard McBride, 325th Iinf., 82nd Div., manuscript, File WWI 2178, World War I Veteran Survey, USAMHI.

41.   MacArthur, *War Bugs,* 22; Pfc. Jonas E. Warrell, 103rd Ammunition Train, 28th Div., unpublished memoir, World War I Veteran Survey, USAMHI, 41; Barber, *Along the Road,* 78–80.

42.   Stringfellow, *Hell! No!* 161.

43.   "The Five Questions Interpreted," undated report, Morale at Army Installations, Box 6, Camp Grant file.

44.   "Our Military System As It Appeared to America's Citizen Soldiers," *INJ* 15, no. 10 (April 1919), 772, 782–83.

45.   Chief, Military Morale Section, "Anonymous Letter of Complaint," memorandum, 6 September 1918, Morale at Army Installations, Box 3, Camp Lee file.

46.   Nancy Gentile Ford, *Americans All: Foreign-born Soldiers in World War I,* 3–15; Ascanio Di Rago, Camp Devens, MA, "Report and Suggestions," 10 August 1918, Morale at Army Installations, Box 3, Camp Devens file.

47.   Intelligence Officer, Camp Devens, MA, "Questions Regarding Morale," report, 20 July 1918, and "Military Morale," report, 9 September 1918, Morale at Army Installations, Box 3, Camp Devens file.

48. Foreign-Speaking Soldier Sub-Section to Captain Perkins, memorandum, 10 September 1918, ibid., Box 5, Camp Gordon file.

49. Quoted in O. E. McKaine, "The Buffaloes: A First-Class Colored Fighting Unit," *Outlook* 119 (22 May 1918), 412.

50. Robert L. Bullard, *Personalities and Reminiscences of the War,* 294.

51. Monroe Mason and Arthur Furr, *The American Negro Soldier with the Red Hand in France,* 41–42; Arthur W. Little, *From Harlem to the Rhine,* 351–52.

52. Chief, Morale Branch, General Staff, "Letters of Complaint from Colored Soldiers, Camp Gordon," memorandum, 27 November 1918, Morale at Army Installations, Box 3, Camp Gordon file (the file contained the original letters between 17 and 25 October); Emmett J. Scott, memorandum, 13 December 1918, ibid., Box 3, Camp Grant file.

53. Arthur E. Barbeau and Florette Henri, *The Unknown Soldiers: Black American Troops in World War I*, 65–66, 78–80.

54. Intelligence Officer, Camp Pike, AR, "Questionnaire Concerning Colored Troops," memorandum, 6 November 1918, Morale at Army Installations, Box 12, Camp Pike file; Mason and Furr, *American Negro Soldier,* 19–20.

55. Intelligence Officer, Camp Pike, AR, "Questionnaire Concerning Colored Troops"; Bullard, *Personalities,* 291, 295.

56. Intelligence Officer, Camp Merritt, NJ, "Captain Timothy Mahoney," memorandum, 15 May 1918, in Records of the American Expeditionary Forces, Entry 186, "Correspondence Related to Officers Suspected of Pro-German Sympathies," Box 6218, RG 120, NARA; Capt. William H. Caldwell, 351st Field Art., 92nd Div., Blois case reports, Box 2291; 1st Lt. Julius Rogovin, 350 FA, 92nd Div., ibid., Box 2312.

57. Fosdick, "Report to the Secretary of War," 17 April 1919; Maury Maverick, *A Maverick American,* 109; Col. Robert Morehead, *The Story of the 139th Field Artillery, American Expeditionary Forces,* 33; Morale Branch Officers' Survey, 63.

58. Will Judy, *A Soldier's Diary,* 25; Hugh Thompson, *Trench Knives and Mustard Gas,* 16, 24.

59. Edwards, *From Doniphan to Verdun,* 115; Col. J. A. Bauer, Advance Section, Inspector General, to Chief of Staff, 1st Army, memorandum, 7 November 1918, Entry 590, Box 2, RG 120, NARA.

60. 2nd Lt. Homer A. Davis, 26th Div., Blois case reports, Box 2294; 1st Lt. Arthur R. Campbell, 26th Div., ibid., Box 2291.

61. Anonymous, *Wine, Women and War: A Diary of Disillusionment,* 44, 52.

62. Intelligence Officer, 31st Division, report, 22 February 1918, Morale at Army Installations, Box 17, Camp Wheeler file.

63. Colby L. McIntyre, *The Old Man of the 103rd: The Biography of Frank M. Hume,* 57.

64. Ben Chastaine, *Story of the 36th,* 15, 55; McIntyre, *Old Man,* 56–58; *The Story of "E" Company, 101st Engineers, 26th Division,* 13.

65. Emerson G. Taylor, *New England in France 1917–1919: A History of the Twenty-Sixth Division,* 23.

66. Ibid.; McIntyre, *Old Man,* 66–67.

67. Edmund A. Grossman, 139th Inf., 35th Div., World War I Veteran Survey, USAMHI; Morale Branch Officers' Survey, 19-b, 61.

68. Morale Branch Officers' Survey, 19-b.

69. Ibid., 27, 31.

70. Paul Fussell, *Wartime: Understanding and Behavior in the Second World War,* 80.

71. Ettinger, *Doughboy,* 13–14, 39–42, 44.

72. Cpl. Frank LaPierre Faulkner, 23rd Inf., 2nd Div., manuscript, World War I Veteran Survey, USAMHI; Archibald Hart, *Company K of Yesterday,* 31–32.

353

73. Cpl. Paul E. Maxwell, Camp Lee, 314th Field Art. and Veterinary Training School, manuscript, File WWI-6694, World War I Veteran Survey, USAMHI, 7, 12.

74. Ibid., 12, 18.

75. *The Battery Book: A History of Battery "A," 306th F.A.*, 28; MacArthur, *War Bugs*, 66; Judy, *Soldier's Diary*, 19.

76. Lucian K. Truscott Jr., *The Twilight of the Cavalry: Life in the Old Army, 1917–1942*, 15–16.

77. Morale Branch Officers' Survey, 19-f.

78. Sirmon, *That's War*, 270.

79. Quoted in, Henry Berry, *Make the Kaiser Dance: Living Memories of a Forgotten War: The American Experience in World War I*, 164.

80. Herman Dacus, 28th Inf., 1st Div., World War I Veteran Survey, USAMHI.

81. *Battery Book*, 174; Sirmon, *That's War*, 193–94; Thompson, *Trench Knives*, 75–76.

82. Sirmon, *That's War*, 145; Colonna, *History of Company B*, 44; Morale Branch Officers' Survey, 19-a, 19-b.

83. S. L. A. Marshall, *Men Against Fire: The Problem of Battle Command*, 44.

Chapter 9

1. Wendell Westover, *Suicide Battalions*, 209.

2. Col. John Parker, "Simplifying the Organization and Tactics of Infantry," *INJ* 16, no. 7 (January 1920), 567.

3. General Headquarters American Expeditionary Force, *Report of Officers Convened by Special Orders No. 98, GHQ AEF, 09 April 1919* (hereafter cited as Lewis Board), USAMHI, Appendix R, 1.

4. Unpublished diary of Maj. Gen. Beaumont Buck, Beaumont Buck Papers, US Army Cavalry Museum Archives, Fort Riley, KS.

5. Mark E. Grotelueschen, *Doctrine under Fire: American Artillery Employment in World War I*, 141–52; Mark E. Grotelueschen, *The AEF Way of War*, 36–38, 343–52.

6. Ernst Otto, *The Battle at Blanc Mont*, 31–33; German 102nd Regt. to the 63rd Inf. Bde., report, Box 200, German Miscellaneous File, RG 165, NARA. I thank D. Scott Stephenson, Department of History, US Army Command and General Staff College, for bringing this report to my attention.

7. General Headquarters, AEF, *Report of Superior Board on Organization and Tactics*.

8. General Headquarters, AEF, *Combat Instructions*, 6.

9. A. Lincoln Lavine, *Circuits of Victory*, 507–8.

10. Ibid., 511.

11. Charles Dienst et al., *They're From Kansas: History of the 353rd Infantry Regiment, 89th Division, National Army*, 261; Hervey Allen, *Towards the Flame*, 220–21.

12. Benedict Crowell, *America's Munitions, 1917–1918: Report of Benedict Crowell, the Assistant Secretary of War, Director of Munitions*, 581; Ashby Williams, *Experiences of the Great War*, 126–27; Robert H. Ferrell, *Five Days in October: The Lost Battalion of World War I*, 25–27, 36.

13. Maj. Gen. C. P. Summerall, memorandum, 25 August 1918, in US Army, 1st Division, *World War Records, First Division, A.E.F. Regular*, Vol. 2, *Field Orders, First Division, June 1, 1918, to Sept. 18, 1918*, not paginated; Joseph D. Lawrence, *Fighting Soldier: The AEF in 1918*, 121.

14. Clair Kenamore, *From Vauquois Hill to Exermont*, 136–38; Evan A. Edwards, *From Doniphan to Verdun: The Official History of the 140th Infantry*, 54–55.

15. Ben Chastaine, *Story of the 36th*, 229–33.

16. Dienst, *They're from Kansas,* 257–59; Lt. Col. James McIlroy, Forward Office, 1st Army Inspector General, "Report of Inspection of 3rd Division, 15 Oct. 18," memorandum, 15 October 1918, Entry 590, Box 4, RG 120, NARA.

17. Lewis Board, Appendix S, 1.

18. General Headquarters, AEF, *Notes on Recent Operations, No. 1,* issued 7 August 1918, and *No. 3,* issued 12 October 1918, 10–11, 16–18.

19. Ibid., *No. 3,* 14.

20. Capt. Charles M. Harrington, Entry 541, Box 2299, "Reclassified Officers National Army and National Guard, Blois," RG 120, NARA; Emil B. Gansser, *History of the 126th Infantry in the War with Germany,* 145–53.

21. Bruce Canfield, *U.S. Infantry Weapons of World War I,* 200–209. **355**

22. Ibid., 191, 211–15.

23. Maj. C. A. Dravo, "Machine Guns: The Offensive in Open Warfare," *INJ* 17, no. 4 (October 1920), 323–25; Capt. A. M. Patch, "Machine Gun Organization," *INJ* 17, no. 2 (August 1920), 145; Edward S. Johnston, "A Study of the Nature of United States Infantry Tactics for Open Warfare on July 18, 1918, and of Their Points of Difference as Contrasted with the United States Army Tactics Taught in 1914," CGSS SIRS, IR-124–1931, Appendix IV, 4.

24. GHQ, AEF, *Notes on Recent Operations, No. 1;* Summerall, memorandum, 25 August 1918.

25. Lewis Board, Appendix S, 9; Sanborn, *The 131st Infantry in the World War,* 191.

26. GHQ, AEF, *Notes on Recent Operations, No. 3:*15; Lawrence, *Fighting Soldier,* 87; Lt. Col. James McIlroy, Forward Office, 1st Army Inspector General, "Report of Inspection of 3rd Division, 15 Oct. 18," memorandum, 15 October 1918, Entry 590, Box 4, RG 120, NARA; "Notes Made by the Inspector General A.E.F., during the Active Operations from 12th September 1918 to 11th November 1918," Entry 588, Box 116, RG 120, NARA.

27. Canfield, *U.S. Infantry Weapons,* 247–60; Col. Robert R. McCormick, *The Army of 1918,* 169–70; Edwards, *From Doniphan to Verdun,* 58.

28. Charles M. DuPuy, *A Machine Gunner's Notebook,* 79; 2nd Lt. Malcolm Helms, 5th Mach. Gun Bn., 2nd Div., World War I Veteran Survey, USAMHI; Ray N. Johnson, *Heaven, Hell, or Hoboken,* 101–2.

29. "Report of Operations Argonne Meuse, Office of the Inspector, 5th Army Corps, Sept. 25–Nov. 11, 18, Extracts of Reports of Div. Inspectors," Entry 588, Box 116, RG 120, NARA.

30. GHQ, AEF, *Notes on Recent Operations, No. 1;* Leslie Baker, *The Company History: The Story of Company B, 106th Machine Gun Battalion,* 47.

31. "Notes Made by the Inspector General A.E.F."; Lt. Col. Oliver Spaulding, "The Tactics of the War With Germany," *INJ* 17, no. 3 (September 1920), 239; McIlroy, "Report of Inspection of 3rd Division"; Col. E. L. Gruber, "Employment of Field Artillery with Infantry," *INJ* 15, no. 12 (June 1919), 969.

32. Harvey L. Harris, *The War as I Saw It: 1918 Letters of a Tank Corps Lieutenant,* 127; GHQ, AEF, *Notes on Recent Operations, No. 3:*5, 18; Dale E. Wilson, *Treat 'Em Rough! The Birth of American Armor, 1917–20,* 220–21.

33. Dienst, *They're from Kansas,* 258, 264–65.

34. US Army Center of Military History, *The United States Army in the World War, 1917–1919,* Vol. 2, *Policies,* 411; Lewis Board, Appendix R, 10.

35. Maj. C. A. Dravo, "Machine Guns: The Offensive in Open Warfare," *INJ* 17, no. 4 (October 1920), 319; Canfield, *U.S. Infantry Weapons,* 147–52, 154; Gerard Demaison and Yves Buffetaut, *Honor Bound: The Chauchat Machine Rifle,* xvii, 131–45.

36. Canfield, *U.S. Infantry Weapons,* 156–60.

37. Ibid., 238–40; Crowell, *America's Munitions,* 208.

38. Summerall, memorandum, 25 August 1918.

39. GHQ, AEF, *Notes on Recent Operations, No. 1;* Diary entries for 28 August 1918 in manuscript "A Tanglefoot's Diary," in John D. McDaniels, 126th Inf., 32nd Div., WWI-456, World War I Veteran Survey, USAMHI.

40. Summerall, memorandum, 25 August 1918.

41. Williams, *Experiences,* 69, 80.

42. Summerall, memorandum, 25 August 1918.

43. Lewis Board, Appendix R, 6.

Chapter 10

1. Quoted in *Infantry in Battle,* 137.

2. Lt. Col. Jennings C. Wise, "The Soldier's Life in Battle," *INJ* 16, no. 11 (May 1920), 929.

3. Brig. Gen. Frank Parker, "Certain Observations on Infantry," lecture, 2 April 1919, to AEF's Army Center of Artillery Studies, AEF, Third Course, in the personal collection of the author.

4. General Headquarters, AEF, *Notes on Recent Operations, No. 1;* Entry for 28 August 1918 in manuscript "A Tanglefoot's Diary," in John D. McDaniels, 126th Inf., 32nd Div., WWI-456, World War I Veteran Survey, USAMHI.

5. General Headquarters, AEF, *Combat Instructions,* 1; *Infantry in Battle,* 1–3.

6. B. A. Colonna, ed. *The History of Company B, 311th Infantry in the World War,* 22–23.

7. Report for 23 July 1918 from the German 7th Army War Diary, in US Army, 1st Division, *World War Records, First Division, A.E.F. (Regular),* Vol. 2, *German Documents: Aisne-Marne (Soissons),* not paginated; Horatio Rogers, *World War I through My Sights,* 182.

8. Beaumont Buck, *Memories of Peace and War,* 210; Charles M. Clement, ed., *Pennsylvania in the World War: An Illustrated History of the Twenty-Eighth Division* 2:395–97, 399.

9. Edward Johnston, "A Study of the Nature of United States Infantry Tactics for Open Warfare on July 18, 1918, and of Their Points of Difference as Contrasted with the United States Army Tactics Taught in 1914," CGSS SIRS, IR-124–1931, Appendix IV, 2–3, 13.

10. Hervey Allen, *Towards the Flame,* 138–39.

11. Grotelueschen, *The AEF Way of War,* 114–22, 240–49; Division Inspector, 26th Division, "Points Noted during the Operations of September 12th, 13th and 14th," report, 30 September 1918, Entry 588, Reports of the Inspector General, Box 108, RG 120, NARA.

12. Justus Owens to "Mamma" (Settie Owens), 14 September 1918, Justus Erwin Owens Scrapbook, folder 2856, Special Collections, Hargrett Rare Book and Manuscript Library, University of Georgia.

13. Pfc. George Loukides, 326th Inf., 82nd Div., File #1547, World War I Veteran Survey, USAMHI; Alvin C. York, *Sergeant York: His Own Life Story and War Diary,* ed. Tom Skeyhill, 208–9.

14. Charles MacArthur, *War Bugs,* 142–43.

15. Hugh Thompson, *Trench Knives and Mustard Gas,* 170, 172.

16. "Lecture Delivered by Colonel Willey Howell on 6 January 1919 to the Assembled General Officers and Chiefs of Staff of the First American Army, Subject: 'The Second Section, G.S., First American Army in the St. Mihiel and Meuse-Argonne Operations,'" in Lectures: [Operations of the First Army]/First Army Staff, Special Collections, Combined Arms Research Library, Fort Leavenworth, KS.

17. Charles Dienst, et al., *They're from Kansas: History of the 353rd Infantry Regiment, 89th Division, National Army,* 261; Society of the Fifth Division, *The Official History of the Fifth Division,* 117, 145, 154.

18. Quoted in Donald Smythe, *Pershing: General of the Armies,* 191.

19. Ernst Otto, *The Battle at Blanc Mont,* 136–38, 155.

20. GHQ, AEF, *Combat Instructions,* 9, 11.

21. Charles M. DuPuy, *A Machine Gunner's Notebook,* 81; General Headquarters, AEF, *Report of Officers Convened by Special Orders No. 98, GHQ AEF, 09 April 1919* (hereafter cited as Lewis Board), Appendix R, 1.

22. Col. E. G. Peyton, "Modern Tactics," *INJ* 17, no. 2 (August 1920), 120; "Notes Made by the Inspector General, A.E.F., during the Active Operations from 12th September 1918 to 11th November 1918," Entry 588, Box 116, RG 120, NARA, 4.

23. "Notes Made by the Inspector General A.E.F."

24. *History of Company "C," 328th Infantry,* 23.

25. AEF Inspector General, "Observations and Investigations Already Made by the Inspectors General in Regard to Straggling and the use of Shelter in the Area Occupied by the First Army," memorandum, 21 October 1918, Entry 588, Box 113, RG 120, NARA.

26. L. V. Jacks, *Service Record by an Artilleryman,* 186–87; Horace Baker, *Argonne Days in World War I,* 54; John L. Barkley, *No Hard Feelings!* 199, 269.

27. Joseph D. Lawrence, *Fighting Soldier: The AEF in 1918,* 86–87; Maury Maverick, *A Maverick American,* 129.

28. Colonna, *History of Company B,* 51.

29. Martin Hogan, *The Shamrock Battalion in the Great War,* 111.

30. Elmer Murphy and Robert S. Thomas, *The Thirtieth Division in the World War,* 194.

31. Charles Heller, *Chemical Warfare in World War I: The American Experience, 1917–1918,* Leavenworth Papers 10:91–93; Rexmond C. Cochran, *Gas Warfare in World War I,* Study 2, *The 78th Division at the Kriemhilde Stellung: October 1918,* 62–75, and Study 3, *The 1st Division in the Meuse-Argonne: 1–12 October 1918,* 51–61; Col. H. L. Gilchrist, *A Comparative Study of World War Casualties from Gas and Other Weapons,* 16, 23–26.

32. 26th Division, "Report of Gas Attack, 10 October 1918," in E. W. Spencer, *The History of Gas Attacks upon the American Expeditionary Forces during the World War,* copy in CARL archives; "Report of Gas Attack, 5–6 October 1918," ibid. (emphasis in the original).

33. US Army Center of Military History, *The United States Army in the World War, 1917–1919* (cited hereafter as *US Army in the World War*), Vol. 14, *Reports,* 403; Society of the Fifth Division, *Official History,* 185.

34. Cpl. Berch Ford, 16th Inf., 1st Div., World War I Veteran Survey, USAMHI.; Lowell Thomas, *Woodfill of the Regulars,* 74–75.

35. Entry for 20 September 1918 in "A Tanglefoot's Diary."

36. Ibid., entries for 1 and 2 October 1918.

37. Forward Office, Inspector General's Department, 1st Army, "Report of Inspection in Sector of 42nd Division," memorandum, Entry 590, Box 2, RG 120, NARA.

38. "Combat Report of the M.G. School of the 3rd Army during Its Employment at Group Argonne from the 26 September to 5th October 1918," in US Army, 1st Division, *World War Records, First Division, A.E.F. (Regular),* Vol. 4, *German Documents: Meuse-Argonne,* not paginated (hereafter cited as *German Documents: Meuse-Argonne*); Otto, *Battle at Blanc Mont,* 79; "Experiences from Last Action," German 1st Battalion, 170th Infantry Regiment, 20 October 1918, in ibid. For examples of German descriptions of mass American attacks, see Otto, *Battle at Blanc Mont,* 76, 112–13, 119, 165; Group Argonne, report, 10 October 1918; Report

of 1st Guard Infantry Division, 30 September 1918; 5th Guards Division Battle Diary entry for 4 October 1918; Report of the 3rd Bn., 150th Infantry, 5 October 1918; and Report of the 2nd Bn., 150th Infantry, 14 October 1918, all in *German Documents: Meuse-Argonne.*

39. Otto, *Battle at Blanc Mont,* 192–94; "Experiences with the American Method of Combat," German 111th Inf. Regiment, 26 October 1918, in *German Documents: Meuse-Argonne.*

40. Lewis Board, Appendix R, 2, 13; General Headquarters, AEF, *Report of Superior Board on Organization and Tactics,* 20–21.

41. George Cornish, "The Twenty-Sixth Infantry (U.S.) in the Meuse-Argonne Offensive," CGSS SIRS, IR-104–1931, 210; War Department, *Offensive Combat of Small Units,* War Department Document No. 802, May 1918, "Supplement to Instructions," Plate 11.

42. Report from the 102nd Regt. to the 63rd Inf. Bde., Box 200, German Miscellaneous File, RG 165, NARA; "Relative to Experiences as to the Method of Attack of the Americans," German 170th Inf. Regt., 15 October 1918, *German Documents: Meuse-Argonne.*

43. "Experiences during Recent Actions," German 37th Division, 13 October 1918, *German Documents: Meuse-Argonne.*; final quote from "Lecture Delivered by Colonel Willey Howell on 6 January 1919 to the Assembled General Officers and Chiefs of Staff of the First American Army, Subject: 'The Second Section, G.S., First American Army in the St. Mihiel and Meuse-Argonne Operations,'" in Lectures: [Operations of the First Army]/First Army Staff, CARL Archives, Fort Leavenworth, KS.

44. Henry Thorn, *History of 313th U.S. Infantry,* 46–48; Allen, *Towards the Flame,* 192.

45. "Experiences with the American Method of Combat," German 111th Inf. Regiment, 26 October 1918, in *German Documents: Meuse-Argonne;* George Viereck, ed., *As They Saw Us: Foch, Ludendorff, and Other Leaders Write Our War History,* 38; "Experiences from Last Action," German 1st Battalion, 170th Infantry Regiment, 20 October 1918, in *German Documents: Meuse-Argonne.*

46. "Experiences Relative the Method of Attack of the Americans," German 169th Inf. Regt., 25 October 1918, *German Documents: Meuse-Argonne;* "Experiences with the American Method of Combat," German 111th Inf. Regiment, 26 October 1918, ibid.

47. "Experiences with the American Method of Combat," German 111th Inf. Regiment, 26 October 1918, ibid.

48. "Notes Made by the Inspector General A.E.F., during the Active Operations from 12th September 1918 to 11th November 1918," Entry 588, Box 116, RG 120, NARA; "Combat Report for the 10th, 11th and 12th October 1918," German 151st Infantry Regiment, 12 October 1918, *German Documents: Meuse-Argonne.*

49. Morale Branch of the War College and War Plans Division to the Chief of Staff, "Replies to Officers' Questionnaires," 5 November 1919, NM-84, Entry 378, Box 6, RG 165, NARA (hereafter cited as Morale Branch Officers' Survey), 26–27.

50. Ibid.

51. General Headquarters, AEF, *Notes on Recent Operations, No. 3:* 8; Donovan quoted in James H. Hallas, *Doughboy War: The American Expeditionary Force in World War I,* 279.

52. "Lecture Delivered by Major General Alexander, 77th Division, on 3 February 1919, Subject: 'Operations of the Division, 26th of September to the 11th of November,'" in Lectures: [operations of the First Army]/First Army Staff, CARL Archives, Fort Leavenworth, KS; *US Army in the World War,* Vol. 15, *Reports of Commander-in-Chief, A.E.F. Staff Sections and Services,* 304–5.

53. Morale Branch Officers' Survey, 21.

54. Ibid.; *US Army in the World War,* 14:403.

55. Lewis Board, Appendix R, 12.

56. *Infantry in Battle,* 92, 131; Capt. J. O. Green, "Operations of 3rd Battalion, 23rd Infantry, from 30 May to 12 July 1918," CGSS SIRS, IR-53–1931, CARL Archives, 12.

57. Otto, *Battle at Blanc Mont,* 140–41, 146.

58. Ben Chastaine, *Story of the 36th,* 94, 98–105.

59. Craig Hamilton and Louise Corbin, eds., *Echoes from Over There,* 205.

60. "Report of Gas Attack, 9–10 October," in Spencer, *History of Gas Attacks* (emphasis in the original).

61. Arthur H. Joel, *Under the Lorraine Cross,* 54–55; 1st Lt. Glen Gardiner, Entry 541, Box 2298, "Reclassified Officers National Army and National Guard, Blois," RG 120, NARA.

62. Jeremiah M. Evarts, *Cantigny: A Corner of the War,* 49–61.

63. Lt. Col. V. M. Elmore, Inspector General, 6th Division, "Recommendations," report, 17 October 1918, Entry 588, Box 108, RG 120, NARA (emphasis in the original).

64. Inspector, 82nd Division, "Daily Report of Past 24 Hours," 12 October 1918, Entry 796, Box 3, RG 120, NARA.

65. Lawrence, *Fighting Soldier,* 109–10.

66. Stanley J. Herzog, *Helmets: Second Battle of the Marne,* 112–19; American Battle Monuments Commission, *American Armies and Battlefields in Europe,* 309.

Chapter 11

1. Quoted in George Viereck, ed., *As They Saw Us: Foch, Ludendorff, and Other Leaders Write Our War History,* 10.

2. G. Edward Buxton Jr., ed., *Official History of 82nd Division, American Expeditionary Forces: "All American" Division, 1917–1919,* 16, 29, 86–87, 213.

3. Arthur C. Havlin, *The History of Company A, 102nd Machine Gun Battalion, Twenty-Sixth Division,* 166–67.

4. War Veteran's Association, *History of Company "E," 107th Infantry,* 105–21.

5. B. A. Colonna, ed., *The History of Company B, 311th Infantry, in the World War,* 58–61.

6. Ibid., 66–68.

7. Leonard Ayres, *The War With Germany: A Statistical Summary,* 121, 130.

8. Edward A. Shils and Morris Janowitz, "Cohesion and Disintegration in the Wehrmacht in World War II," *Public Opinion Quarterly* 12, no. 2 (Summer, 1948), 280–99.

9. Joseph D. Lawrence, *Fighting Soldier: The AEF in 1918,* 60–65; Henry Thorn, *History of 313th U.S. Infantry,* 36, 40.

10. L. Wardlaw Miles, *History of the 308th Infantry,* 315–22. Also see Kerr Rainsford, *From Upton to the Meuse with the Three Hundred and Seventh Infantry,* 294–99.

11. John L. Barkley, *No Hard Feelings!* 187, 251; *History of the Three Hundred and Twelfth Infantry,* 73; Pvt. Milton B. Sweningsen, 138th Inf., 35th Div., unpublished memoirs, World War I Veteran Survey, USAMHI, 2.

12. Pfc. Charles W. Flacker, 112th Inf., 28th Div., World War I Veteran Survey, USAMHI; Robert R. McCormick, *The Army of 1918,* 171–74; Forward Office, Inspector General's Department, First Army, "Inspection 5th Division," memorandum, 22 Oct. 1918, Entry 590, Box 8, RG 120, NARA.

13. Pvt. Herman Dacus, 28th Inf., 1st Div., World War I Veteran Survey, USAMHI.

14. Entries for 31 August, 29 September, and 9 and 18 October 1918 in manuscript "A Tanglefoot's Diary," John D. McDaniels, 126th Inf., 32nd Div., WWI-456, World War I Veteran Survey, USAMHI.

359

15. Report of effective strength of 77th Division, 24 October 1918, Entry 590, Box 1, RG 120, NARA; General Headquarters, AEF, *Report of Officers Convened by Special Orders No. 98, GHQ AEF, 09 April 1919* (hereafter cited as Lewis Board), 9–10.

16. Miles, *History of the 308th,* 173; Lawrence, *Fighting Soldier,* 84, 124.

17. Horace Baker, *Argonne Days in World War I,* 108, 109, 112–13.

18. Craig Hamilton and Louise Corbin, eds., *Echoes from Over There,* 105; Pvt. Duncan M. Kemerer, 111th Inf., 28th Div., World War I Veteran Survey, USAMHI.

19. Kemerer, World War I Veteran Survey.

20. Shils and Janowitz, "Cohesion and Disintegration," 297; Pvt. Fred Takes, 325th Inf., 82nd Div., file WWI-1760, World War I Veteran Survey, USAMHI.

21. Wilbert F. Stambaugh, 2nd Field Signal Battalion, 1st Div., World War I Veteran Survey, USAMHI; William S. Triplet, *A Youth in the Meuse-Argonne,* 246; Clair Kenamore, *From Vauquois Hill to Exermont,* 206, 240.

22. Capt. John K. Taylor, Commanding C Company, to Col. Whitman, Commander 325th Infantry, "Impressions and Recollections of Operations, C Co., 325 In.," report, 26 December 1918, contained in unpublished "History of the 325th INF, Letters from Company Commanders," in Brig. Gen. Walter Whitman, 325th Inf., 82nd Div., WWI 6052, World War I Veteran Survey, USAMHI.

23. Sgt. Charles Strikell, 5th Field Art., 1st Div., World War I Veteran Survey, USAMHI.

24. C. V. L. to Maj. Charles D. Gentsch, G-2, 83rd Division, "Classification Camps Report This Date," memorandum, 3 November 1918, Entry 195 (one box), "Reports Related to the Morale of American Troops 1917–1918," RG 120, NARA.

25. Leonard V. Smith, *Between Mutiny and Obedience: The Case of the French Fifth Infantry Division during World War I,* 11–19.

26. "Notes Made by the Inspector General, A.E.F., during the Active Operations from 12th September 1918 to 11th November 1918," Entry 588, Box 116, RG 120, NARA.

27. Stanley J. Herzog, *Helmets: Second Battle of the Marne,* 81.

28. Capt. Joseph H. W. Hinkson, 311th Machine Gun Battalion, statement, 5 November 1918, and Lt. Col. C. M. Dowell to Chief of Staff, 26th Division, confidential memorandum, 7 November 1918, Entry 540, Box 2273, "Gondracourt Reclassification Depot, Investigation of BG Charles Cole," RG 120, NARA; Colby L. McIntyre, *The Old Man of the 103rd: The Biography of Frank M. Hume,* 84–98. Also see Commanding General, 1st Army, to Commander in Chief, AEF, "Recommendations Regarding 26th Division," memorandum, 7 November 1918, Entry 540, Box 2273, "Gondracourt Reclassification Depot, Investigation of BG Charles Cole"; "Brief History of the Case of Major General Clarence R. Edwards," Entry 464, Box 2267, "Personal Files of Brigadier General R. C. Davis"; and Maj. Elson A. Hosford, 103rd Inf., 26th Div., Entry 541, Box 2301, "Reclassified Officers National Army and National Guard, Blois," all in RG 120, NARA.

29. Reports of self-inflicted wounds for August 1918, Entry 588, Box 108, RG 120, NARA; Lt. Col. V. M. Elmore, Inspector General, 6th Division, "Recommendations," report, 17 October 1918, ibid. In all but 15 of the 179 cases, the soldiers were in infantry or machine gun units.

30. Robert L. Bullard, *Personalities and Reminiscences of the War,* 251; Hunter Liggett, *A.E.F.: Ten Years Ago in France,* 207.

31. "Notes Made by the Inspector General, A.E.F., during the Active Operations from 12th September 1918 to 11th November 1918," Entry 588, Box 116, RG 120, NARA; Inspector, V Corps, "Extract Report of Division Inspector, 91st Division," report, 5 October 1918, Entry 590, Box 1, RG 120, NARA; Lt. Col. J. C. McIlroy, Advance Section, GHQ Inspector General, "Stragglers," memorandum, 10 November 1918, Entry 590, Box 8, RG 120, NARA; Brig. Gen. Henry H.

Bandholtz, *History of the Provost Marshal General's Department, American Expeditionary Forces,* 6; Commanding Officer, 2nd Provisional M.P. Bn., 2nd Army, "Reports of Arrests by Months up to Dec. 31, 1918," report, Entry 918, Box 84, RG 120, NARA.

32. "Report of Inspection of the 37th Division," 2 October 1918, Entry 590, Box 2, RG 120, NARA; "Lecture Delivered by Colonel A. W. Foreman on 27 January 1919 at Fifth Army Corps Headquarters, Subject: 'Administration and Supply of the 5th Army Corps during the St. Mihiel Offensive and Meuse-Argonne Offensive 1918,'" in Lectures: [Operations of the First Army]/First Army Staff, CARL Archives, Fort Leavenworth, KS.

33. "Lecture Delivered by Maj. Gen. George B. Duncan, Commanding 82nd Division, on 3 February 1919, Subject: 'General Missions of the 82nd Division in the Argonne-Meuse Offensive,'" in Lectures: [Operations of the First Army]/First Army Staff, CARL Archives, Fort Leavenworth, KS; Bandholtz, *Provost Marshal General's Department,* 8; AEF Inspector General, "Observations and Investigations Already Made by the Inspectors General in Regard to Straggling and the Use of Shelter in the Area Occupied by the First Army," memorandum, 21 October 1918, Entry 588, Box 113, RG 120, NARA.

34. Provost Marshal, Second Army, "Memorandum to Corps Provost Marshals and Division A.P.Ms," 9 November 1918, Entry 55, Box 83, RG 120, NARA; Reports of Stragglers Apprehended on October 28–31 and 1 November 1918 by Company A, 313th MP, Assistant Provost Marshal, Souilly, France, A Company, 117th MP, B, C, and D Companies, 1st Army MP Bn., and H Troop, 2nd Cavalry, Entry 865, Box 269, RG 120, NARA. The MPs apprehended 108 stragglers on 28 October, 140 on 29 October, 193 on 30 October, 75 on 31 October, and 97 on 1 November 1918.

35. Wendell Westover, *Suicide Battalions,* 250.

36. First Army Inspector General, "Memorandum from Observations and Investigations Already Made by the Inspector General in Regard to Straggling and the Use of Shelter in the Area Occupied by the First Army," 21 October 1918, Entry 588, Box 113, RG 120, NARA.

37. Inspector, 82nd Division, to Inspector General, AEF, "Report of Past 24 Hours," 10 October 1918, Entry 590, Box 8, RG 120, NARA; Ray N. Johnson, *Heaven, Hell, or Hoboken,* 95.

38. Entries for 26–30 September 1918 in the dairy of Capt. Clarence J. Minick, Clarence J. Minick Papers, Liberty Memorial Archives, Kansas City, MO.

39. Lowell Thomas, *Woodfill of the Regulars,* 284–85.

40. Jennings C. Wise, "The Soldier's Life in Battle," *INJ* 16, no. 11 (May 1920), 931.

41. Barkley, *No Hard Feelings!* 126–27; Inspector, 82nd Division, to Inspector General, AEF, "Daily Report of Past 24 Hours," report, 12 October 1918, Entry 796, Box 3, RG 120, NARA.

42. Dillman quoted in Ernest Fisher, Jr., *Guardians of the Republic: A History of the Noncommissioned Officer Corps of the U.S. Army,* 201; Officer of the 119th Infantry quoted in Elmer Murphy and Robert S. Thomas, *The Thirtieth Division in the World War,* 194.

43. John W. Nell, *The Lost Battalion: A Private's Story,* 87–88; Thomas Barber, *Along the Road,* 82–83.

44. Bandholtz, *Provost Marshal General's Department,* 2.

45. Thorn, *History of the 313th,* 33, 35; Entries for 11–16 October 1918 in "A Tanglefoot's Diary"; L. V. Jacks, *Service Record by an Artilleryman,* 209.

46. Shils and Janowitz, "Cohesion and Disintegration," 281, 291–92.

47. Inspector General, 82nd Division, "Condition of Troops from 78th Division," report, 9 October 1918, Entry 590, Box 1, RG 120, NARA.

48. Baker, *Argonne Days,* 39; Cpl. Fred Takes, 325th Inf., 82nd Div., World War I Veteran Survey, USAMHI; Barber, *Along the Road,* 71–72, 83.

49. Barber, *Along the Road,* 82 "Lecture presented by Lt. Col. Troup Miller on 20 January 1919 at Fifth Army Corps Headquarters, Subject: 'Plan of Communication, Supply and Evacuation, 1st Corps, for St. Mihiel Offensive and Meuse-Argonne Offensive,'" in Lectures: [Operations of the First Army]/First Army Staff, CARL Archives, Fort Leavenworth, KS.

50. Lawrence, *Fighting Soldier,* 114–15.

51. Barber, *Along the Road,* 71–72, 83; Maj. Merritt Olmstead, "A Critical Analysis of Troop Leading within the 5th Division during the Second Phase of the Meuse-Argonne Offensive," CGSS SIRS, 1933, CARL Archive, Fort Leavenworth, KS.

52. Lewis Board, Appendix P, 16.

53. Leslie Langille, *Men of the Rainbow,* 158; Ben Chastaine, *Story of the 36th,* 134; L. Wardlaw Miles, *History of the 308th Infantry,* 132; Carl E. Haterius, *Reminiscences of the 137th Infantry,* 153.

54. Forward Office, Inspector General's Department, 1st Army, "Combined Reports on 91st Division Made Up from Report of Lt. Col. McKenney, 16 Oct., Lt. Col. J. G. McIlroy, Oct. 16 and 18," memorandum, 19 October 1918, Entry 590, Box 1, RG 120, NARA.

55. Baker, *Argonne Days,* 79; *The Service Record: Atlanta's Military Weekly,* 5 June 1919, 57; Col. Edward Carpenter, IV Corps Inspector General, "Observations on the Offensive for the Reduction of the St. Mihiel Salient," report, 27 September 1918, Entry 588, Box 108, RG 120, NARA; US War Department, *War Department Annual Report, 1919,* 1, part 3:3374–76; "Report of MAJ Oliver Q. Melton, Commander K Company, 325th Infantry, to COL Whitman, Commander, 325th Infantry," contained in unpublished "History of the 325th INF, Letters from Company Commanders," in Brig. Gen. Walter Whitman, 325th Inf., 82nd Div., WWI 6052, World War I Veteran Survey, USAMHI.

56. Col. A. C. Read, 1st Army Inspector General, "Ammunition Supply, Morale, Roads, etc.," memorandum, 2 October 1918, Entry 590, Box 4, RG 120, NARA.

57. Inspector General, 3rd Division, "Report of Inspection, 3rd Division, 15 and 16 Oct 18," Entry 590, Box 4, RG 120, NARA.

58. John S. Stringfellow, *Hell! No!* 251–52.

59. Inspector, 82nd Division, "Daily Report of Past 24 Hours," 12 October 1918, Entry 796, Box 3, RG 120, NARA; Forward Officer, Inspector General's Department, 1st Army, "Investigation of Straggling in 82nd Division," memorandum, Entry 590, Box 8, RG 120, NARA.

60. William Wright, *Meuse-Argonne Diary: A Division Commander in World War I,* 23; "Report of Operations Argonne Meuse, Officer of the Inspector, 5th Army Corps, Sept. 25–Nov. 11, Extracts of Reports of Div. Inspectors," Entry 588, Box 116, RG 120, NARA.

61. Inspector, 32nd Division, report, 14 October 1918, Entry 590, Box 2, RG 120, NARA; Commander, H Troop, 2nd Cavalry, "Report of Stragglers Apprehended," 1 November 1918, Entry 856, Box 269, RG 120, NARA.

62. Archibald Hart, *Company K of Yesterday,* 108–9.

63. Baker, *Argonne Days,* 118.

64. Barber, *Along the Road,* 82; Ernesto Bisogno, Pvt., 328th Inf., 82nd Div., World War I Veteran Survey, USAMHI; Lawrence, *Fighting Soldier,* 88, 102, 114–15.

65. Col. J. A. Bauer, memorandum, 18 October 1918, Entry 590, Box 4, RG 120, NARA; Olmstead, "Critical Analysis of Troop Leading."

66. Westover, *Suicide Battalions,* 219–20.

67. Hart, *Company K of Yesterday,* 92–93.

68. Lawrence, *Fighting Soldier,* 121–23; Baker, *Argonne Days,* 120.

69. Shils and Janowitz, "Cohesion and Disintegration," 279; William D. Henderson, *Cohesion: The Human Element in Combat,* 109.

70. "Notes Made by the Inspector General, A.E.F., during the Active Operations from 12th September 1918 to 11th November 1918," Entry 588, Box 116, RG 120, NARA; Forward Office, Inspector General's Department, 1st Army, "Report of Inspection in Sector of 42nd Division," memorandum, Entry 590, Box 2, RG 120, NARA; Thomas M. Johnson, *Without Censor,* 315–16, 318.

71. Diary entries for 28 August 1918 in "A Tanglefoot's Diary."

72. Field agents to Capt. (later Maj.) Charles Gentsch, G-2, 83rd Division, reports for the dates 12 August 1918, 7 September 1918, 16 September 1918, 3 October 1918, 29 October 1918, and 1 November 1918, Entry 195 (one box), "Reports Related to the Morale of American Troops 1917–1918," RG 120, NARA; Hugh Thompson, *Trench Knives and Mustard Gas,* 160.

73. Testimony of Lt. Col. Eugene H. Houghton in Capt. Albert Rich, Asst. Inspector General, 1st Army, to Inspector General, 1st Army, "77th Division Cutting Off of Seven Companies and One Machine Gun Company, October 3rd, 1918," report, 8 October 1918, Entry 590, Box 1, RG 120, NARA.

74. Du Cane quoted in Robert C. Walton, ed., *Over There: European Reaction to Americans in World War I*, 195; Viereck, *As They Saw Us,* 302; US Army Center of Military History, *United States Army in the World War, 1917–1919,* Vol. 14, *Reports,* 310.

Chapter 12

1. George C. Marshall, *The Papers of George Catlett Marshall,* Vol. 2, *We Cannot Delay, July 1939–December 6, 1941,* ed. Larry Bland and Sharon Ritenour Stevens, 511–13; Henry L. Stimson, *On Active Service in Peace and War,* 348–50; Forrest Pogue, *George C. Marshall: Ordeal and Hope 1939–1942,* 102–3.

2. *Infantry in Battle,* first page of the Introduction.

3. Daniel Bolger, "Zero Defects: Command Climate in the First U.S. Army, 1944–1945," *Military Review* 71 (May 1991), 61–73.

4. *Infantry in Battle,* 199.

5. General Headquarters, American Expeditionary Force, *Report of Officers Convened by Special Orders No. 98, GHQ AEF, 09 April 1919,* 9–10, USAMHI Library.

6. John T. Carlton and John F. Slinkman, *The ROA Story: A Chronicle of the first 60 Years of the Reserve Officers Association of the United States,* 20.

7. Robert W. Palmer, et al., *The Procurement and Training of Combat Ground Troops,* 93–95, 264–73, 335–45; Peter R. Mansoor, *The GI Offensive in Europe: The Triumph of American Infantry Divisions, 1941–1945,* 11–15, 20–21, 24–28, 40–46.

8. Michael S. Neiberg, *Making Citizen Soldiers: ROTC and the Ideology of American Military Service,* 112–50.

9. Quoted in Samuel Lipsman and Edward Doyle, *Fighting for Time,* 96–101, 112; Shelby Stanton, *The Rise and Fall of an American Army: U.S. Ground Forces in Vietnam, 1965–1973,* 271–72.

Bibliography

Archival and Unpublished Primary Sources
COMBINED ARMS RESEARCH LIBRARY, SPECIAL COLLECTIONS, FORT LEAVENWORTH, KS

Bell, J. Franklin. "Reflections and Suggestions: An Address by General J. Franklin Bell," 17 March 1906.

Burdett, Allen M. "A Critical Analysis of the Operations of the American II Corps from September 20 to October 2, 1918." GGSC SIRS, IR-17–1934.

Chambliss, T. M. "A Study of the Operations of the 30th Infantry, 3rd Division, in the Second Battle of the Marne on July 15th 1918." GGSC SIRS, IR-87–1930.

Cornish, George. "The Twenty-Sixth Infantry (U.S.) in the Meuse-Argonne Offensive." CGSS SIRS, IR-104–1931.

Emery, Frank. "What Should Be the Plan of Education for Officers of the Army." CGSS SIRS, IR-85–1930.

Green, J. O. "Operations of 3rd Battalion, 23rd Infantry from 30 May to 12 July 1918." CGSS SIRS, IR-53–1931.

Johnston, Edward S. "A Study of the Nature of United States Infantry Tactics for Open Warfare on July 18, 1918, and of Their Points of Difference as Contrasted with the United States Army Tactics Taught in 1914." CGSS SIRS, IR-124–1931.

Joseph J. Koch Collection.

"Lecture Delivered by Colonel A. W. Foreman on 27 January 1919 at Fifth Army Corps Headquarters. Subject: 'Administration and Supply of the 5th Army Corps during the St. Mihiel Offensive and Meuse-Argonne Offensive 1918.'"

"Lecture Delivered by Colonel Willey Howell on 6 January 1919 to the Assembled General Officers and Chiefs of Staff of the First American Army. Subject: 'The Second Section, G.S., First American Army in the St. Mihiel and Meuse-Argonne Operations.'"

"Lecture Delivered by Major General Alexander, 77th Division, on 3 February 1919. Subject: 'Operations of the Division, 26th of September to the 11th of November.'"

"Lecture Delivered by Major General George B. Duncan, Commanding, 82nd Division, on 3 February 1919. Subject: 'General Missions of the 82nd Division in the Argonne-Meuse Offensive.'"

"Lecture Presented by LTC Troup Miller on 20 January 1919 at Fifth Army Corps Headquarters. Subject: 'Plan of Communication, Supply and Evacuation, 1st Corps, for St. Mihiel Offensive and Meuse-Argonne Offensive.'"

Olmstead, Merritt. "A Critical Analysis of the Troop Leading within the 5th U.S. Division during the 2d Phase of the Meuse-Argonne Offensive." CGSS SIRS, GR-74–1933.

Bibliography

Rehman, Edward. "An Analysis of Casualties in the 3rd Division in the Second Battle of the Marne, July 15–20, 1918." CGSS SIRS, GR-107–1930.

Waltz, W. P. "Personal Experiences of a Machine Gun Officer at Cantigny, 28th–30th May." CGSS SIRS, IR-6–1933.

HARGRETT RARE BOOK AND MANUSCRIPT LIBRARY, UNIVERSITY OF GEORGIA

Justus Erwin Owens Scrapbook.

LIBERTY MEMORIAL AND NATIONAL WORLD WAR I MUSEUM ARCHIVES, KANSAS CITY, MO

Clarence J. Minick Papers.

Ernest F. McKeighan Papers.

William R. Phillips Papers.

NATIONAL ARCHIVES AND RECORDS ADMINISTRATION

Records of the American Expeditionary Forces (World War I), Record Group 120

Correspondence Relating to American Officers Suspected of Pro-German Sympathies.

Personal Files of Brig. Gen. R. C. Davis.

Records of the Combat Officers' Depot.

Records of the Gondracourt Reclassification Depot.

Records of the Inspector General.

Records of the Judge Advocate General.

Records of the Provost Marshal.

Records of Reclassified Officers National Army and National Guard, Blois.

Reports Related to the Morale of American Troops, 1917–19.

Records of the Office of the Judge Advocate General (Army), Record Group 153

"Record of the Trial by General Courts-Martial of Captain Charles G. Clement, 328th Infantry," 15 July 1918, U.S., Box 5977, Docket no. 120515.

Records of the War Department General Staff, Record Group 165

Letters, Memorandums, Reports, etc., of the Citizen Training Camps, Officers' Training Camps, Central Officers' Training Schools, and Student Army Training Corps.

Reports by US Military Attaché in France, 1914–1917.

Reports on Military Education at Colleges and Universities.

Reports of the Morale Branch of the War College and War Plans Division.

Reports of Morale at Army Installations.

Reports of Army War College Historical Branch, G5 Schools.

PERSONAL COLLECTION OF THE AUTHOR

Knox Alexander Papers.

Albert Carmoody Papers.

Lecture Given by Brig. Gen. Frank Parker entitled "Certain Observations on Infantry," 2 April 1919, to AEF's Army Center of Artillery Studies, AEF, Third Course.

P. Benson Oakley Papers.

PERSONAL COLLECTION OF CONRAD CRANE
Crane, C. L. "The Great War: 1917–1918–1919." Unpublished diary.

ROBERT W. WOODRUFF LIBRARY SPECIAL COLLECTIONS, EMORY UNIVERSITY
Adelaide Bowen Papers.

F. L. Miller Papers.

UNITED STATES ARMY CAVALRY MUSEUM ARCHIVES, FORT RILEY, KS
Beaumont Buck Papers.

UNITED STATES ARMY MILITARY HISTORY INSTITUTE, CARLISLE BARRACKS, PA
Fiske, Harold, "Report of Training in the American Expeditionary Forces." Timberman-Fiske Papers.

Goddard, Calvin H. "Relations between American and British Expeditionary Forces, 1917–1920." USAMHI AWC #57200E, Part I, Number 5 (June 1942).

Humber, Robert C. "Absences and Desertions during the First World War." USAMHI AWC Report 36 (1942).

Spinks, M. G. "Major Problems of the Inspector General, AEF, and Their Solutions." Transcript of lecture presented to Army War College, 9 October 1933. USAMHI File AWC 401-A-5.

Transcript of interview of Gen. John E. Hull by Lt. Col. James W. Wurman, 22 October 1973, Washington, DC. USAMHI Oral History Collection.

Transcript of oral history Interview of John G. Oechsner by CSM Erwin H. Koehler, 19 January 1982. Noncommissioned Officer Oral History Program.

USAMHI World War I Veteran Survey. The following files were used in this work:

Adams, Charles D., 5th Inf., Maryland National Guard, Mexican Border.

Aamoth, Walter J., Officers' Training Schools, Camp Pike.

Bernet, Milton E., 89th Div.

Bisogno, Ernesto, Pvt., 328th Inf., 82nd Div.

Briggs, William McKinley, Camp Zachary Taylor, Ky., 159th Depot Bde.

Burke, Mervyn F., Headquarters Troops, 1st Div.

Burton, John McNab, Camp Jackson, 156th Depot Bde.

Campbell, Charles G., Evacuation Ambulance Company No. 1.

Carroll, Tom, 16th Inf., 1st Div.

Chayes, Edward, Officers' Training Schools, Camp Johnston.

Cristo, Pandelis, 327th Inf., 82nd Div.

Dacus, Herman, 28th Inf., 1st Div.

Dongarra, George, 2nd Ammunition Train, 2nd Div.

Engleman, Edwin Earl, Camp Dodge, 163rd Depot Bde.

Faulkner, Frank L., 23rd Inf., 2nd Div.

Flacker, Charles W., 112th Inf., 28th Div.

Ford, Berch, 16th Inf., 1st Div.

Grossman, Edmund A., 139th Inf., 35th Div.

Hagen, Fendell A., 1st Sgt., 140th Inf., 35th Div.

Harter, Arthur G., Camp Hancock, 1st Provisional Rgt.

Heath, Benjamin, 328th Inf., 82nd Div.

Helms, Malcolm, 5th Mach. Gun Bn., 2nd Div.

Honaker, H. A., AEF North Russia and 142nd Mach. Gun Bn.

Hook, Hugh L., 353rd Inf., 89th Div.

Hopkins, Claude A., Camp Gordon, File.

House, Harry, 320th Mach. Gun Bn., 82nd Div.

Kemerer, Duncan M., 111th Inf., 28th Div.

King, Henry P., 23rd Inf., 2nd Div.

Kyler, Donald, 16th Inf., 1st Div.

LaVenture, Alonzo M., 111th Inf., 28th Div.

Loukides, George, 326th Inf., 82nd Div.

Maxwell, Paul E., Camp Lee, 314th Field Art., 80th Div.

McBride, Richard, 325th Inf., 82nd Div.

McDaniels, John D., 126th Inf., 32nd Div. This file contains the manuscript "A Tanglefoot's Diary," compiled by 1st Sgt. Harold C. Woehl.

O'Hair, Robert, Student Army Training Corps, Indiana University.

Osgood, Charles G., Student Army Training Corps, Michigan State University.

Shaw, Frederick, 18th Inf., 1st Div.

Stambaugh, Wilbert F., 2nd Field Signal Btn., 1st Div.

Stovall, Elmer, 1st Ammunition Train, 1st Div.

Strikell, Charles, 5th Field Art., 1st Div.

Supple, Howard, 137th Inf., 35th Div.

Sweningsen, Milton B., 138th Inf., 35th Div.

Takes, Fred, 325th Inf., 82nd Div.

Warrell, Jonas E., 103rd Ammunition Train, 28th Div.

Whitman, Walter, 325th Inf., 82nd Div.

Williams, George O., Student Army Training Corps, Washington State College.

Wolf, Walter L., 129th Inf., 33rd Div.

UNITED STATES MILITARY ACADEMY LIBRARY SPECIAL COLLECTIONS, WEST POINT, NY

John W. Castles Papers.

Captain John H. Horton Papers.

Julian L. Schley Papers.

Published Primary Sources

Adams, Myron, ed. *The Officer's Responsibility for His Men.* Fort Sheridan, IL: locally published, 1917.

Adjutant General's Office. *Official Army Register,* 1897, 1905, 1915. Washington, DC: Government Printing Office.

———. *General Orders.* Washington, DC: Government Printing Office, 1878–1919.

———. *The Personnel System of the United States Army.* Vol. 1, *History of the Personnel System.* Washington, DC: Government Printing Office, 1919.

———. *The Personnel System of the United States Army.* Vol. 2, *The Personnel Manual.* Washington, DC: Government Printing Office, 1919.

Albertine, Connell. *The Yankee Doughboy.* Boston: Branden Press, 1968.

Alexander, Robert. *Memories of the World War, 1917–1918.* New York: MacMillan Company, 1931.

Allen, Hervey. *Towards the Flame.* New York: Grosset and Dunlap, 1934.

Amerine, William H. *Alabama's Own in France.* New York: Eaton and Gettinger, 1919.

Andrews, Lincoln C. *Fundamentals of Military Service.* Philadelphia: J. B. Lippincott, 1916.

———. *Leadership and Military Training.* Philadelphia: J. B. Lippincott, 1918.

———. *Military Manpower.* New York: E. P. Dutton, 1920.

Annual Report of the Commandant, the Army Service Schools, for the years 1908–20. Fort Leavenworth, KS: Army Service Schools Press.

Anonymous [Bob Casey]. *The Cannoneers Have Hairy Ears.* New York: J. H. Sears, 1927.

Anonymous. *Wine, Women, and War: A Diary of Disillusionment.* New York: J. H. Sears [1926].

Armstrong, F. C. *The Story of the Sixteenth Infantry in France.* Frankfurt am Main, Germany: Martin Flock, 1919.

Army War College. *Lectures on Discipline and Training by Colonel Applin of the British General Staff and Hints to Young Officers by a British Officer.* Washington, DC: War College Press, February 1918.

———. *Manual for Commanders of Infantry Platoons.* Washington, D.C.: Government Printing Office, 1917.

———. *Notes on Recent Operations, No. 3.* Washington, D.C.: Government Printing Office, 1917.

———. *Statement of a Proper Military Policy for the United States, September 1915.* Washington, DC: Government Printing Office, 1916.

Arps, George F. "Science as Applied to the Selection of Noncommissioned Officers." *INJ* 15, no. 7 (January 1919).

Ayers, Leonard. *The War with Germany: A Statistical Summary.* Washington, DC: Government Printing Office, 1919.

Azan, Paul. *The War of Positions.* Cambridge: Harvard University Press, 1917.

Bach, C. A. "Leadership." *INJ* 14, no. 8 (February 1918).

Bach, Christian A., and Henry N. Hall. *The Fourth Division: Its Services and Achievements in the World War.* New York: Country Life Press, 1920.

Bairnsfather, Bruce. *Fragment from France, Part Six.* New York: G. P. Putnam's Sons, 1918.

Baker, Chester E. *Doughboy's Diary.* Shippensburg, PA: Burd Street Press, 1998.

Baker, Horace. *Argonne Days in World War I.* Columbia: University of Missouri Press, 2007.

Baker, Leslie. *The Company History: The Story of Company B, 106th Machine Gun Battalion.* New York: privately printed, 1920.

Baltzell, George. "The Proper Training of an Infantry Company." *INJ* 5, no. 5 (March 1909).

Bandholtz, Henry H. *History of the Provost Marshal General's Department, American Expeditionary Forces.* Chaumont, France: AEF Provost Marshal, 15 April 1919.

Barber, Thomas H. *Along the Road.* New York: Dodd, Mead and Company, 1924.

Barbusse, Henri. *Under Fire.* London: J. M. Dent & Sons, 1926.

Barkley, John L. *No Hard Feelings!* New York: Cosmopolitan Book Corporation, 1930.

The Battery Book: A History of Battery "A," 306th F.A. New York: The DeVinne Press, 1921.

Bennett, Arnold. *Over There.* New York: George H. Doran Company, 1915.

Bisogno, Ernesto. "The Life and Death of Charles Clement." *American Legion Magazine,* March 1938, 50–51.

Bjornstad, A. W. "Infantry Combat." *INJ* 8, no. 6 (May–June 1912).

Black, William. "The Education and Training of Army Officers." *JMSIUS* 32 (1903).

Braddan, William S. *Under Fire with the 370th Infantry (8th I.N.G.) A.E.F.* Chicago: privately published, no date.

Brenner, Leo. "The American 'Million Army.'" *INJ* 13, no. 5 (February 1917).

Briggs, Allan L. "Training in Morale." *INJ* 14, no. 4 (October 1917).

Brown, William. *The Adventures of an American Doughboy.* Tacoma, WA: Smith-Kenney Co., 1919.

Browne, George. *An American Soldier in World War I.* Ed. David L. Snead. Lincoln: University of Nebraska Pres, 2006.

Buck, Beaumont. *Memories of Peace and War.* San Antonio: The Naylor Company, 1935.

Bullard, Robert L. "The Military Study of Men." *INJ* 8, no. 3 (November–December 1911).

———. *Personalities and Reminiscences of the War.* New York: Doubleday, Page and Company, 1925.

Burtt, W. B. "Tactical Instruction of Officers." *INJ* 10, no. 2 (September–October 1913).

Buxton, G. Edward, Jr., ed. *Official History of 82nd Division, American Expeditionary Forces: "All American" Division, 1917–1919.* Indianapolis: Bobbs-Merrill, 1919.

Camp, Charles W. *History of the 305th Field Artillery.* New York: Country Life Press, 1919.

Camp MacArthur Central Infantry Officers Training Camp. *Farewell Book, CIOTS, 1918.* Camp MacArthur, TX, n.p. [1918].

Camp Sherman Officer Training Camp. *The Gold Bar.* Indianapolis: The Art Press, 1918.

Carter, William H. "Our Defective Military System." *North American Review,* March 1917.

Chastaine, Ben. *Story of the 36th.* Oklahoma City: Harlow Publishing Company, 1920.

Cherfils, General. "Infantry Fire in the Present War." *INJ* 12, no. 3 (November 1915).

Clarke, J. L. J. "Infantry in the Attack." *INJ* 14, no. 11 (May 1918).

———. "Notes on Infantry Work on Western Front." *INJ* 14, no. 11 (May 1918).

Clarke, William F. *Over There with O'Ryan's Roughnecks.* Seattle: Superior, 1966.

Clement, Charles M. ed. *Pennsylvania in the World War: An Illustrated History of the Twenty-Eighth Division.* 2 Vols. Pittsburgh: States Publication Society, 1921.

Colonna, B. A., ed. *The History of Company B, 311th Infantry, in the World War.* Freehold, NJ: Transcript Printing House, 1922.

Colson, W. N., and A. B. Nutt. "The Failure of the Ninety-Second Division." *The Messenger,* September 1919.

Company C, 328th Infantry. *History of Company "C," 328th Infantry.* N.p.: privately published, 1919.

Crawford, Charles. *Weapons and Munitions of War,* Part I, *Infantry Weapons.* Fort Leavenworth: Staff College Press, 1907.

Crowell, Benedict. *America's Munitions, 1917–1918: Report of Benedict Crowell, the Assistant Secretary of War, Director of Munitions.* Washington, DC: Government Printing Office, 1919.

Cutchins, John, and George Stewart. *History of the 29th Division, 1917–1919.* Philadelphia: MacCalla & Co., 1921.

Deckard, Percy. *List of Officers Who Served with the 371st Infantry and Headquarters 186th Infantry Brigade during the World War.* Allegany, MD: Allegany Citizen, 1929.

Dickinson, John. *The Building of an Army.* New York: Century, 1922.

Dickman, J. T. *Modern Improvements in Fire Arms and Their Tactical Effects.* General Service and Staff College Lectures Number Ten. Fort Leavenworth: General Service and Staff College Press, 26 September 1902.

Dienst, Charles, et al. *They're from Kansas: History of the 353rd Infantry Regiment, 89th Division, National Army.* Wichita, KS: The Eagle Press, 1921.

Dittmar, Gus. *They Were First.* Austin, TX: Steck-Warlick Company, 1969.

Dougherty, A. J. "The Making of a Soldier." *INJ* 7, no. 5 (March 1911).

Dravo, C. A. "Machine Guns: The Offensive in Open Warfare." *INJ* 17, no. 4 (October 1920).

Drennan, L. H. "Psychology of the Bayonet." *INJ* 11, no. 2 (September–October 1914).

DuPuy, Charles M. *A Machine Gunner's Notebook.* Pittsburgh: Reed & Witting, 1920.

Eames, Henry E. *The Rifle in War.* Fort Leavenworth: US Cavalry Association, 1908.

Edwards, Evan A. *From Doniphan to Verdun: The Official History of the 140th Infantry.* Lawrence, KS: The World Company, 1920.

Edwards, William. "The Squad Leader and His Squad." *Journal of the United States Cavalry Association* 23, no. 4 (January 1913).

Eighth Company, Central Officers' Training School, Camp Lee, Virginia. *Take His Name.* Richmond, VA: Garrett and Massie, 1918.

Eisenhower, Dwight D. *At Ease: Stories I Tell to Friends.* New York: Doubleday, 1967.

———. *Eisenhower: The Prewar Diaries and Selected Papers, 1905–1941.* Ed. Daniel D. Holt. Baltimore: Johns Hopkins University Press, 1998.

Ellis, O. O. "Hints for Service at Training Camps." *INJ* 13, no. 6 (March 1917).

———, and E. B. Garey. *The Plattsburg Manual: A Handbook for Military Training.* New York: Century, 1917.

English, George H., Jr. *History of the 89th Division, U.S.A.* Denver: Smith-Brooks Printing Company, 1920.

Ettinger, Albert M. *A Doughboy with the Fighting 69th.* Shippensburg, PA: White Mane Publishing Company, 1992.

Evans, R. K. "Infantry Fire in Battle." *INJ* 5, no. 6 (May 1909).

Evarts, Jeremiah M. *Cantigny: A Corner of the War.* N.p.: privately published, 1938.

Farley, J. P. "Military Service for College Men." *JMSI* 50 (1912).

Fell, Edgar. *History of the Seventh Division, 1917–1919.* Philadelphia: Seventh Division Officer's Association, 1927.

Fort Devens OTC Yearbook Committee. *The Pick: 3rd O.T.C., Camp Devens, Mass.* Boston: George H. Dean, 1918.

Fort Sheridan Association. *The History and Achievements of the Fort Sheridan Officers' Training Camps.* Chicago: Hawkins & Loomis Company, 1920.

French, Allen. *At Plattsburg.* New York: Charles Scribner's Sons, 1917.

Frothingham, Francis. "Plattsburgh Lessons." *Journal of the United Service Institution of the United States* 58 (1916).

371

Gallagher, D. B. *The Battle of Bolts and Nuts in the Sector of Cognac Hill.* Fort Worth: Gallagher-Crosby Company, 1931.

Gansser, Emil B. *History of the 126th Infantry in the War with Germany.* Grand Rapids, MI: Dean Hicks Company, 1920.

General Headquarters, AEF. *Bulletin for Field Officers.* Chaumont, France: AEF GHQ, 8 September 1918.

———. *Combat Instructions.* Chaumont, France: AEF GHQ, 5 September 1918.

———. *Manual of the Chief of Platoon of Infantry.* Paris: Imprimerie Nationale, August 1917.

———. *Notes on Recent Operations, No. 1.* Chaumont, France: AEF GHQ, 7 August 1918.

———. *Notes on Recent Operations, No. 2.* Chaumont, France: AEF GHQ, 8 September 1918.

———. *Notes on Recent Operations, No. 3.* Chaumont, France: AEF GHQ, 12 October 1918.

———. *Report of Officers Convened by Special Orders No. 98, GHQ AEF, 09 April 1919* (the Lewis Board).

———. *Report of Superior Board on Organization and Tactics.* Chaumont, France: AEF GHQ, 1919.

———. *Supplement to Instructions for the Offensive Combat of Small Units.* Nancy, France: Berger-Levarault, April 1918.

General Service Schools *Studies in Minor Tactics.* Fort Leavenworth, KS: Army Service Schools Press, 1908.

Greene, Francis. "Important Improvements in the Art of War during the Past Twenty Years and Their Probable Effect on Future Military Operations." *Journal of the Military Service Institution of the United States* 4, no. 13 (1883).

Gruber, E. L. "Employment of Field Artillery with Infantry." *INJ* 15, no. 12 (June 1919).

Guild, G. R. "The Company in Combat Exercises." *INJ* 11, no. 1 (July–August 1914).

H. R. H. "Why Is It? An Answer." *Cavalry Journal* 28, no. 117 (January 1918).

Hale, H. C. "Military Instruction in Colleges." *INJ* 10, no. 5 (March–April 1914).

Haltzell, George. "The Proper Training of an Infantry Company." *INJ* 5, no. 5 (March 1909).

Hamilton, Craig, and Louise Corbin, eds. *Echoes from Over There.* New York: Soldier's Publishing Company, 1919.

Harbord, James G. *Leaves from a War Diary.* New York: Dodd, Meade and Company, 1925.

———. *The American Army in France, 1917–1919.* Boston: Little Brown and Co., 1936.

Harris, Harvey L. *The War as I Saw It: 1918 Letters of a Tank Corps Lieutenant.* Saint Paul, MN: Pogo Press, 1998.

Hart, Archibald. *Company K of Yesterday.* New York: Vantage Press, 1969.

Haterius, Carl E. *Reminiscences of the 137th Infantry.* Topeka, KS: Carne and Company, 1919.

Havlin, Arthur C. *The History of Company A, 102nd Machine Gun Battalion, Twenty-Sixth Division.* Boston: privately printed, 1928.

Herzog, Stanley J. *Helmets: Second Battle of the Marne.* Stamford, CT: The Bell Press, 1930.

History of Company "C," 328th Infantry. New York: Hunter Collins, no date.

History of the 305th Field Artillery. Garden City, NJ: The Country Life Press, 1919.

History of the Three Hundred and Twelfth Infantry. New York: privately published, 1919.

Hitchcock, C. H. "A Letter to a Training Camp Student." *INJ* 14, no. 1 (July 1917).

Hocking, William E. "Fundamentals of Military Psychology." *INJ* 14, no. 10 (April 1918).

———. *Morale and Its Enemies*. New Haven: Yale University Press, 1918.

Hogan, Martin. *The Shamrock Battalion in the Great War*. Columbia: University of Missouri Press, 2007.

Holden, Frank A. *War Memories*. Athens, GA: Athens Book Company, 1922.

Howard, James M. *The Autobiography of a Regiment*. New York: privately printed, 1920.

Howe, Lucien. "A Brief for Military Education in Our Schools and Universities." *Journal of the United Service Institution of the United States* 57 (1915).

Hoyt, Charles B. *Heroes of the Argonne*. Kansas City, MO: Franklin Hudson, 1919.

Hughes, W. N. "A Practical Plan for Infantry Training." *INJ* 14, no. 1 (July 1917).

Huidenkopfer, Frederick L. *The History of the 33rd Division A.E.F.* Springfield: Illinois State Historical Library, 1921.

Hungerford, Edward. *With the Doughboy in France*. New York: MacMillian, 1920.

Hunton, Addie, and Kathryn M. Johnson. *Two Colored Women with the American Expeditionary Forces*. Brooklyn: Brooklyn Eagle Press, 1920.

"Infantry Training." *INJ* 12, no. 4 (December 1915).

"Information on Trench Warfare." *JUSCA* 27, no. 111 (July 1916).

Jacks, L. V. *Service Record by an Artilleryman*. New York: Charles Scribner's Sons, 1928.

Jackson, Rhees. "Revision of Our System of Military Education." *INJ* 8, no. 3 (November–December, 1911).

Jacobson, Gerald F. *History of the 107th Infantry U.S.A.* New York: The De Vinne Press, 1920.

Joel, Arthur H. *Under the Lorraine Cross*. East Lansing, MI: privately published, 1921.

Johnson, Ray N. *Heaven, Hell, or Hoboken*. Cleveland: O. S. Hubbell Printing Company, 1919.

Johnson, Thomas M. *Without Censor*. Indianapolis: Bobbs-Merrill Company, 1927.

Joint War History Commissions of Michigan and Wisconsin. *The 32nd Division in the World War*. Madison: Wisconsin Printing Company, 1920.

Judy, Will. *A Soldier's Diary*. Chicago: Judy Publishing Company, 1930.

Kenamore, Clair. *From Vauquois Hill to Exermont*. Saint Louis: Guard Publishing, 1919.

Kilbourne, Charles, ed. *The National Service Library*. 4 Vols. New York: F. P. Collier and Son, 1917.

Laffargue, André. "A Study on the Attack in the Present Period of the War." *INJ* 13, no. 2 (September–October 1916).

———. "Advice to the Infantry Soldier." *INJ* 13, no. 3 (November–December 1916).

Lafferty, Frederick. *Roster of Officers of the First Provisional Class, 1916*. Leavenworth, KS: privately published, 1929.

Langer, William L. *Gas and Flame in World War I*. New York: Alfred E. Knopf, 1965.

Langille, Leslie. *Men of the Rainbow*. Hamond, IN: W. B. Coakley Company, 1933.

Lavine, A. Lincoln. *Circuits of Victory*. Garden City, NJ: Doubleday, Page & Company, 1921.

Lawrence, Joseph D. *Fighting Soldier: The AEF in 1918*. Boulder: Colorado Associated University Press, 1985.

Lazelle, Henry. "Important Improvements in the Art of War during the Past Twenty Years and Their Probable Effect on Future Military Operations." *Journal of the Military Service Institution of the United States* 3, no. 11 (1882).

"The Lessons of Trench Warfare." *INJ* 12, no. 4 (December 1915).

Lewis, R. W. "The Soldier Versus Energy, Ambition, and Initiative." *Journal of the United States Cavalry Association* 25, no. 4 (October 1914).

Liggett, Hunter. *A.E.F.: Ten Years Ago in France*. New York: Dodd, Meade and Company, 1927.

———. *Commanding an American Army: Recollections of the World War*. New York: Houghton Mifflin, 1925.

Little, Arthur W. *From Harlem to the Rhine*. New York: Covici Friede, 1936.

Luby, James, ed. *One Who Gave His Life: War Letters of Quincy Sharpe Mills*. New York: G. P. Putnam's Sons, 1923.

Lyon, C. C. *Experience of a Recruit in the United States Army*. Washington, DC: Government Printing Office, 1916.

MacArthur, Charles. *War Bugs*. New York: Doubleday, Doran and Company, 1929.

Machine Gun Officers' Training School. *Four Months of Sand*. Augusta, GA: Phoenix Printing Company, 1918.

MacIntyre, W. Irwin. *Colored Soldiers*. Macon, GA: J. W. Burke, 1923.

Mackin, Elton E. *Suddenly We Didn't Want to Die*. Novato, CA: Presidio Press, 1993.

Malstrom, George M. *The 131st Infantry in the World War*. Chicago: privately published, 1919.

March, Peyton. *The Nation at War*. New York: Doubleday, Doran & Company, 1932.

Marshall, George C. *Memories of My Service in the World War: 1917–1918*. New York: Houghton Mifflin Company, 1976.

———. *The Papers of George Catlett Marshall*. Vol. 1, *The Soldierly Spirit, December 1880–June 1939*. Ed. Larry Bland and Sharon Ritenour Stevens. Baltimore: Johns Hopkins University Press, 1981.

———. *The Papers of George Catlett Marshall*. Vol. 2, *We Cannot Delay, July 1939–December 6, 1941*. Ed. Larry Bland and Sharon Ritenour Stevens. Baltimore: Johns Hopkins University Press, 1981.

Mason, Monroe, and Arthur Furr. *The American Negro Soldier with the Red Hand in France*. Boston: The Cornhill Company, 1920.

Maverick, Maury. *A Maverick American*. New York: Covici Friede Publishers, 1937.

McAndrews, J. W. *Address to the Second Class of Provisional Second Lieutenants*. Fort Leavenworth: Army Service Schools Press, April 17, 1917.

McArthur, John C. *What a Company Officer Should Know*. New York: Harvey Press, 1918.

McCormick, Robert R. *The Army of 1918.* New York: Harcourt, Brace and Howe, 1920.

McDonald, Dale. "Training and Promotion of Second Lieutenants." *Journal of the Military Service Institution of the United States* 59 (August 1916).

McIntyre, Colby L., *The Old Man of the 103rd: The Biography of Frank M. Hume.* Houlton, ME: Aroostock Print Shop, 1940.

McKaine, O. E. "The Buffaloes: A First-Class Colored Fighting Unit." *Outlook* 119 (22 May 1918).

Merrill, Dana. "Infantry Training." *INJ* 9, no. 1 (July–August 1912).

Miles, L. Wardlaw. *History of the 308th Infantry.* New York: G. P. Putnam's Sons, 1927.

375

Military Training Camp Association. *Roster of Attendants at Federal Military Training Camps, 1913–1916.* New York: Anderson & Ruwe, 1916.

"Memorandum on Infantry Instruction." *INJ* 16, no. 7 (January 1920).

Miller, Charles. *The Customs of the Service: Also Some Suggestions and Advice.* Fort Leavenworth: Army Service School Press, 1917.

Minder, Charles. *This Man's War.* New York: Pevensey Press, 1931.

Morehead, Robert. *The Story of the 139th Field Artillery, American Expeditionary Forces.* Indianapolis: Bobbs-Merrill Company, 1920.

Morrison, John F. *Infantry Training.* Fort Leavenworth, KS: US Cavalry Association, 1914.

Moss, James A. *Manual of Military Training.* Menasha, WI: George Banta, 1914.

———. *The Noncommissioned Officer's Manual.* Menasha, WI: George Banta, 1917.

———. *Officer's Manual.* Menasha, WI: George Banta Publishing Co, 1917.

Murphy, Elmer, and Robert S. Thomas. *The Thirtieth Division in the World War.* Lepanto, AR: Old Hickory Publishing, 1936.

Murrin, James A. *With the 112th in France.* Philadelphia: J. B. Lippencott, 1919.

Nell, John W. *The Lost Battalion: A Private's Story.* San Antonio: Historical Publishing Network, 2001.

Nichols, Vernon. "Our Battle of the Argonne." *INJ* 15, no. 15 (September 1919).

91st Division Publication Committee. *The Story of the 91st Division.* San Francisco: H. S. Crocker Company, 1919.

O.N.E. [John B. Barnes]. *Letters of a Plattsburg Patriot.* Washington, DC: United States Infantry Association, 1917.

"An Officer Abroad: Notes on the European War." *JUSCA* 26, no. 107 (July 1915).

"An Officer of High Rank, Berlin, February 1915: What Has the World's War Taught Us up to the Present Time That Is New in a Military Way." *JUSCA* 26, no. 107 (July 1915).

One of Them [pseud.]. "Why Is It? Another Answer." *Cavalry Journal* 28, no. 118 (April 1918).

O'Ryan, John F. "The Psychology of Discipline." *INJ* 10, no. 5 (March–April 1914).

Otto, Ernst. *The Battle at Blanc Mont.* Annapolis: United States Naval Institute, 1930.

"Our Military System As It Appeared to America's Citizen Soldiers." *INJ* 15, no. 10 (April 1919).

Parker, James. *The Old Army: Memories, 1872–1918.* New York: Stackpole Books, 2003.

Parker, John. "Simplifying the Organization and Tactics of Infantry." *INJ* 16, no. 7 (January 1920).

Parker, William E. "The Company Officer and His Work." *INJ* 4, no. 1 (July–August 1912).

Patch, A. M. "Machine Gun Organization." *INJ* 17, no. 2 (August 1920).

Peck, R. H. *Infantry in Attack*. General Service and Staff College Lectures No. 14. Fort Leavenworth: General Service and Staff College, 13 December 1904.

Peixotto, Ernest. *The American Front*. New York: Charles Scribner's Sons, 1919.

Perry, Ralph B. *The Plattsburg Movement: A Chapter of America's Participation in the World War*. New York: E. P. Dutton and Company, 1921.

Pershing, John J. *My Experiences in the World War*. 2 Vols. New York: Frederick Stokes Company, 1931.

Peyton, E. G. "Modern Tactics." *INJ* 17, no. 2 (August 1920).

The Plattsburger. New York: Wynkoop Hallenbeck Crawford Co. [1917].

Pollard, James E. *The Forty-Seventh Infantry: A History 1917–1918–1919*. Saginaw, MI: Press of Seeman & Peters, 1919.

Pratt, Walter M. *Tin Soldiers: The Organized Militia and What It Really Is*. Boston: The Gorham Press, 1912.

"The Present Problem." *INJ* 6, no. 2 (September, 1909).

"Probationary Officers." *JUSCA* 27, no. 112 (November 1916).

Proctor, H. G. *The Iron Division in the World War*. Philadelphia: John C. Winston Company, 1919.

"The Products of the Training Camps." *INJ* 14, no. 1 (July 1917).

Rainsford, Kerr. *From Upton to the Meuse with the Three Hundred and Seventh Infantry*. New York: D. Appleton, 1920.

Reeves, Ira L. *Military Education in the United States*. Burlington, VT: Free Press, 1914.

Reilly, Henry J. *Why Preparedness?* Chicago: Daughaday and Company, 1916.

"Reserve Officers' Training Camps." *INJ* 14, no. 6 (December 1917).

"Reveries of an Old Field Officer." *JUSCA* 27, no. 113 (January 1917).

Richards, John. "Some Experiences with Colored Soldiers." *Atlantic Monthly*, August 1919.

Rodney, George B. *As a Cavalryman Remembers*. Caldwell, ID: The Caxton Printers, 1944.

Rogers, Horatio. *World War I through My Sights*. San Rafael, CA: Presidio Press, 1975.

Ross, Warner A. *My Colored Battalion*. Chicago: self-published, 1920.

Ruschke, Egmont W., ed. *Lieuie VI: Being the Chronicle of the Battle of Camp Lee as Fought by the Deathless Sixth Battalion, Central Officers' Training School, Camp Lee, Virginia*. Petersburg, VA: privately published, 1919.

Sanborn, Joseph. *The 131st Infantry in the World War*. Chicago: privately printed, 1919.

Schmidt, Paul. *Co. C, 127th Infantry, in the World War*. Sheboygan, WI: Press Publishing Company, 1919.

Schofield, John M. *Forty-Six Years in the Army*. New York: Century Publishing Company, 1897.

"A School for Noncommissioned Officers." *INJ* 12, no. 8 (April 1916).

Scott, Emmett J. *Scott's Official History of the American Negro in the World War*. New York: n.p., 1919.

Scott, Hugh L. *Some Memories of a Soldier*. New York: The Century Company, 1928.

Shanks, David C. *Management of the American Soldier*. New York: Thomas Ryan [1917].

"Shortages of Officers on the Border." *INJ* 13, no. 2 (September–October 1916).

Sibley, Frank. *With the Yankee Division in France*. Boston: Little, Brown, and Company, 1919.

Sirmon, W. A. *That's War: An Authentic Diary*. Atlanta: The Linmon Company, 1929.

Society of the Fifth Division. *The Official History of the Fifth Division*. Washington, DC: privately published, 1919.

"The Soldier at School." *Cavalry Journal* 27, no. 3 (January, 1917).

Spaulding, Oliver. "The Tactics of the War with Germany." *INJ* 17, no. 3 (September 1920).

Spears, Francis, et al, eds. *Damitall: Twentieth Company, Central Officers' Training Camp, Camp Gordon, Georgia*. Atlanta: privately published, 1918.

Spencer, E. W. *The History of Gas Attacks upon the American Expeditionary Forces during the World War*. Edgewood Arsenal, MD: Chemical Warfare Service, US War Department, 1928.

"The Spirit of Training." *INJ* 14, no. 7 (January 1918).

Stewart, E. M. *Handbook for Noncommissioned Officers of Infantry*. Kansas City, MO: Franklin Hudson Publishing Company, 1903.

Stewart, M. B. "The Military Training Camps." *INJ* 13, no. 3 (November–December 1916).

Stewart, Worth P. *The History of Company "K" 117th Infantry in the Great War*. N.p.: privately published, 1919.

Stimson, Henry L. *On Active Service in Peace and War*. New York: Harper and Brothers, 1948.

Stockton, Richard. "Military Schools and the Nation." *INJ* 11, no. 1 (July–August 1914).

———. "Military Training—Valuable and Valueless." *JMSI* 59, no. 202 (July–August 1916).

Story of "E" Company, 101st Engineers, 26th Division. Boston: privately published, 1919.

Straub, Elmer F. *A Sergeant's Diary in the World War*. Indiana History Collections 10. Indianapolis: Indiana History Commission, 1923.

Stringfellow, John S. *Hell! No!* Boston: Meador Publishing Company, 1936.

Sutherland, S. J. *The Reserve Officers' Handbook*. Boston: Houghton Mifflin Company, 1917.

Sutliffe, Robert S. *Seventy-First New York in the World War*. New York: Seventy-First Regiment Association, 1922.

Swan, Carroll J. *My Company*. New York: Houghton Mifflin Company, 1918.

Swann, Thomas. "The Top-Sergeant." *INJ* 15, no. 12 (June 1919).

Taylor, Emerson G. *New England in France, 1917–1919: A History of the Twenty-Sixth Division*. Boston: Houghton Mifflin Company, 1920.

Tebbetts, Frank. "War and Emotionalism." *JUSCA* 25, no. 104 (October 1914).

———. "Leadership." *JUSCA* 27, no. 111 (July 1916).

Thompson, Hugh. *Trench Knives and Mustard Gas*. College Station: Texas A&M University Press, 2004.

Thorn, Henry. *History of 313th U.S. Infantry.* New York: Wynkoop Hallenbeck Crawford Company, 1920.

328th Infantry Historical Committee. *History of the Three Hundred and Twenty-Eighth Infantry Regiment.* N.p.: privately published, 1922.

Tiebout, Frank B. *A History of the 305th Infantry.* New York: Wynkoop Hallenbeck Crawford Company, 1919.

Tips, Charles R. "Selecting and Training Military Leaders." *INJ* 15, no. 7 (January 1919).

Tompkins, Raymond S. *The Story of the Rainbow Division.* New York: Boni & Liveright, 1919.

"Training Negro Officers." *Literary Digest,* no. 55 (21 July 1917).

Triplet, William S. *A Youth in the Meuse-Argonne.* Columbia: University of Missouri Press, 2000.

Truscott, Lucian K., Jr. *The Twilight of the Cavalry: Life in the Old Army, 1917–1942.* Lawrence: University Press of Kansas, 1989.

US Army Center of Military History. *Order of Battle of the United States Land Forces in the World War: American Expeditionary Forces.* 2 vols. Washington, DC: Government Printing Office, 1988.

———. *The United States Army in the World War, 1917–1919.* 17 vols. Washington, DC: Government Printing Office, 1988–1992.

US Army, 1st Division. *World War Records, First Division A.E.F. (Regular).* 25 vols. Washington, DC: Army War College, 1930.

US Army, 2nd Division. *World War Records, Second Division A.E.F. (Regular).* 10 vols. Washington, DC: Army War College, 1924.

US Congress. House. Committee on Military Affairs. *To Increase the Efficiency of the Military Establishment of the United States.* 64th Cong., 1st sess., January–February 1916.

US National Guard Bureau. *Report on Mobilization of the Organized Militia and National Guard of the United States, 1916.* Washington, DC: Government Printing Office, 1916.

US War Department. *The Army as a Life Occupation for Enlisted Men.* Washington, DC: War Department, 1907.

———. *Field Service Pocket Book.* Washington, DC: Government Printing Office, 1917.

———. *Field Service Regulations of the United States Army, with Corrections to May 21, 1913.* Washington, DC: Government Printing Office, 1913.

———. *Infantry Drill Regulations (Provisional).* Washington, DC: Government Printing Office, 1919.

———. *Infantry Training.* Document no. 656. Washington, DC: Government Printing Office, 1917.

———. *Instructions for the Training of Platoons for Offensive Action.* Washington, DC: Government Printing Office, 1917.

———. *Instructions for the Offensive Combat of Small Units.* War Department Document No. 802. Washington, DC: Government Printing Office, May 1918.

———. *Instructions on the Offensive Conduct of Small Units.* War Department Document No. 583. Washington, DC: Government Printing Office, 1917.

———. *Manual for Noncommissioned Officers and Privates of Infantry.* Washington, DC: Government Printing Office, 1917.

———. *The National Defense Act Approved June 3, 1916, with Updates to June 1924.* Washington, DC: Government Printing Office, 1924.

———. *Notes on Infantry, Cavalry, and Field Artillery.* Three parts. Washington, DC: Government Printing Office, 1917.

———. *Regulations for the Army of the United States, 1913, Corrected to April 15, 1917.* Washington, DC: Government Printing Office, 1917.

———. *Report of the Acting Chief of the Militia Bureau, 1916.* Washington, DC: Government Printing Office, 1916.

———. *Reports of Military Observers Attached to the Armies in Manchuria during the Russo-Japanese War.* Five parts. Washington, DC: Government Printing Office, 1906–1907.

———. *Soldier's Hand Book.* Rev. ed. Washington, DC: Government Printing Office, 1913.

———. *Special Regulations No. 49: Training Camps for Reserve Officers and Candidates for Appointment As Such, May 15–August 11, 1917.* Washington, DC: Government Printing Office, 5 May 1917.

———. *Study on Educational Institutions Giving Military Training as a Source for a Supply of Officers for a National Army.* Washington, DC: Government Printing Office, 1916.

———. *Three Year Enlistment for the Army.* Washington, DC: Government Printing Office, 1912.

———. *Training Circular No. 12: Combined Training of a Division.* Washington, DC: Government Printing Office, 10 October 1918.

———. *War Department Annual Report.* Washington, DC: Government Printing Office, 1901–1921.

Viereck, George, ed. *As They Saw Us: Foch, Ludendorff, and Other Leaders Write Our War History.* New York: Doubleday and Doran, 1929.

Walters, Raymond, et al. *F.A.C.O.T.S.: The Story of the Field Artillery Central Officers' Training School, Camp Zachary Taylor, Kentucky.* New York: Knickerbocker Press, 1919.

War Veteran's Association. *History of Company "E," 107th Infantry.* New York: privately published, 1920.

Ward, Frank H., ed. *The Camp Sherman Souvenir.* Cincinnati: Lambertson, 1918.

Westover, Wendell. *Suicide Battalions.* New York: G. P. Putnam's Sons, 1929.

Wharton, James B. "A Battalion in Action," *INJ* 16, no. 6 (December 1919).

"Why Is It?" *Cavalry Journal* 28, no. 116 (October 1917).

Wilhelm, Carl, et al. *Pass in Review: The Book of the Fourth Officers' Training School, Camp Dodge, Iowa, 1918.* Camp Dodge, IA: privately published, 1918.

Williams, Ashby. *Experiences of the Great War.* Roanoke, VA: Stone Printing, 1919.

Williams, Charles. *Sidelights on Negro Soldiers.* Boston: B. J. Brimmer, 1923.

Wise, Jennings C. "Organization and Initial Training of a Company." *INJ* 14, no. 3 (September 1917).

———. "The Soldier's Life in Battle." *INJ* 16, no. 11 (May 1920).

Wright, William. *Meuse-Argonne Diary: A Division Commander in World War I.* Columbia: University of Missouri Press, 2004.

Yerkes, Robert M., ed. *Psychological Examining in the United States Army.* Memoirs of the National Academy of Sciences 15. Washington, DC: Government Printing Office, 1921.

Yoakum, Clarence, and Robert Yerkes. *Army Mental Tests.* New York: Henry Holt and Company, 1920.

York, Alvin C. *Sergeant York: His Own Life Story and War Diary*. Ed. Tom Skeyhill. New York: Doubleday, Doran, 1928.

Secondary Sources

American Battle Monuments Commission. *American Armies and Battlefields in Europe*. Washington, DC: Government Printing Office, 1938.

Armstrong, David. *Bullets and Bureaucrats: The Machine Gun and the United States Army*. Westport, CT: Greenwood Press, 1982.

Ashworth, Tony. *Trench Warfare, 1914–1918: The Live and Let Live System*. London: Macmillan Press, 1980.

Barbeau, Arthur F., and Florette Henri. *The Unknown Soldiers: Black American Troops in World War I*. Philadelphia: Temple University Press, 1974.

Baynes, John. *Morale: A Study of Men and Courage: The Second Scottish Rifles at the Battle of Neuve Chappelle, 1915*. New York: Frederick A. Praeger, 1967.

Berlin, Ira, ed. "A Wisconsinite in World War I: Reminiscences of Edwin P. Arpin, Jr." *Wisconsin Magazine of History* 51, nos. 1–3 (Autumn–Spring 1962).

Berry, Henry. *Make the Kaiser Dance: Living Memories of a Forgotten War: The American Experience in World War I*. New York: Doubleday, 1978.

Bolger, Daniel. "Zero Defects: Command Climate in the First U.S. Army, 1944–1945," *Military Review* 71 (May 1991).

Braim, Paul F. *The Test Of Battle*. 2nd ed. Shippensburg, PA: White Mane Books, 1998.

Brands, H. W. *TR: The Last Romantic*. New York: Basic Books, 1997.

Brereton, T. R. *Educating the U.S. Army: Arthur L. Wagner and Reform, 1875–1905*. Lincoln: University of Nebraska Press, 2000.

Bruce, Robert. *A Fraternity in Arms: America and France in the Great War*. Lawrence: University Press of Kansas, 2003.

Byerly, Carol R. *The Fever of War: The Influenza Epidemic in the U.S. Army during World War I*. New York: New York University Press, 2005.

Canfield, Bruce. *U.S. Infantry Weapons of the First World War*. Lincoln, RI: Andrew Mowbray Publishers, 2000.

Carlton, John T., and John F. Slinkman. *The ROA Story: A Chronicle of the first 60 Years of the Reserve Officers Association of the United States*. Washington, DC, n.p., 1982.

Chambers, John W. *To Raise an Army*. New York: Free Press, 1987.

Chodoff, Elliot P. "Ideology and Primary Groups," *Armed Forces & Society* 9, no. 4 (Summer 1983).

Clifford, John G. *The Citizen Soldiers*. Lexington: University Press of Kentucky, 1972.

Cochran, Rexmond C. *Gas Warfare in World War I*. 20 vols. Washington, DC: U.S. Army Chemical Corps Historical Office, 1959–60.

Coffman, Edward M. *The Hilt of the Sword: The Career of Peyton C. March*. Madison: University of Wisconsin Press, 1966.

———. *The Old Army: A Portrait of the American Army in Peacetime, 1784–1898*. New York: Oxford University Press, 1986.

———. *The Regulars: The American Army, 1898–1941*. Cambridge: The Belknap Press of Harvard University Press, 2004.

————. *The War to End All Wars.* Madison: University of Wisconsin Press, 1968.

Cooke, James J. *Pershing and His Generals.* Westport, CT: Praeger, 1997.

Cooper, Jerry. *The Rise of the National Guard: The Evolution of the American Militia, 1865–1920.* Lincoln: University of Nebraska Press, 1997.

Cosmas, Graham. *An Army for Empire.* Shippensburg, PA: White Mane, 1994.

Davis, Henry B. *Generals in Khaki.* Raleigh, NC: Pentland Press, 1998.

Demaison, Gerard, and Yves Buffetaut. *Honor Bound: The Chauchat Machine Rifle.* Cobourg, Ontario: Collector Grade Publications, 1996.

Doughty, Robert A. *Pyrrhic Victory: French Strategy and Operations in the Great War.* Cambridge: The Belknap Press of Harvard University Press, 2005.

Echevarria, Antulio J. *After Clausewitz: German Military Thinkers before the Great War.* Lawrence: University Press of Kansas, 2000.

Farwell, Byron. *Over There: The United States in the Great War, 1917–1918.* New York: W. W. Norton, 1999.

Faulkner, Richard S. "Hard Knocks, Hubris, and Dogma: Leader Competence in the American Expeditionary Forces." In *Leadership: The Art of the Warrior.* Ed. Chris Kolinda. Carlisle, PA: Army War College Association Press, 2001.

————. "Our Patriotic Duty at Home and Abroad: The University of Georgia in World War I." *Georgia Historical Quarterly* 79, no. 4 (Winter 1995).

————. "Up in the Argonne: The Tragedy of Lieutenant Justus Owens and the 82nd Division in the First World War." *Georgia Historical Quarterly* 80, no. 2 (Summer 1996).

Ferrell, Robert H. *America's Deadliest Battle: Meuse-Argonne, 1918.* Lawrence: University Press of Kansas, 2007.

————. *Collapse in the Meuse-Argonne: The Failure of the Missouri-Kansas Division.* Columbia: University of Missouri Press, 2004.

————. *Five Days in October: The Lost Battalion of World War I.* Columbia: University of Missouri Press, 2005.

Fisher, Ernest, Jr. *Guardians of the Republic: A History of the Noncommissioned Officer Corps of the U.S. Army.* New York: Ballantine Books, 1994.

Foner, Jack. *The United States Soldier between Two Wars.* New York: Humanities Press, 1970.

Ford, Nancy Gentile. *Americans All!: Foreign-born Soldiers in World War I.* College Station: Texas A&M University Press, 2001.

Fussell, Paul. *Wartime: Understanding and Behavior in the Second World War.* New York: Oxford University Press, 1989.

Gilchrist, H. L. *A Comparative Study of World War Casualties from Gas and Other Weapons.* Washington, DC: Government Printing Office, 1928.

Graff, Alan D. *Blood in the Argonne.* Norman: University of Oklahoma Press, 2005.

Gregory, Stanford W. "Toward a Situated Description of Cohesion and Disintegration in the American Army." *Armed Forces & Society* 2, no. 3 (May 1976).

Grotelueschen, Mark E. *The AEF Way of War.* New York: Cambridge University Press, 2007.

————. *Doctrine under Fire: American Artillery Employment in World War I.* Westport, CT: Greenwood Press, 2001.

Gruber, Carol S. *Mars and Minerva: World War I and the Uses of Higher Learning in America.* Baton Rouge: Louisiana State University Press, 1975.

Bibliography

Gudmundsson, Bruce. *Stormtroop Tactics: Innovation in the German Army, 1914–1918.* Westport, CT: Praeger, 1989.

Hallas, James H. *Squandered Victory: The American First Army at St. Mihiel.* Westport, CT: Praeger, 1995.

———. *Doughboy War: The American Expeditionary Force in World War I.* Boulder, CO: Lynne Rienner, 2000.

Hamburger, Kenneth E. *Learning Lessons in the American Expeditionary Forces.* CMH Publication 24–1. Washington, DC: Government Printing Office, 1997.

Harries, Meirion, and Susie Harries. *The Last Days of Innocence: America at War, 1917–1918.* New York: Random House, 1997.

Harris, Stephen L. *Duffy's War.* Washington, DC: Brassey's, 2006.

———. *Duty, Honor, Privilege.* Washington, DC: Brassey's, 2001.

Heller, Charles. *Chemical Warfare in World War I: The American Experience, 1917–1918.* Leavenworth Papers No. 10. Fort Leavenworth, KS: Combat Studies Institute, 1984.

Henderson, William D. *Cohesion: The Human Element in Combat.* Washington, DC: National Defense University Press, 1985.

Howard, Michael. "Men against Fire: The Doctrine of the Offensive in 1914," in *Makers of Modern Strategy.* Ed. Peter Paret. Princeton: Princeton University Press, 1986.

Infantry in Battle. Washington, DC: Infantry Journal, 1939.

James, D. Clayton. *The Years of MacArthur.* Vol. 1, *1880–1941.* Boston: Houghton Mifflin, 1970.

Jamieson, Perry, D. *Crossing the Deadly Ground: United States Army Tactics, 1865–1899.* Tuscaloosa: University of Alabama Press, 1994.

Johnson, Douglas V., and Rolfe L. Hillman, Jr. *Soissons 1918.* College Station: Texas A&M University Press, 1999.

Johnson, Thomas M., and Fletcher Pratt. *The Lost Battalion.* Indianapolis: Bobbs-Merrill, 1936.

Keene, Jennifer D. *Doughboys, the Great War, and the Remaking of America.* Baltimore: Johns Hopkins University Press, 2001.

Kennedy, David M. *Over Here.* New York: Oxford University Press, 1980.

Kennett, Lee. "The AEF through French Eyes." *Military Review* 52, no. 11 (November 1972).

Kipling, Rudyard. "Only a Subaltern," in *Soldiers Three: A Collection of Stories.* New York: John W. Lovell Company, 1890.

Lane, Jack. *Armed Progressive: General Leonard Wood.* San Rafael, CA: Presidio, 1978.

Lengle, Edward G. *To Conquer Hell.* New York: Henry Holt and Company, 2008.

Lerwill, Leonard L., ed. *The Personnel Replacement System in the United States Army.* Washington, DC: Government Printing Office, 1954.

Lipsman, Samuel, and Edward Doyle. *Fighting for Time.* The Vietnam Experience. Boston: Boston Publishing Company, 1983.

Lupfer, Timothy. *The Dynamics of Doctrine: The Changes in German Tactical Doctrine during the First World War.* Fort Leavenworth, KS: Combat Studies Institute, 1981.

Lyons, Gene M., and John W. Massland. "The Origins of the ROTC." *Military Affairs* 23, no. 1 (Spring 1959).

MacHoian, Ronald G. *William Harding Carter and the American Army*. Norman: University of Oklahoma Press, 2006.

Mansoor, Peter R. *The GI Offensive in Europe: The Triumph of American Infantry Divisions, 1941–1945*. Lawrence: University Press of Kansas, 1999.

Marshall, S. L. A. *Men Against Fire: The Problem of Battle Command*. Gloucester, MA: Peter Smith, 1978.

Miller, Nathan. *Theodore Roosevelt: A Life*. New York: William Morrow, 1992.

Millett, Allan R. *The General: Robert L. Bullard and Officership in the United States Army, 1881–1925*. Westport, CT: Greenwood Press, 1975.

———, and Williamson Murray, eds. *Military Effectiveness*. Vol. 1, *The First World War*. Boston: Allen and Unwin, 1988.

Moran, Lord. *Anatomy of Courage*. London: Constable and Company, 1945.

Neiberg, Michael S. *Making Citizen Soldiers: ROTC and the Ideology of American Military Service*. Cambridge: Harvard University Press, 2000.

Nenninger, Timothy K. "The Army Enters the Twentieth Century." In *Against All Enemies*. Ed. Kenneth Hagen and William Roberts. Westport, CT: Greenwood Press, 1986.

———. "John J. Pershing and the Relief for Cause in the American Expeditionary Forces, 1917–1918." *Army History*, Spring 2005.

———. *The Leavenworth Schools and the Old Army*. Westport, CT: Greenwood Press, 1978.

———. "Tactical Dysfunction in the AEF, 1917–1918." *Military Review* 51, no. 4 (October 1987).

Owen, Peter. *To the Limits of Endurance: A Battalion of Marines in the Great War*. College Station: Texas A&M University Press, 2007.

Painter, Nell. *Standing at Armageddon*. New York: W. W. Norton, 1987.

Palmer, Robert W.; Bell I. Wiley, and William R. Keast. *The Procurement and Training of Combat Ground Troops*. United States Army in World War II: The Army Ground Forces. Washington, DC: US Army Center of Military History, 1948.

Patton, Gerald W. *War and Race: The Black Officer in the American Military 1915–1941*. Westport, CT: Greenwood Press, 1981.

Pearlman, Michael. *To Make Democracy Safe for America: Patricians and Preparedness in the Progressive Era*. Urbana: University of Illinois Press, 1984.

Pogue, Forrest. *George Marshall: Education of a General, 1880–1939*. New York: Viking, 1963.

———. *George Marshall: Ordeal and Hope, 1939–1942*. New York: Viking, 1963.

Rainey, James W. "Ambivalent Warfare: The Tactical Doctrine of the AEF in World War I." *Parameters* 13 (September 1983).

———. "The Questionable Training of the AEF in World War I." *Parameters* 22, no. 4 (Winter 1992–93).

Ramsay, M. A. *Command and Cohesion: The Citizen Soldier and Minor Tactics in the British Army, 1870–1918*. Westport, CT: Praeger, 2002.

Rickey, Don. *Forty Miles a Day on Beans and Hay*. Norman: University of Oklahoma Press, 1963.

Rush, Robert S. "A Different Perspective: Cohesion, Morale, and Operational Effectiveness in the German Army, Fall 1944." *Armed Forces & Society* 25, no. 3 (Spring 1999).

Samuels, Martin. *Command or Control: Training and Tactics in the British and German Armies, 1888–1918*. London: Frank Cass, 1995.

383

Savage, Paul L., and Richard A. Gabriel. "Cohesion and Disintegration in the American Army: An Alternative Perspective." *Armed Forces & Society* 2, no. 3 (May 1976).

Sheffield, G. D. *Leadership in the Trenches: Officer-Man Relations, Morale, and Discipline in the British Army of the First World War.* New York: St. Martin's Press, 2000.

Shephard, Ben. *A War of Nerves: Soldiers and Psychiatrists in the Twentieth Century.* Cambridge: Harvard University Press, 2001.

Shils, Edward, and Morris Janowitz. "Cohesion and Disintegration in the Wehrmacht in World War II." *Public Opinion Quarterly* 12, no. 2 (Summer, 1948).

Slotkin, Richard. *Lost Battalions.* New York: Henry Holt and Company, 2005.

Smith, Leonard V. *Between Mutiny and Obedience: The Case of the French Fifth Infantry Division during World War I.* Princeton: Princeton University Press, 1994.

Smythe, Donald. *Pershing: General of the Armies.* Bloomington: Indiana University Press, 1986.

Stallings, Laurence. *The Doughboys: The Story of the AEF, 1917–1918.* New York: Harper and Row, 1963.

Stanton, Shelby. *The Rise and Fall of an American Army: U.S. Ground Forces in Vietnam, 1965–1973.* Novato, CA: Presidio Press, 1985.

Stephenson, Donald Scott. "Frontschwene and Revolution: The Role of Front-Line Soldiers in the German Revolution of 1918." PhD dissertation, University of Kansas, 2007.

Thomas, Lowell. *Woodfill of the Regulars.* Garden City, NJ: Doubleday, Doran & Company, 1929.

Trask, David. *The AEF and Coalition Warmaking, 1917–1918.* Lawrence: University Press of Kansas, 1993.

Travers, Tim. *The Killing Ground.* London: Unwin Hyman, 1987.

Vandiver, Frank. *Black Jack: The Life and Times of John J. Pershing.* College Station: Texas A&M University Press, 1977.

White, Lonnie J. *The 90th Division in World War I.* Manhattan, KS: Sunflower University Press, 1996.

———. *Panthers to Arrowheads: The 36th Division in World War I.* Austin, TX: Presidial Press, 1982.

Whitehorn, Joseph. *The Inspectors General of the United States Army, 1903–1939.* Washington, DC: Government Printing Office, 1998.

Wiebe, Robert H. *The Search for Order, 1877–1920.* New York: Hill and Wang, 1967.

Wiengartner, Steven, ed. *Cantigny at Seventy-Five: A Professional Discussion.* Weaton, IL: Robert R. McCormick Tribune Foundation, 1993.

Wilson, Dale E. *Treat 'Em Rough! The Birth of American Armor, 1917–20.* Novato, CA: Presidio, 1989.

Winter, Dennis. *Death's Men.* London: Penguin Books, 1978.

Index

391